LOG DATA ACQUISITION AND QUALITY CONTROL

Second Edition

FROM THE SAME PUBLISHER

Dictionary of Drilling and Boreholes. *English-French/French-English*
    M. MOUREAU, G. BRACE

Dictionary of Seismic Prospecting. *English-French/French-English*

Formation Imaging by Acoustic Logging   J.L. MARI, Ed.

Peri-Tethyan Platforms   F. ROURE, Ed.

Petroleum and Tectonics in Mobile Belts   J. LETOUZEY, Ed.

Rock Mechanics   Ph. A. CHARLEZ
   1. Theoretical Fundamentals
   2. Petroleum Applications

Seismic Surveying and Well Logging   J.L. MARI, S. BOYER

Seismic Well Surveying   J.L. MARI, F. COPPENS

Wave Separation   F. GLANGEAUD, J.L. MARI

Cement Evaluation Logging Handbook   D. ROUILLAC

Cementing Technology and Procedures   J. LECOURTIER, U. CARTALOS Eds.

Drilling. Oil and Gas Field Development Techniques   J.P. NGUYEN

Drilling Data Handbook. *Seventh edition*   G. GABOLDE, J.P. NGUYEN

Encyclopedia of Well Logging   R. DESBRANDES

Progressing Cavity Pumps   H. CHOLET

Well Testing: Interpretation Methods   G. BOURDAROT

Philippe THEYS
Graduate Engineer from École Centrale de Paris
Data Quality Manager, Anadrill-Schlumberger

Foreword by
Farouk Al-Kasim
Previously Director of Resources
Norwegian Petroleum Directorate

LOG DATA ACQUISITION AND QUALITY CONTROL

Second Edition
Completely revised and expanded

1999

t ÉDITIONS TECHNIP   25 rue Ginoux 75015 PARIS FRANCE

© 1999, Éditions Technip, Paris

All rights reserved. No part of this publication may be reproduced or transmitted in any form or by any means, electronic or mechanical, including photocopy, recording, or any information storage and retrieval system, without the prior written permission of the publisher.

ISBN 978-2-7108-0748-3

In memory of Paulette and Jean

Table of contents

Acknowledgements xix

Foreword xxi

Preface xxiii

I Premises

1 Introduction 3
   1.1 Acquiring reservoir data in the eighties 3
   1.2 Changes in the nineties 4
   1.3 What can be expected from log data? 5
   1.4 What can be expected from this book? 5
   1.5 What cannot be found in this book? 6
       1.5.1 Physics of logging measurements 6
       1.5.2 Log interpretation 6

2 Evaluation of hydrocarbon volume 7
   2.1 Inputs to evaluation of hydrocarbon volume 9
       2.1.1 Coring 10
       2.1.2 Conventional logging 10
       2.1.3 Deep-investigation logging 11
       2.1.4 Well testing 12
       2.1.5 Seismics 14
       2.1.6 Summary 17
   2.2 Continuity of geological formations 19
   2.3 Combining information 21

3 Data collection and decision-making ... 23
3.1 Requirements of decision-making ... 24
3.1.1 Timeliness ... 24
3.1.2 Accuracy ... 24
3.1.3 Cost-effectiveness ... 25
3.1.4 Planning ... 25
3.2 Data quality objectives method ... 25
3.2.1 Description of the method ... 26
3.2.2 Example of DQO in the field of environmental protection ... 26
3.2.3 Hydraulic separation with a cement job ... 27
3.2.4 Calculating the length of the producing drain ... 27
3.3 Other decision-making examples ... 28
3.3.1 Quality of the data required for quantitative evaluation ... 28
3.3.2 Hydraulic continuity between two wells ... 28
3.3.3 Geosteering ... 30

4 Elements of metrology I: error analysis ... 33
4.1 True value and errors ... 33
4.1.1 True value of a physical characteristic ... 33
4.1.2 Error ... 34
4.1.3 Types of error ... 34
4.1.4 Error analysis ... 35
4.2 Evaluating systematic error ... 36
4.3 Estimating random error ... 37
4.3.1 General ... 37
4.3.2 Binomial distribution ... 39
4.3.3 Poisson distribution ... 39
4.3.4 Gaussian distribution ... 40
4.3.5 One-σ, two-σ, three-σ ... 41
4.4 Precision and accuracy ... 42
4.4.1 Definitions ... 42
4.4.2 Importance of precision and accuracy ... 43
4.5 Repeatability and reproducibility ... 45
4.5.1 Repeatability ... 45
4.5.2 Reproducibility ... 45
4.6 Uncertainty ... 46

5 Elements of metrology II: volume considerations ... 47
5.1 Geometrical factor ... 48
5.1.1 Definition ... 49
5.1.2 Elemental geometrical factor ... 50
5.1.3 Successive integrations of the geometrical factor ... 51
5.1.4 Pseudogeometric factor ... 53
5.2 Depth of investigation ... 56
5.2.1 Conventional definition of depth of investigation ... 56

		5.2.2	Ambiguity of the conventional definition of the depth of investigation	58
		5.2.3	Modern definition of depth of investigation	59
		5.2.4	Variation of depth of investigation with formation and environment	60
		5.2.5	Electric diameter	60
	5.3	Vertical resolution		61
		5.3.1	Conventional definitions	61
		5.3.2	Modern definition of vertical resolution	66
		5.3.3	Reduction or enhancement of vertical resolution	67
		5.3.4	Combination of measurements with different vertical resolutions	67
		5.3.5	Consonant sensors	68
	5.4	Relation between depth of investigation and vertical resolution		69

6 Elements of metrology III: other attributes — 71

	6.1	Sensitivity to the environment		71
		6.1.1	Effect of temperature	71
		6.1.2	Effect of pressure	71
		6.1.3	Effects related to magnetism	72
		6.1.4	Effect of flux of other particles	72
	6.2	Time response		73
		6.2.1	Hysteresis	73
		6.2.2	Dead time	74
	6.3	Effect of the measuring apparatus on the measurement		75
	6.4	Sensitivity to shocks and vibrations		76
	6.5	Downhole access		76
	6.6	Reliability		77

7 Mathematical preliminary: propagation of errors — 79

	7.1	Derivatives		79
		7.1.1	Function of one variable	79
		7.1.2	Function of several variables	80
	7.2	Setting the problem of error propagation		81
	7.3	Analytical quantification of propagated error		83
	7.4	Applications		84
		7.4.1	Sums or differences	84
		7.4.2	Multiplication by a constant	85
		7.4.3	Combination of independent measurements	86
	7.5	Examples of propagation of errors		87
	7.6	Propagation of errors derived from a numerical method		90
		7.6.1	Theory	90
		7.6.2	Examples	90
		7.6.3	Practical use of numerical methods	94

7.7	Applications of error propagation		95
	7.7.1 Ranking of critical parameters		95
	7.7.2 Worth of a porosity unit		97

II Data acquisition

8 Data acquisition . . . 101
  8.1 The data cycle . . . 101
  8.2 Causes of error . . . 103

9 Sensor and source technology . . . 107
  9.1 Logging sources . . . 107
    9.1.1 Nuclear sources . . . 107
    9.1.2 Sound sources . . . 108
    9.1.3 Electromagnetic transmitters . . . 108
    9.1.4 Natural sources . . . 109
  9.2 Logging sensors . . . 109
    9.2.1 Geiger-Müller detector . . . 109
    9.2.2 Scintillation detectors . . . 109
    9.2.3 Semiconductor detectors . . . 110
    9.2.4 Other detectors . . . 111
    9.2.5 Helium detectors . . . 111
    9.2.6 Sonic transducers . . . 111
    9.2.7 Pressure gauges . . . 112
    9.2.8 Magnetometers and accelerometers . . . 113
  9.3 Optimization of sensor design . . . 113
    9.3.1 Compensation of effects influencing the measurement . . . 113
    9.3.2 Detector geometry . . . 114
    9.3.3 Optimization of detector design for the density tool . . . 114

10 Effect of measurement duration on precision . . . 117
  10.1 Duration of the basic interactions experienced in logging . . . 117
    10.1.1 Induction, laterolog, microresistivity, electromagnetic propagation . . . 117
    10.1.2 Sonic and ultrasonic . . . 117
    10.1.3 Nuclear . . . 118
    10.1.4 Nuclear magnetic resonance . . . 119
    10.1.5 Measurement duration in logging . . . 119
  10.2 Data transmission . . . 120
    10.2.1 Analog transmission with a cable . . . 120
    10.2.2 Digital transmission with cable . . . 120
    10.2.3 Mud-pulse transmission . . . 121

       10.2.4 Electromagnetic propagation 123
       10.2.5 Memory dump 123
  10.3 Conversion of time to depth 124
       10.3.1 Relating time to depth in wireline logging 124
       10.3.2 Relating time to depth in MWD 125
       10.3.3 Merging data and depth information 126
  10.4 Effect of logging speed/rate of penetration and sampling rate on
       measurement precision 130
       10.4.1 Induction, laterolog, imaging 130
       10.4.2 Nuclear tools 130
       10.4.3 NMR tools 132

11 Signal processing: filtering 133
  11.1 Precision of a measurement over a thick homogeneous interval 133
  11.2 Combination of successive measurements 134
       11.2.1 Precision of a weight-averaged measurement 134
       11.2.2 Reduction of vertical resolution 135
  11.3 Optimal filtering 135
       11.3.1 Incidence of sampling interval 137
  11.4 How is filtering controlled in the logging environment? 139
  11.5 Median averaging 140
  11.6 Stacking 140
  11.7 Active filtering 142
       11.7.1 Spectral gamma ray logging 142
  11.8 Azimuthal averaging 144

12 Enhancement of vertical resolution through processing 147
  12.1 Effect of vertical resolution 147
       12.1.1 Hydrocarbon volume estimation 147
       12.1.2 Recovery factor 150
  12.2 High-resolution sensors 151
  12.3 Signal processing 153
       12.3.1 Deconvolution 153
       12.3.2 Combination of measurements 153

13 Tool response 159
  13.1 Physical simulation 160
       13.1.1 The resistivity network 160
  13.2 Experimental measurements 161
       13.2.1 Test formation design considerations 163
  13.3 Scaled experiments 164
       13.3.1 Resistivity 164
       13.3.2 Sonic 164
  13.4 Mathematical modeling 167
       13.4.1 General 167

    13.4.2 Sonic modeling 168
    13.4.3 Resistivity modeling 169
    13.4.4 Nuclear modeling 170

14 Environmental corrections 175
14.1 Definition 175
14.2 Correcting for the environmental effects 176
14.3 Determination of environmental corrections 177
    14.3.1 Anticipating the need for corrections 177
    14.3.2 Methods of determination of correction algorithms 177
    14.3.3 Challenges linked to environmental corrections 178
    14.3.4 Modeling 179
    14.3.5 How can environmental corrections be checked? 180
14.4 Error propagation from the input to environmental corrections 180
    14.4.1 Induction borehole correction 182
    14.4.2 Thermal neutron correction 185
14.5 Importance of the description of the environment 186
    14.5.1 The role of sensors describing the environment 186
    14.5.2 A note of concern: caliper accuracy 186
    14.5.3 A note of optimism 187
    14.5.4 MWD and the well environment 187
    14.5.5 Are environmental corrections really useful? 188
    14.5.6 How can environmental corrections be improved? 188

15 The real environment 191
15.1 Temperature 191
15.2 Borehole shape 192
    15.2.1 Distortion of hole shape 192
    15.2.2 Distortion from the drilling mode 196
    15.2.3 Effects of hole shape and tool positioning on logs 196
    15.2.4 Borehole alteration 198
15.3 Borehole trajectory 199
    15.3.1 Anisotropy 201
15.4 The mud 203
    15.4.1 Mud characteristics that impact logs 203
    15.4.2 Oil-base mud 203
    15.4.3 Effect of mud solids 204
    15.4.4 Mud homogeneity 205
    15.4.5 Invasion 206
15.5 Dielectric constant variation 211
15.6 Putting it all together 212

16 Density logging 217
- 16.1 The first density algorithms 217
  - 16.1.1 Single detector 217
  - 16.1.2 Dual-detector device 218
  - 16.1.3 Non-Compton gamma rays 219
- 16.2 Spectrometric detectors 220
  - 16.2.1 Compensating the upper windows for photoelectric information 221
  - 16.2.2 The heavy mud algorithm 221
  - 16.2.3 Experimental establishment of the response of a spectrometric tool 222
  - 16.2.4 Experimental evaluation of environmental effects 224
- 16.3 MWD density 227
  - 16.3.1 Nonazimuthal density 227
  - 16.3.2 Azimuthal density while drilling 227
- 16.4 PLATFORM EXPRESS* density 227
- 16.5 Later developments and summary 228

17 Calibration 233
- 17.1 Definitions 233
  - 17.1.1 Misuse of the word calibration 234
- 17.2 Evolution of calibration methods 234
  - 17.2.1 Logging calibrations in precomputer times 234
  - 17.2.2 Surface system alignment 236
- 17.3 Importance of calibration 236
  - 17.3.1 Variation between tools 237
  - 17.3.2 Variation during the life of a tool 237
- 17.4 Calibration requirements 238
  - 17.4.1 Calibrators 238
  - 17.4.2 Procedures 238
  - 17.4.3 Calibration environment 239
  - 17.4.4 Control of calibration conditions 240
- 17.5 Presentation of calibration 242
  - 17.5.1 Evolution of the calibration tail 242
  - 17.5.2 Guidelines for calibration tails 242
  - 17.5.3 Calibration records in digital format 246
  - 17.5.4 Meaning of tolerances 246
- 17.6 Error introduced by calibrations 248
  - 17.6.1 Caliper calibration 249
  - 17.6.2 Natural gamma ray calibration 250
  - 17.6.3 Induction tool calibration 251
  - 17.6.4 Two-height sonde error determination 252
  - 17.6.5 Neutron calibration 252
  - 17.6.6 Density calibration 256

18 Monitoring of tool behavior 259
18.1 Verification 259
18.1.1 Before-survey verification 260
18.1.2 After-survey verification 260
18.2 Need for downhole performance monitoring 262
18.3 Individual downhole checks 264
18.3.1 Thermal neutron quality control diagrams 264
18.3.2 Litho-Density tool quality control curves 265
18.3.3 Natural gamma ray spectrometry control curves 265
18.3.4 Induction control curves 269
18.4 Consistency checks 269
18.4.1 Array induction tool 270
18.4.2 Multidetector nuclear tool 271
18.4.3 Coherence plots in the frequency domain 272
18.5 Integrated control monitoring 273
18.6 Conclusions and recommendations 274

19 Measurement of depth 275
19.1 Importance of depth 275
19.1.1 Standards of depth control 276
19.2 How is depth measured? 277
19.2.1 Wireline depth measurement 277
19.2.2 Driller's depth 284
19.2.3 MWD depth 284
19.3 Causes and effects of depth mismatch on logs and interpretation . . . 286
19.3.1 Effect of depth matching errors 286
19.3.2 Apparent gas effect caused by depth mismatch 287
19.3.3 Mismatch introduced by depth-derived compensation 290
19.3.4 Apparent depth mismatch from high apparent dip 290
19.4 Speed correction 292
19.4.1 Speed electrode 293
19.4.2 Downhole measurement of acceleration 293
19.5 Depth matching 293
19.6 Guidelines of depth control 294
19.6.1 Depth policy 295
19.6.2 Documenting depth acquisition 296
19.6.3 Other recommendations 296
19.7 Future developments 297

20 Directional surveys 299
20.1 Introduction 299
20.1.1 Potential errors in directional surveys 300
20.2 Magnetic surveys 301
20.2.1 General 301
20.2.2 Earth's magnetic field 302

| 20.2.3 Misalignment errors 303
| 20.3 Gyroscope surveys 304
| 20.3.1 Gyroscope technologies 304
| 20.3.2 Sources of errors in gyroscope measurements 304
| 20.4 Importance of accurate directional surveys 305
| 20.4.1 Collision avoidance 305
| 20.4.2 Undulations in well trajectories 305

III Data quality control

21 Data quality plan 309
21.1 Definition of quality 309
21.2 Data quality plan 310
       21.2.1 Job planning and preparation 311
       21.2.2 Job execution 314
       21.2.3 Review after the job 314
21.3 Log quality control 314
       21.3.1 Log quality control requirements 316
       21.3.2 Log quality control reporting 317
       21.3.3 Example of LQC form 319
       21.3.4 Log quality data base 320
21.4 And, if a log does not conform? 321

22 Completeness of information 325
22.1 Units 325
       22.1.1 Unit systems 326
       22.1.2 Abbreviating units 326
       22.1.3 Conversion from one system to another 326
22.2 References 326
       22.2.1 Concentration 327
       22.2.2 Depth 327
       22.2.3 Location 328
       22.2.4 Porosity 328
       22.2.5 Pressure 329
22.3 Key information 329
       22.3.1 Client and product name 329
       22.3.2 Field and well name, run number 329
       22.3.3 Date 330
22.4 Other general information 330
       22.4.1 Tracking software 330
       22.4.2 Other information 331

	22.5 Remarks	331
	22.6 Logging parameters	331
	22.6.1 Mud data	331
	22.6.2 Environmental data	332
	22.6.3 Signal processing characteristics	332
	22.6.4 Temperature	332
	22.7 Tool description	333
	22.8 Indexed data	334
	22.8.1 Guidelines	334
	22.8.2 Time-indexed data	334
	22.8.3 Caveat	335
	22.8.4 Auxiliary indexed information	336
	22.9 Additional quality features	337
	22.9.1 Repeat sections and additional passes	337
	22.9.2 Calibration parameters	337
	22.9.3 Directional surveys	337
	22.9.4 Quality control tail	337
	22.9.5 Time marks	337
	22.9.6 Depth editing	339
	22.9.7 Dialogue between the logging engineer and the surface system	339
	22.10 Housekeeping recommendations	339
	22.10.1 Be consistent	339
	22.10.2 Don't throw anything away	339
	22.10.3 Avoid cosmetic surgery	340
	22.10.4 Maintain experts' presence at the wellsite	340
	22.11 Consequences of incomplete information	340
23	**Data management**	**343**
	23.1 Introduction	343
	23.2 Data transfers	344
	23.2.1 Reality	345
	23.2.2 Digital transfer of data	346
	23.2.3 Sorting and archiving data	346
	23.3 Data storage media	346
	23.3.1 Historical summary	346
	23.3.2 Current data storage media	347
	23.3.3 Archiving of data	347
	23.3.4 Tape quality checks	348
	23.4 Data models	348
	23.4.1 Data objects	349
	23.5 Data formats	349
	23.5.1 ASCII	350
	23.5.2 LAS	350
	23.5.3 LIS	350
	23.5.4 DLIS and RP66	351

| 23.5.5 Encapsulation 352
            23.5.6 Graphics 352
      23.6 Attempts of standardization 353
      23.7 Prints and films 353
            23.7.1 Standard presentation 355

24 Log quality checks 359
      24.1 Logging speed/rate of penetration 359
      24.2 Tool rotation 361
            24.2.1 Wireline logging 361
            24.2.2 MWD 361
      24.3 Tool configuration 362
            24.3.1 Selection of the correct auxiliary equipment 362
            24.3.2 Combination of tools 362
      24.4 Importance, advantage and limitations of repeatability checks 364
            24.4.1 Current checks of repeatability 364
            24.4.2 Repeatability standards 364
            24.4.3 Normalized repeatability check 367
            24.4.4 Practical hints for the repeatability check 368
            24.4.5 Numerical application of the normalized repeatability method . 370
            24.4.6 Repeatability standards for nuclear logging tools 370

25 Data quality evaluation 373
      25.1 First look 373
            25.1.1 Quick interpretation to discriminate zones of interest 374
      25.2 Looking at a larger picture 374
      25.3 Data quality control system 375
      25.4 Noise 376
            25.4.1 Spontaneous potential 376
            25.4.2 Sonic 377
            25.4.3 Dipmeter 378
      25.5 Anomalous response 380
            25.5.1 Hole conditions 380
      25.6 Tool limitations 385
            25.6.1 Geometrical limits 385
            25.6.2 Pressure and temperature 386
      25.7 Anomalous but correct readings: a formation trouble-shooting 386
            25.7.1 Delaware and anti-Delaware effects 386
            25.7.2 Squeeze and antisqueeze 387
            25.7.3 Groningen effect 387
            25.7.4 Conductive anomalies 390
            25.7.5 Cycles 391
            25.7.6 Lithology 392
            25.7.7 Pyrite 393

26 Images and nuclear magnetic resonance 395
26.1 Borehole imaging 395
26.1.1 Job preparation 395
26.1.2 Acquisition issue: tool movement 396
26.1.3 Processing guidelines 396
26.1.4 General guidelines for image quality control 397
26.2 NMR 398
26.2.1 Behavior of hydrogen proton spins 398
26.2.2 Relaxation times and mechanisms 398
26.2.3 NMR acquisition sequence 399
26.2.4 Other NMR specifications 401
26.2.5 Job planning 401
26.2.6 Quality control curves and parameters 403

27 Comparison of logged data with other information 407
27.1 Response in known conditions 407
27.1.1 Casings 407
27.1.2 Geological formations 410
27.1.3 Muds and formation fluids 412
27.2 Comparison of data acquired on the same well 413
27.2.1 Caveat 413
27.3 Multiwell analysis 413
27.3.1 Choice of markers 413
27.4 Comparison with cores 416
27.4.1 Comparison of logs with drilled cores 416
27.4.2 Comparison with sidewall cores 419
27.5 Comparison with production data 420
27.5.1 Capillary pressure curves 420

28 Optimum logging and uncertainty management 423
28.1 Optimal logging 423
28.1.1 Case study 1 424
28.1.2 Case study 2 426
28.2 Quantification of uncertainties 428
28.3 Using uncertainties in interpretation 428

Appendix 1: abbreviations, unit symbols and acronyms 431

Appendix 2: metrological definitions 433

Appendix 3: cumulative Gaussian probability 437

Appendix 4: solution of exercises 439

Bibliography 447

Index 449

Acknowledgements

I wish to acknowledge the help of Jim Spurlin for his multiple contributions to the text and his frequent attempts to revive this project. The ideas and findings of Jeff Phelps are at the basis of the section on invasion in Chapter 15.

My thanks go to Philip Bevington and Glenn Knoll. Their books on propagation of errors and nuclear detectors have been written in such a manner that it was difficult to take much distance from these references when I dealt with the same topics.

Tom Barber and Charlie Flaum have confirmed privately a number of technical details, the former in the fields of induction calibration and environmental corrections, the latter on high resolution and nuclear logging. Francis Gras has developed the fundamental ideas of the computerized log quality data base. Jean-Rémy Olesen has communicated his experience on artificial formations farms.

Christian Clavier and Oberto Serra have brought their extensive and multidisciplinary knowledge to comment, edit and correct the original manuscript and Jim Kent has toiled through the original text to lift ambiguities and correct the syntax.

My thanks also to the Norwegian oil industry at large for their strong commitment to quality assurance/quality control, commitment that has motivated publication of this book. In particular *Statoil*, *Elf* and *Total* have given their kind approval for release of the examples. I am also indebted to the authors and publishing companies who have accepted to have their figures reproduced in this work. This applies particularly to *Schlumberger Oilfield Services*, which has allowed the use of many figures originating from *The Technical Review* and the *Oilfield Review*.

The second edition could not have been completed without the support of Ivanna Albertin, Robert Freedman, Peter Ireland, Glenda de Luna, Laurent Moinard, Jeff Prilliman, Kirby Schrader, Hugh Scott and Stephen Whittaker.

Foreword

It was fortuitous that the first North Sea discoveries coincided in time with substantial technological developments in the oil industry. Improvements in seismic acquisition, processing and interpretation during the sixties were key factors in locating and defining North Sea discoveries. At the same time, substantial breakthroughs were taking place in various geoscience disciplines including a more fundamental understanding of basin tectonics and sedimentary processes, combined with the development of better tools to identify and map these features in the subsurface.

Nevertheless, errors in reservoir management were by no means eliminated. The large size of the initial discoveries, combined with the sharp rise in oil prices, helped absorb some otherwise serious setbacks to individual projects, caused either by budget overruns or by disappointing reservoir performance. In the wake of the price fall in the mid-eighties, the industry cannot afford serious reservoir misjudgements. This challenge is not likely to be of short duration. Predictability in the oil prices is yet to be achieved, while the remaining petroleum resources are progressively becoming more and more demanding in terms of economic and technical development.

There is therefore an increasing emphasis within the industry on reservoir description. Perhaps the greatest challenge in this respect is to quantify reservoir uncertainty as the first step in managing this uncertainty in the field development plan. To do this requires an intimate assessment of the individual input parameters which in turn determine the composite reservoir uncertainty. Not only is data quality important here, but also the way it interacts with other reservoir parameters.

By and large, our traditional approach to reservoir description has been rather qualitative. The focus must now be turned towards assessing the quantitative properties of the reservoir as a means of avoiding risky investments. This is indeed a continuous process that must go on throughout all phases from delineation to abandonment. It requires not only a highly selective attitude to data acquisition but also a multidisciplinary interpretation of these data. The importance of this new mind-set is evident in ongoing challenges to maximize recovery from producing fields.

In the past couple of years I have on several occasions had the pleasure of discussing these broad aspirations with Philippe Theys. His understanding of the challenges involved has been profound and dedicated. This book exemplifies this dedication. Not only is this book important as a professional contribution to the understanding of log acquisition and quality control, it is, indeed, a timely addition to the all-embracing challenge of appreciating and managing reservoir uncertainty.

Farouk Al-Kasim

Preface

After working for one year on the mechanical behavior of long molecules in a laboratory environment, it was with some admiration that I found that logging companies were attempting similar measurements in a less controlled and comfortable environment—several miles underground, at high pressure and temperature. The quality of the data is certainly a tribute to the ingenuity and adaptability of logging tools designers.

This achievement is surrounded by a large degree of mystery. Few documents describe the characteristics of the measuring devices, and much less so the uncertainty of their measurements. This paucity of information has helped develop the reputation of logging as an esoteric art, if not a branch of black magic open only to the cognoscenti. Logging procedures are often qualified as ritual and the data so derived is used following traditions that disregard technologic evolution and give little attention to the underlying physical processes that control data collection.

This is why my early inclination toward log interpretation has been gradually complemented by a concern for the meaning and relevance of the input data, later directed toward the development of methods to control log quality.

While many books deal with log interpretation, few address data acquisition. Little or no information from other sources is available to validate the data just acquired. In addition, it is necessary to judge the value and relevance of the data under extreme time pressure: each hour dedicated to log data acquisition is made at great expense, whereas subsequent log interpretation can be performed at a more leisurely pace. At the wellsite, the decision on accepting or refusing a log should be correct and fast.

Unlike other documents dealing with log quality control, this one excludes an exhaustive list of checks to be performed on specific tools. Such lists are available only from logging companies and need frequent updating. Once this information is obtained from the logging contractor, this book allows the building of a log quality control system.

This book summarizes information collected during numerous meetings with oil company representatives, many log quality audits and several seminars. Although some paragraphs would find application in other earth sciences, this writing concentrates on electrical wireline logging. Examples originate mostly from the field of formation evaluation, which performs the static description of a reservoir through the combination of logging and seismic surveys, cuttings and core analysis, and outcrop studies.

Philippe Theys

Part I

Premises

1
Introduction

1.1 Acquiring reservoir data in the eighties

A foremost problem in acquiring relevant data on reservoir rocks has been the conflicting motives of management and staff. Management focuses its energy on lowering costs and maximizing short-term return on investment, whereas technical staff is concerned with acquiring data that will have important implications five, ten or twenty years hence.[1]

In the cost-cutting environment of the eighties, justifying the value of what could be considered an extravagant luxury with little return was an even more difficult task. Under the pressure of the oil/gas companies, service companies lowered prices to maintain a presence on the market. Falling revenues stimulated strong control of costs, often with cuts in personnel, equipment, maintenance, and research and development. All these components impacted data quality and service efficiency.

Low service standards were evidenced directly by the failure rate of measuring equipment. Unfortunately, it was difficult to pinpoint problems in equipment performance and data quality, since data acquisition companies used to be secretive about the specifications of their equipment as well as their data quality control methods. Many oil and gas companies had a confidential internal system, but generally, these systems included little information on the specifications of the measuring devices.

---

[1] Refer to *Slip* Slider [XVII]. Each chapter contains a section citing relevant books and articles. They are referred to by Arabic numbers, for instance [12]. A general bibliography is located at the end of the book. These fundamental documents are denoted by Roman numerals, for instance [XVII].

1.2 Changes in the nineties

At the start of the nineties, the increasing number of technical papers on logging tool response, mathematical modeling, environmental effects and calibrations is showing that the past trend of secretiveness is gradually being replaced by a more open attitude of the logging companies. Reference manuals that describe the detailed quality features of the logs have been made available by most log suppliers.

Several conferences[2] have addressed the data quality issues. Workgoups[3] have been established to leverage the efforts necessary to understand logging measurements better. In addition, many multidisciplinary organizations have been formed to establish standards for the industry. Several methods (e.g., value analysis and data quality objectives[4]) have been imported from other business segments and have been further refined to yield a systematic and cost-effective approach to designing logging programs.

This cultural change has been paralleled by a technological explosion. Classical wireline logging in vertical wells is complemented by similar techniques in wells with complex trajectories, but now undertaken with drillpipe or coil-tubing conveyance.

The routine logs such as the "triple-combo" are being replaced by multidetector, multiarray devices. The logging suite is augmented by complex tools that provide imaging, azimuthal and nuclear magnetic resonance measurements, to quote a few modern techniques. It is fair to say that the intuitive understanding of logging that prevailed in the oil industry in the early days is now missing (*Fig. 1.1*).

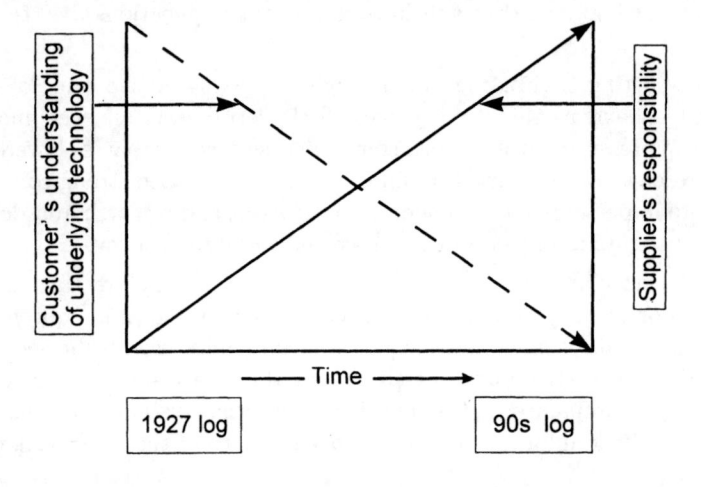

FIG. 1.1 : Evolution of customer and supplier behavior.
For all technology-intensive products, the understanding of the underlying technology by end-users has decreased with time, which leaves the onus on the suppliers to deliver controlled products.

---

[2] London, October 1993; Taos, October 1997.
[3] Among them, the Log Characterization Consortium.
[4] Refer to Chapter 3 for a summary of the Data Quality Objectives method (DQO).

The garbage-in, garbage-out syndrome fully applies to data collected in the field (from logs, cores, drilling reports, directional surveys, etc.) as input and to oil-in-place, production forecasts, return-on-investment as outputs. It is therefore critical to optimize data acquisition and to control the information it delivers.

1.3 What can be expected from log data?

All logs look the same. Although two logs may appear similar, the relevance and validity of the information they contain could differ vastly. The following questions address the qualification of data:

1. How does the borehole environment affect reservoir data? (192)[5]
2. How does precision vary with the range of the measurement? (118)
3. Is it useful to run a repeat section? (364)
4. Does a valid wellsite calibration record guarantee that the logging tool is functioning? (262)
5. Is the recorded depth reliable? (277)
6. How accurate is it? (278)
7. How reliable is the tool response in the range of interest? (159)
8. Is data filtered? How? Is signal processing affecting log analysis? (139)
9. How does measurement-while-drilling (MWD)[6] data compare with wireline data? (187)

Trivial questions also arise:

10. Is rain affecting tool calibration? (256)
11. Where are the centralizers located on the tool string? (333)
12. What is the size of the centralizers? (333)
13. Will we get any information out of the tape? (347)

1.4 What can be expected from this book?

This book dwells on quality control of log data and how quality control methods are related to the uncertainty of pore volume computation and to the level of confidence given to the log information used in decision-making. It begins (Part I) with a description of the jargon of the measuring industry and of the mathematical tools used to evaluate the way errors propagate. The components of data acquisition, from raw data to the beginning of the interpretation process are then listed exhaustively

---

[5]Questions 1 to 13 are answered on pages indicated by numbers in parentheses at the end of the questions.

[6]In this book, MWD relates to all measurements—directional surveys and formation evaluation data—acquired while drilling.

(Part II). Finally, methods of log quality control are exposed in detail (Part III). Exercises are scattered in the book to complement the practical chapters.

Many examples are extracted from *Schlumberger* literature. This does not mean the text only applies to *Schlumberger* logs or that quality control is a problem specific to *Schlumberger* data. It was easiest to obtain released information from *Schlumberger*. This textbook is in fact applicable to data acquired by all logging companies.

1.5 What cannot be found in this book?

1.5.1 Physics of logging measurements

The reader will not find details on the physics of the logging measurements. For details on logging detectors, refer to the listed literature:

- Knoll, G. F., *Radiation detection and measurement*, John Wiley and sons, New York, 1979.

- Serra, O., *Diagraphies différées: acquisition des donnés diagraphiques.*, Elf Aquitaine, Pau, 1983.

- Tittman, J., *Geophysical well logging*, Academic Press, Orlando, Florida, 1986.

1.5.2 Log interpretation

This book is not intended to be used as a reference on log interpretation. Recommended reading on log interpretation includes the following:

- Dewan, J. T., *Essentials of modern openhole log interpretation*, PennWell, Tulsa, 1983.

- Ellis, D. V., *Well logging for earth scientists*, Elsevier Science Publishing Company, New York, 1987.

- *Schlumberger log interpretation principles/applications*, Schlumberger Oilfield Services, Houston, 1995.

- Serra, O., *Fundamentals of well log interpretation*, Elsevier, Amsterdam, 1984.

- Serra, O., *Interprétation des données diagraphiques*, Elf Aquitaine, Pau, 1985.

2
Evaluation of hydrocarbon volume

> Information on the external world is acquired through observation and measurement. It is bound to be affected by errors.
>
> *Helstroem: Quantum detection and estimation*

In most countries, the volume of gasoline containers (*Fig. 2.1a*) is inspected on a regular basis by special regulatory departments. The expected error in this volume is of the order of 0.1%. Certificates (*Fig. 2.1b*) are then issued to authorize the use of such tanks.

It is impossible to hope for the same range of error for the volume of reservoir rocks. The traditional formula used to define hydrocarbon pore volume is

$$\text{Volume} = V_R \times \phi \times (N/G) \times (1 - S_w),$$

in which V_R is the volume of the reservoir, ϕ is the average porosity and S_w is the average water saturation. N/G is the net-to-gross ratio, the usable thickness of the reservoir as a fraction of the total thickness.

The prediction of V_R is based on a limited number of observations in wells, on measurements of seismic reflection times and on geological considerations. Because knowledge of the reservoir is limited, the prediction will be associated with some uncertainty. Evaluation of this uncertainty should be part of the prediction. A number of research projects [5, 21 and 22] have established a relationship between the expected uncertainty and the number of wells where observations are available (*Fig. 2.2*). However, very few papers have considered the inherent uncertainty of the measurements themselves.

8   2. Evaluation of hydrocarbon volume

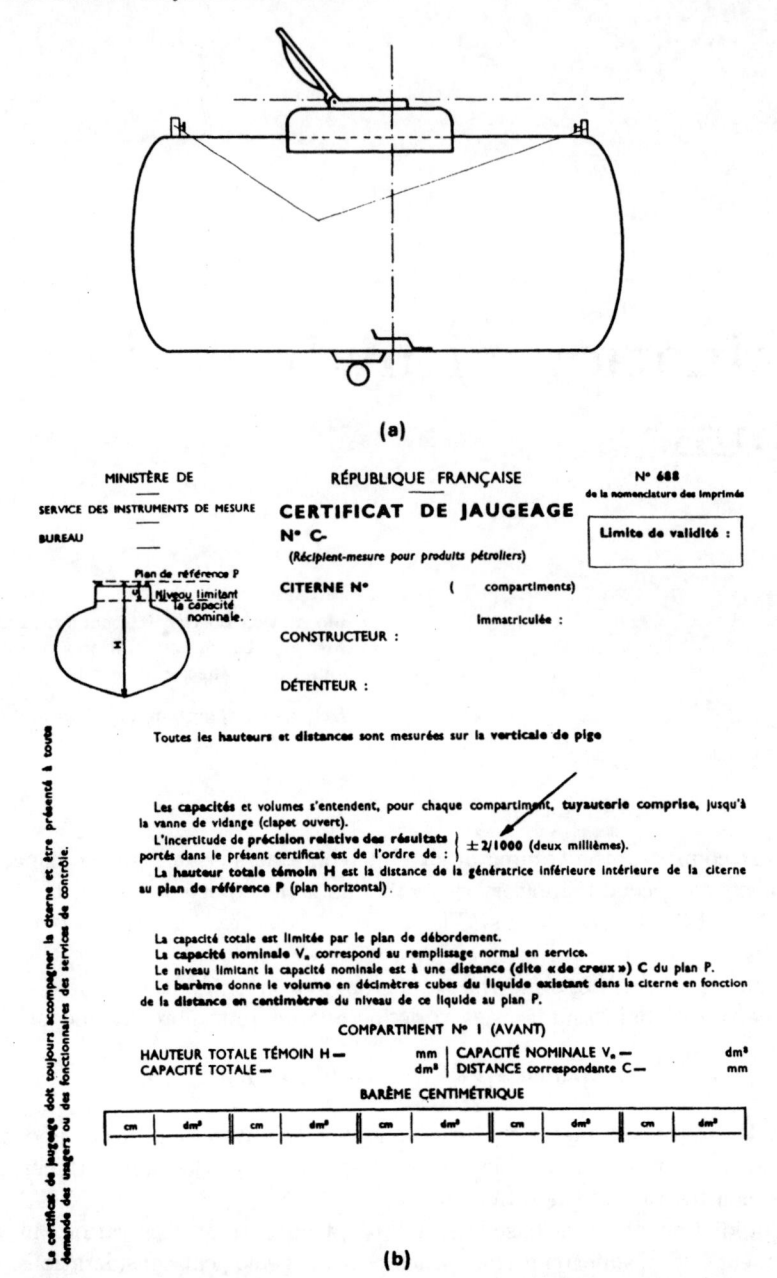

FIG. 2.1 : Volumetric characteristics of a gasoline container in France.
a. Drawing of the container.
b. Certificate for the container delivered by the French administration. Note that the volume uncertainty is reported in one-thousandths.

Courtesy of Technip [6]

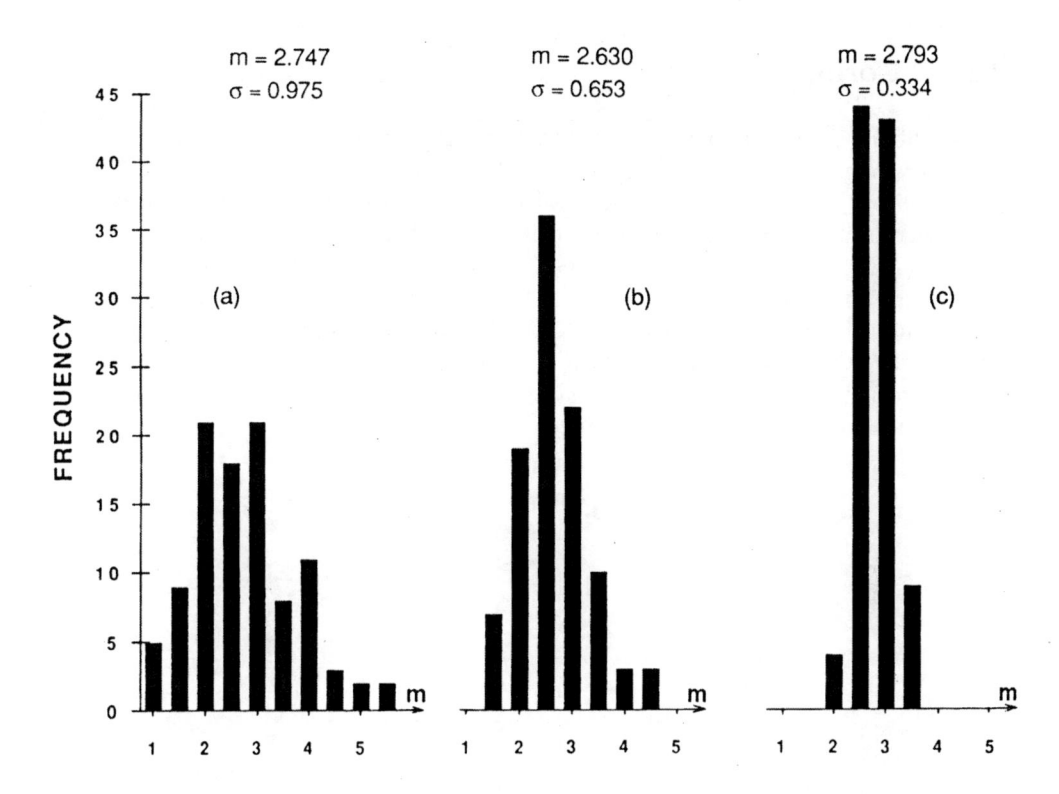

FIG. 2.2 : Reduction of hydrocarbon pore volume uncertainty. m is the oil-in-place estimate expressed in $10^7 \times m^3$. σ is the standard deviation expressed in the same unit. It is reduced three-fold as the number of wells goes from one (a) to six (c). The estimated volume derived from three wells in shown in the center (b).

Courtesy of the Norwegian Computing Center [21]

2.1 Inputs to evaluation of hydrocarbon volume

Logging, coring, well testing and seismic methods sample earth characteristics in a selective manner. To give some idea of the scarcity of information available on reservoir rocks, take as an example a reservoir with a simple configuration. Assume a field with a constant well spacing. One vertical well drains the equivalent of a cylinder with a radius of 0.5 km[1] (1640 ft). The reservoir is 100 m (330 ft) thick. The total volume of the reservoir drained by the well, including all solids and fluids, is 78.5×10^6 m^3 (2800 MMcf) and is supposed to be traversed by a 20.3-cm (8.5-in.) vertical borehole.

---

[1] A list of abbreviations is given in Appendix 1.

2.1.1 Coring

A fullbore, or "conventional," core drilled in a 20.3-cm (8.5-in.) hole has a diameter no larger than 10 cm (4 in.). Over 100 m (330 ft) this yields a volume of 0.785 m^3 (28 ft^3) of rock given an optimistic total recovery. This represents 10^{-6} percent of the reservoir. In most cases, plugs are drilled out of the fullbore core at a constant spacing (*Fig. 2.3*). Special analysis is performed and the formation characteristics of these plugs are measured. A generous assumption is 600 plugs taken over the 100-m (330-ft) interval. Plugs are generally 1.27 cm (0.5 in.) in diameter and 2.54 cm (1 in.) long, so that the total volume of plugs is 1.93×10^{-3} m^3 [0.068 ft^3]. If all the plugs were crunched together, they would make a cube with sides of 12.4 cm (5 in.). This corresponds to 0.25×10^{-10} times the volume of the reservoir.

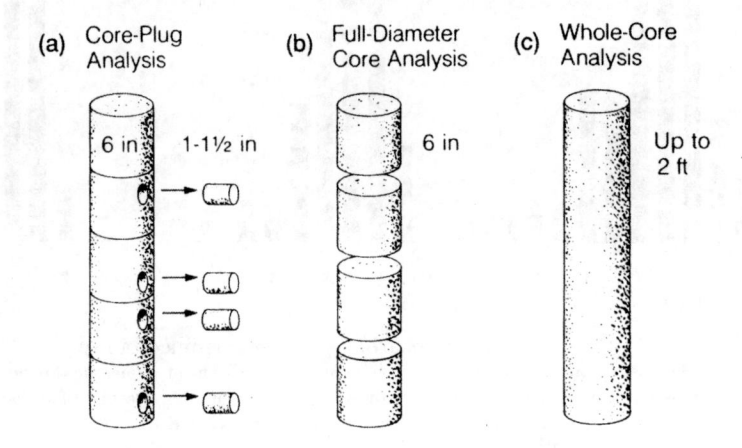

FIG. 2.3 : Core distribution in a well.
  a. Core plugs. Plugs are selected within intervals of 6 in. (15 cm). Most of the time, the distance between plugs is not constant.
  b. Full-diameter core.
  c. Whole core.
Courtesy of The Technical Review [3]

2.1.2 Conventional logging

One of the logging tools with the deepest investigation,[2] the dual laterolog, has a radius of investigation of 1.25 m (50 in.). The tool has a vertical resolution of 0.61 m (2 ft). In a single measurement, it investigates 3 m^3 (105 ft^3). Over 100 m (330 ft), the section surveyed corresponds to 490 m^3 (17×10^3 ft^3), or 6.25×10^{-6} times the volume of the reservoir.

---
[2] Detailed definitions of depth of investigation and vertical resolution are found in Chapter 5.

Similarly, the neutron tool is considered to investigate the formation 25.4 cm (10 in.) from the borehole wall, which corresponds to 0.12 m^3 (4.3 ft^3) if a 0.31-m (1-ft) vertical resolution is accounted for. Over the reservoir section, 40 m^3 (1460 ft^3) or 0.5×10^{-6} times the volume of the reservoir is scanned.

2.1.3 Deep-investigation logging

2.1.3.1 Deep electromagnetics

The only commercial deep electrical probing measurement is the *Ultra Long Spacing Electrical Log (ULSEL)*, a *Schlumberger* service. The spacing of the electrodes is variable with an upper limit at 1000 m (3300 ft). The *ULSEL* service is used to map electrical anomalies, for instance, a salt dome or the casing of a blowout well as seen from a relief well [12 and 23]. The maximum depth of investigation of this survey is 500 m (1600 ft) and the resolution of the device is poor, from 100 m (330 ft) to 1000 m (3300 ft).

Studies of crosswell tomography by electromagnetic propagation have been made at frequencies from 1 to 30 MHz. Long range electromagnetics have the potential to bring complementary information on the presence and location of geological features such as faults, fractures, karsts and lumps [26].

2.1.3.2 Borehole gravimetry

A gravimetry survey [10 and 11] measures formation bulk density over a radius of 15 m (50 ft) at given stations. Considering the large volume of investigation, borehole, mudcake and rugosity effects are small and can be neglected. The vertical resolution is mediocre, probably 3 m (10 ft).

2.1.3.3 Sonic imaging

Sonic imaging [13] uses a sonic array with one transmitter and many receivers. It can be schematized as a small scale seismic streamer moving in a well. Sophisticated signal processing allows imaging of events up to 15 m (50 ft) away from the well. *Fig. 2.4* shows the acoustic ray path that delineates beds of different acoustic impedances. When the tool is below the impedance boundary, the acoustic energy returning to the receivers of the tool originates only from the downdip portion of the boundary. Conversely, with the tool above, the energy comes back only from the updip part. *Fig. 2.5* is an image produced by the tool. Acoustic impedance changes 15 m (50 ft) from the borehole axis are visible.

In fractured formations, low-frequency acoustic waves, typically at about 500 Hz, generate signals representative of volumes of formation 10 m (30 ft) from the borehole. The actual depth of penetration depends on fracture size, fluid viscosity and fracture roughness.

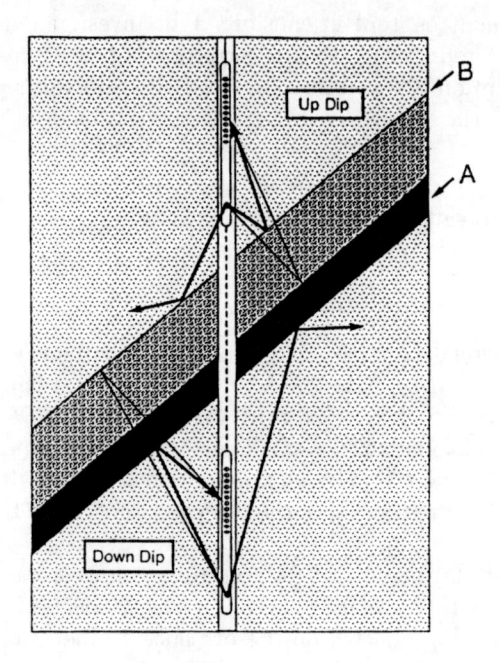

FIG. 2.4 : Sonic imaging tool.
Acoustic ray paths from and to the sonic imaging tool.
Courtesy of the Geophysics [13]

2.1.3.4 Relief well detection: magnetic surveys

Observation of magnetic anomalies caused by tubulars can be performed up to 30 m (100 ft) from a downhole magnetometer located in the openhole section of a well. This technique, explained in detail in Chapter 5.5 of reference [VI], is mostly used to detect the casing of a blowout well from a relief well.

2.1.4 Well testing

The main objective of well testing is to determine whether a formation can produce hydrocarbons economically [17]. Essential information, such as formation permeability, degree of borehole damage, reservoir pressure, presence of reservoir boundaries and heterogeneities, can be derived concurrently from the continuous recording of bottomhole pressure during a well test. In the context of this chapter, well testing can be perceived as a powerful way to scan the formation at a large distance from the well.

The radius of investigation of a well test is used to design a test that will ascertain the presence of formation features (for instance, faults or massive heterogeneities) and characterize them. The a priori evaluation of the radius of investigation provides a guide on test duration and also on the successive sequences of flows and shut-ins.

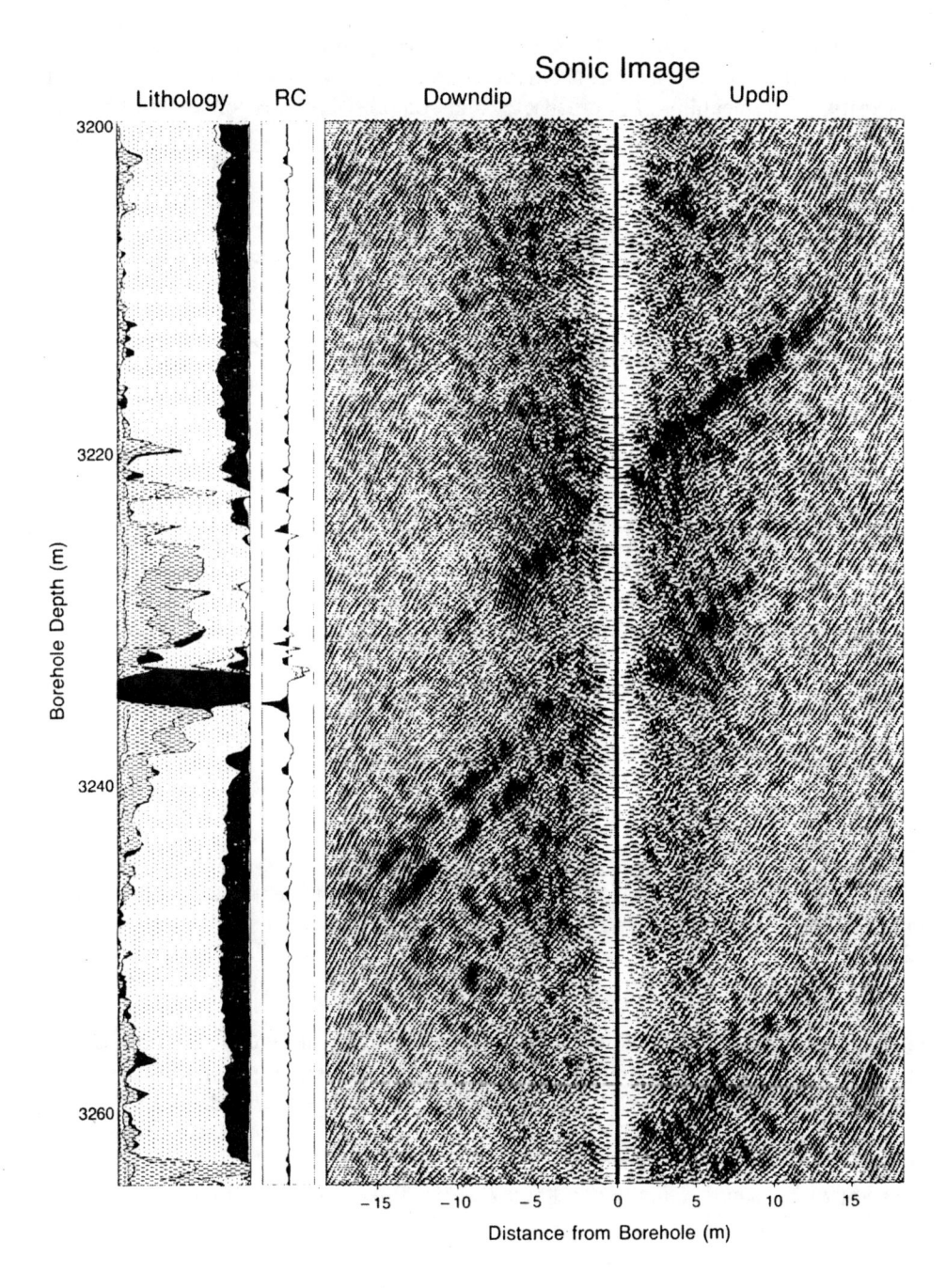

FIG. 2.5 : Sonic imaging tool.
Image of the formation obtained from the sonic imaging tool.
Courtesy of the Geophysics [13]

The diffusivity equation [XVII] shows that a pressure disturbance created at the well reaches a distance r_i, the radius of investigation, after a time t. t and r_i are linked to viscosity, μ, permeability, k, porosity, ϕ, and compressibility, c_t, of the fluid by the equation:[3]

$$r_i = \sqrt{\frac{k \times t}{a \times \phi \times \mu \times c_t}}$$

Numerical application

The radius of investigation of an 80-h test in a formation with the following characteristics is 2000 ft (600 m):

$k = 100$ mD
$\phi = 0.2$ (20 pu)
$c_t = 2 \, 10^{-5}$ psi^{-1}
$\mu = 0.5$ cP

The equation above shows that only t can be controlled by the operating company. Therefore, in principle, it could be possible to investigate the formation at any distance, if the well test duration is sufficiently long. In practice, however, two additional parameters come into play, the flow rate and the pressure gauge accuracy.

2.1.5 Seismics

2.1.5.1 Surface seismics

Surface seismics can scan large volumes of underground formations [8 and 19]. The depth of investigation is limited by the source energy but is generally adequate for shallow- to medium-depth reservoirs. The weak point of seismic data is its resolution. For most purposes, the resolution is taken as one-quarter of the wavelength of the field produced by the seismic source. Because attenuation of seismic energy increases exponentially with frequency and distance, only low frequencies will reach the target at a large depth. The same attenuation can be observed for a 20-Hz wave down to 7010 m (23,000 ft) and to a 2000-Hz wave to 70 m (230 ft). In addition, the earth behaves as an effective filter for high frequencies so that an optimistic upper limit of the frequency range of the signal received is about 100 Hz. This means that for a formation with a sound velocity of 6000 m/s (20,000 ft/s), the maximum resolution is 15 m (50 ft) (*Fig. 2.6*).

---

[3] a is a constant with a value of 948 if k is in mD, t in h, μ in cP, c_t in psi^{-1}, and ϕ such that ϕ equals 1 for 100 pu.

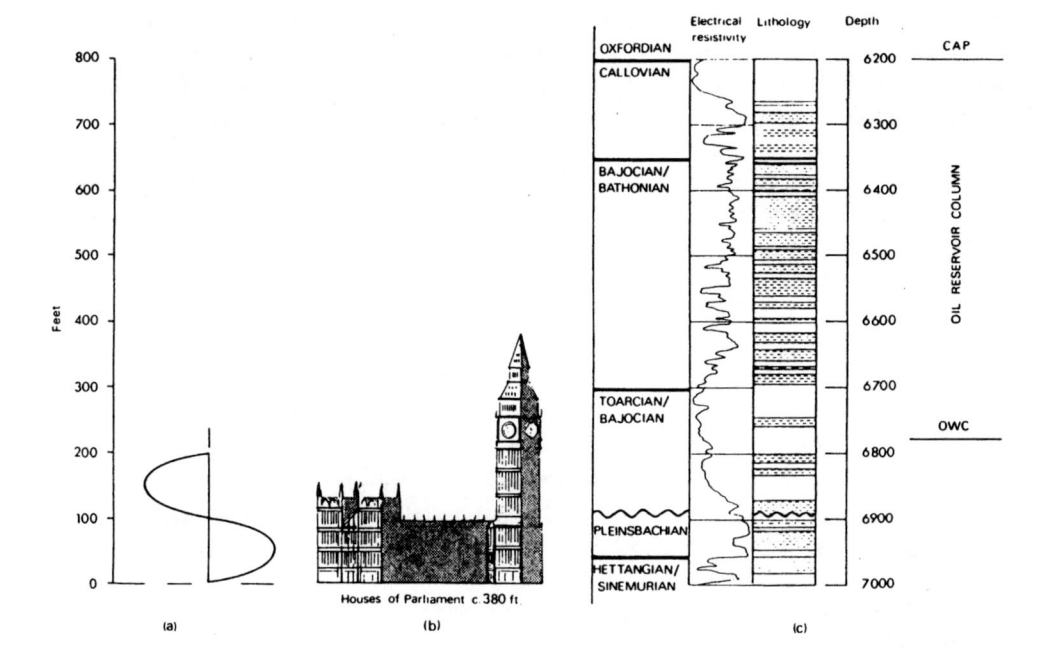

FIG. 2.6 : Vertical resolution of surface seismics.
  a. A single sine wave of 30 Hz in medium of velocity 6000 ft/s (or 60 Hz-12,000 ft/s).
  b. Big Ben, London.
  c. Geophysical logs through the Beatrice oil field.
  Courtesy of Graham and Trotman [19]

Horizontal resolution should also be carefully analyzed. Because of the vibratory nature of the seismic phenomenon, a reflection does not originate from a point on a reflector but from a zone. Individual contributors from within this zone are smeared together and cannot be separated. The radius of this Fresnel zone is computed as

$$\text{Fresnel radius} = \frac{v}{4} \times \sqrt{\frac{t}{f}},$$

in which v is the average velocity to the reflector, t is the two-way time and f is the center of the frequency content of the reflection.

Numerical application

At a depth of 3000 m (10,000 ft), assuming a center frequency of 40 Hz and a sound velocity of 6000 m/s (20,000 ft/s), the Fresnel radius is 240 m (800 ft).

Surface seismics is therefore of little use to map a field with a 16-ha (40-acre) cell, which corresponds to a well spacing of 400 m (1300 ft). Limits are set when the vertical resolution is thicker than the pay zone and the horizontal resolution is wider than the well spacing. For this reason, surface seismics is a great tool for exploration but not for reservoir development unless the reservoir is shallow, big and homogeneous.

Another fundamental problem with surface seismics resolution relates to the distance from the source and the frequency of data. This may be a serious limitation because attenuation is higher at high frequency. Shallow formations, especially weathered rocks, are strong attenuators and introduce large uncertainties in the depth determination of the deeper reflectors.

2.1.5.2 Borehole seismics

In a vertical seismic profile (VSP) [20], an acoustic source is triggered close to surface and the resulting vibration is measured downhole by one geophone or an array. This technique has several advantages over surface seismics:

(a) The energy detected and analyzed has only traveled one way—down. This results in less attenuation and preservation of a higher frequency content.
(b) The depth of the geophone(s) is precisely controlled and correlated with other logs so that an accurate time-depth conversion is possible.
(c) Multiples can be positively identified as well as the layers generating them.
(d) From the downgoing wave field, attenuation of seismic energy can be modeled, and the model used to guide the design of deconvolution filters applied to the upgoing wave field.

The surface source can be moved to optimize the way the reflectors are illuminated by the technique. Several source locations are used in offset VSP and walkaway surveys. They allow the completion of a local seismic survey with the advantages mentioned above. *Fig. 2.7* shows one way to integrate borehole and surface seismic data, along with core and log data.

2.1.5.3 Crosswell seismics

In a crosswell seismics setup [1, 14 and 15], sources and geophones are located downhole in separate wells. A large portion of the high-frequency content of the source is preserved, which implies that the source must have a larger spectrum width than conventional sources.

The important attributes required for downhole sources are directivity, high energy and the feature of causing no damage to the hole. The seismic resolution loss is related to the dispersion of the high-frequency content of the incident seismic energy. This occurs to a large extent in shallow formations and more particularly in the weathered zone. An advantage of crosswell seismics is that it allows placing the source and receiver under these highly absorbing zones. For most practical purposes, the resolution of crosswell seismics is 1.5 m (5 ft). It is hoped that, with this technique, information on zones 500 m (1500 ft) from the observation well will be collected.

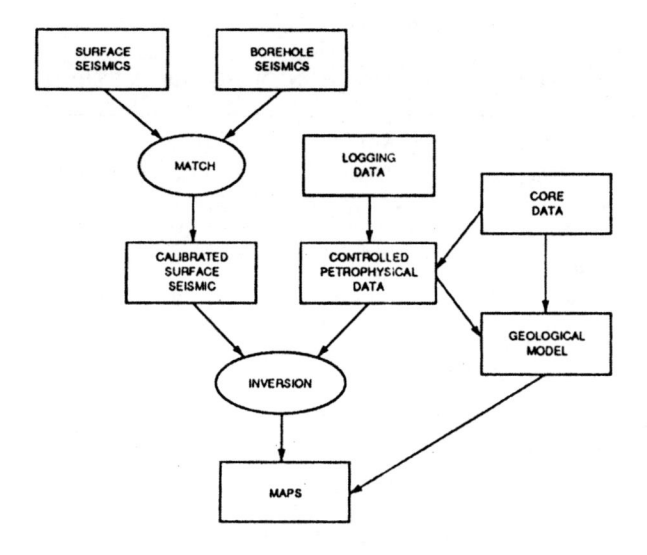

FIG. 2.7 : A possible scheme of integrated borehole/surface seismics.

2.1.6 Summary

The volumes of investigation of the different seismic methods are illustrated in *Fig. 2.8* and listed in *Table 2.1*.

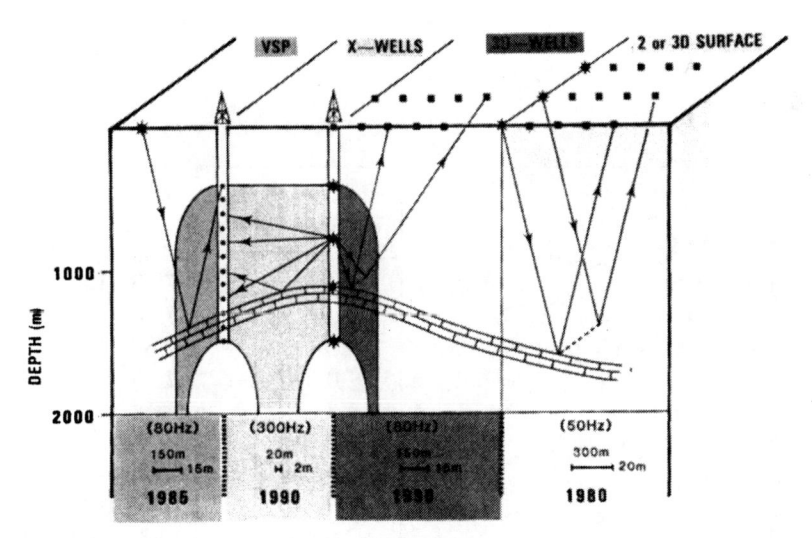

FIG. 2.8 : Volume considerations of the different seismic methods.
Courtesy of P. Simandoux [25]

TABLE 2.1
VOLUMES OF INVESTIGATION OF SEISMIC METHODS

Seismic method	Maximum frequency (Hz)	Vertical resolution (m)	Vertical resolution (ft)	Lateral resolution (m)	Lateral resolution (ft)
Surface	50	30	90	300	1000
Borehole	100	15	50	150	500
Crosswell	1000	2	6	20	60

Depth of investigation and resolution of earth surveying techniques are gathered in *Table 2.2*. The respective volumes investigated are listed in *Table 2.3*.

TABLE 2.2
COMPARISON OF RESOLUTION AND DEPTH OF INVESTIGATION

Method	Vertical resolution (m)	Vertical resolution (ft)	Depth of investigation (m)	Depth of investigation (ft)
Cores	10^{-6}	10^{-6}	10^{-1}	0.3
Logs	10^{-2} to 2.5	10^{-2} to 10	10^{-2} to 3	10^{-2} to 10
Deep-investigation logs	0.5 to 5	1 to 15	20	60
Well tests	Perforated interval	Perforated interval	10^3	$3\cdot 10^3$
Seismics	2 to 30	6 to 90	10^3	$3\cdot 10^3$

TABLE 2.3
COMPARISON OF EARTH SURVEYING TECHNIQUES

	Portion of the reservoir volume	Example (m^3)
Reservoir	1.0	$78.5\cdot 10^6$
Surface seismics	1.0	$78.5\cdot 10^6$
Vertical seismic profile	0.1	$7.85\cdot 10^6$
Gravimetry log	$1.0\cdot 10^{-3}$	$78.5\cdot 10^3$
Dual laterolog	$6.3\cdot 10^{-6}$	490
Neutron porosity log	$5.0\cdot 10^{-7}$	40.0
Cuttings	$4.1\cdot 10^{-8}$	3.24
Fullbore core	$1.0\cdot 10^{-8}$	$7.85\cdot 10^{-1}$
Plugs	$2.5\cdot 10^{-11}$	$1.93\cdot 10^{-3}$
One plug	$4.2\cdot 10^{-14}$	$3.20\cdot 10^{-6}$

2.2 Continuity of geological formations

After adding the information provided by the different geophysical methods, it could be felt that not much data is available to assist reservoir development. Nature is softening this limitation, however, as the shape of the reservoir is not totally random. In fact, fewer than a dozen families of reservoir shapes [24] are found. They depend mostly on the environment of deposition, on diagenesis and on subsequent tectonism.

For a given environment, there are strong constraints on the relation between the dimensions of rock bodies. For instance, in gross terms, it could be said that in a given geological province, sand channels have width, height, and length linked [2] by

$$\text{Length} = a \times \text{Width} = b \times \text{Height},$$

where a and b are constant. This relation is depicted in *Fig. 2.9* with $a = 10$ and $b = 100$. The previous equation is by no means general. Where turbidity currents have taken place, the reservoir would be thinner but might occupy several square kilometers (*Fig. 2.10*). This type of environment of deposition can be easily recognized by a specific signature on cores and logs: the column of thin sediments is arranged in a characteristic way, called the Bouma sequence (*Fig. 2.11*). Data collected at the well location could therefore be diagnostic. Tiny volumes of the formation, as small as a tenth of a billionth part of the reservoir such as a core plug, may bring as diverse information as depth and precise environment of deposition through observation of fossils.

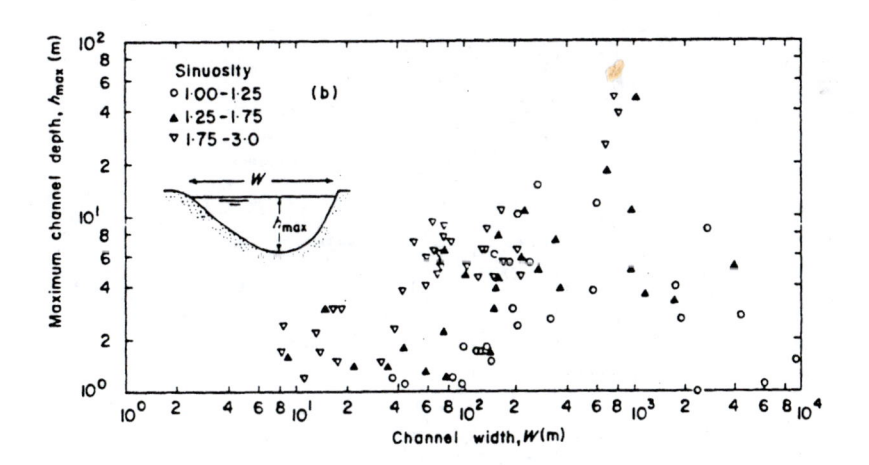

FIG. 2.9 : Geometrical characteristics of a channel.
Courtesy of Elsevier [2]

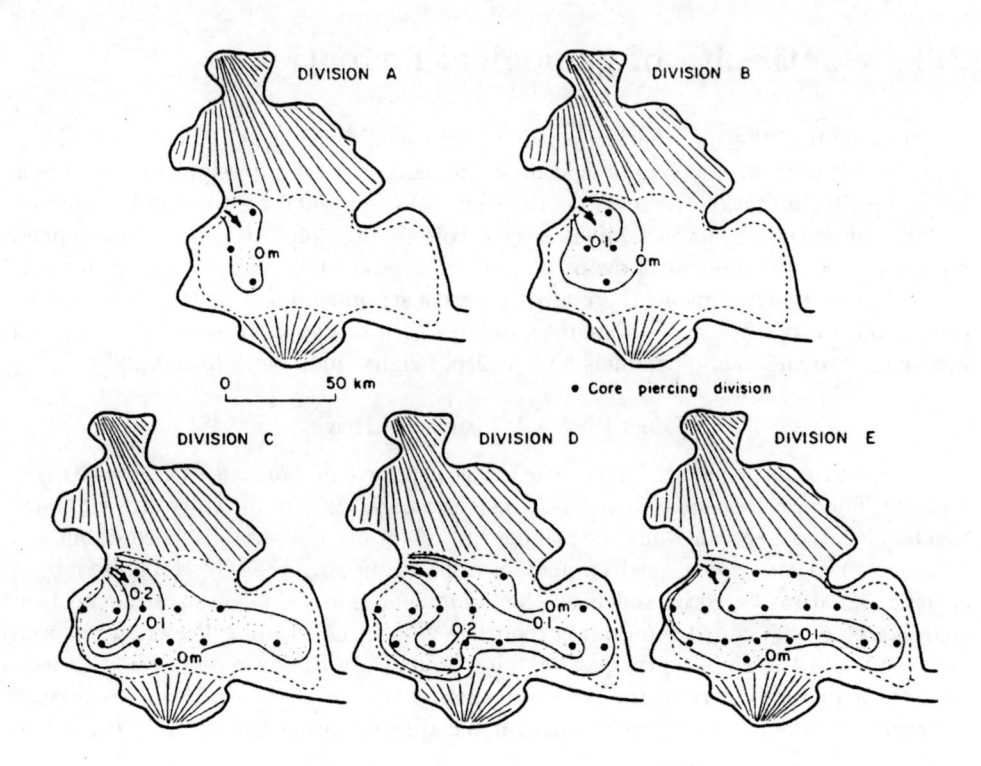

FIG. 2.10 : Lateral extension of a turbidite.

Courtesy of Elsevier [2]

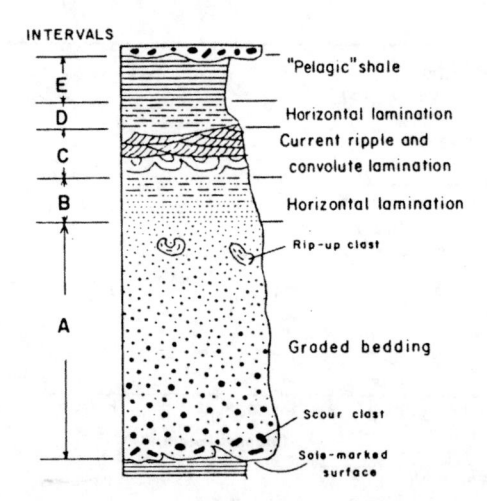

FIG. 2.11 : Vertical development of a Bouma sequence.

Courtesy of Journal of Sedimentary Petrology [27]

The core is like a chromosome, which characterizes a living being though its volume may be less than a billionth of the total volume of the total body. The same way as the recognition of a cat chromosome leaves no ambiguity on the type of animal to which it belongs, the recognition of the dinoflagellate Areoligera senonensis diagnoses the environment of deposition as Early Tertiary marine [4].

Environments of deposition establish strong constraints on reservoir shape. In the same time, the observation of outcrops shows that homogeneity is not the rule. Finding the right balance between assuming too much continuity and integrating too many details is the challenge of the explorationist [7, 9 and 16].

2.3 Combining information

When combining information from different sources [18], there is often some overlap, generally used for calibration of lower resolution information. Crosschecking between methods is advocated but an attempt to reconcile data too exactly can be detrimental. For instance, core and log information are often compared, yet both are subject to a number of uncertainties. Forcing one data set to fit the other exactly may lead to incorrect evaluation of the formation.

References

1 Albright, J. N., Terry, D. A., Bradley, C. R., "Pattern recognition and tomography using crosswell acoustic data," SPE 13854, *Trans.* SPE/DOE low permeability gas reservoirs symposium, Denver, 1985.

2 Allen, J. R. L., *Developments in sedimentology, sedimentary structures, their characteristics and physical basis*, Elsevier, Amsterdam, 1982.

3 Basan, P., Hook, J., Hughes, K., Rathmell, J., Thomas, D., "Measuring porosity, saturation and permeability from cores: an appreciation of the difficulties," pp. 22-36, *The Technical Review*, Vol. 36, No. 4, 10-1988.

4 Bell, D. G., Bjorøy, M., Kilenyi, T., Grogan, P., Selnes, H., Trayner, P., "Better prospect evaluation with organic geochemistry, biostratigraphy and seismics," pp. 24-42, *Oilfield Review*, Vol. 2, No. 1, 1-1990.

5 Berteig, V., Halvorsen, K. B., Omre, H., Hoff, A. K., Jorde, K., Steilein, O. A., "Prediction of hydrocarbon pore volume with uncertainties," SPE 18325, *Trans.* SPE 63rd annual technical conference, Houston, 1988.

6 Defix, A., *Mesurages des volumes des carburants et combustibles liquides*, Éditions Technip, Paris, 1975.

7 Delfiner, P., Delhomme, J. P., Pélissier-Combescure, J., "Application of geostatistical analysis to the evaluation of petroleum reservoirs with well logs," paper WW, *Trans.* SPWLA 24th annual logging symposium, 1983.

8 Dobrin, M. B., Savit, C. H., *Introduction to geophysical prospecting*, McGraw-Hill, New York, 1988.

9 Edmundson, H. N., "Trends in reservoir management," *Oilfield Review*, Vol. 4, No. 1, 1-1992.

10 Gournay, L. S., "Application of borehole gravimetric techniques to determine oil saturation," U.S. Patent 4399693, *Mobil Oil Corp.*, 23-08-1983.

11 Gournay, L. S., Lyle, W. D., "Determination of hydrocarbon saturation and porosity using a combination borehole gravimeter (BHGM) and deep investigating electric log," paper WW, *Trans.* SPWLA 25th annual logging symposium, New Orleans, 1984.

12 Haanshoten, G. W., "Ulsel logging in blow-out relief wells," pp. 23-27, *The Log Analyst*, Vol. 18, No. 1, 1-1976.

13 Hornby, B. E., "Imaging of near-borehole structure using the full-waveform sonic data," pp. 747-757, *Geophysics*, Vol. 54, No. 6, 6-1989.

14 Iverson, W. P., "Crosswell logging for acoustic impedance recovery," pp. 75-82, *Journal of Petroleum Technology*, Vol. 40, No. 1, 1-1988.

15 Kretzchmar, J., Kibbe, K. L., Witterholt, E. J., "Tomographic reconstruction techniques for reservoir monitoring," SPE 10990, *Trans.* SPE 57th annual technical conference, 1982.

16 Lasseter, T., "The reservoir: heterogeneities and modeling," pp. 39-49, *The Technical Review*, Vol. 32, No. 1, 9-1984.

17 Lee, J., *Well Testing*, SPE textbook series, Vol. 1, New York, 1982.

18 Lorenz, J., "Net value of our information," pp. 499-503, *Journal of Petroleum Technology*, Vol. 40, No. 4, 4-1988.

19 Mc Quillin, R., Bacon M., Barclay W., *An introduction to seismic interpretation*, Graham and Trotman, London, 1979.

20 Mons, F., Babour, K., *Vertical seismic profile recording, processing and applications*, Schlumberger, Paris, 1981.

21 Omre, H., "Quantifying the uncertainty in estimates of the in-situ oil/gas volume in a reservoir - outline of methods," Norsk Regnesentral, 1985.

22 Omre, H., Halvorsen, K. B., Berteig, V., "Prediction of hydrocarbon pore volume in petroleum reservoir," Norsk Regnesentral, 1987.

23 Runge, R. J., Worthington, A. E., Lucas, D. R., "Ultra long spaced electric log (Ulsel)," paper H, *Trans.* SPWLA 10th annual logging symposium, 1969.

24 Selley, R. C., *Ancient sedimentary environments*, Chapman and Hall, London, 1970.

25 Simandoux, P., "Future trends in oil and gas recovery," Vista conference on fundamental oil and gas research, Stavanger, 1988.

26 Spies, B. R., Habashy, T. M., "Sensitivity analysis of crosswell electromagnetics," *Geophysics*, Vol. 60, No. 3, 5-1995.

27 Stanley, D. J., *Journal of Sedimentary Petrology*, Vol. 33, 1963.

3

Data collection and decision-making

> The object of taking data is to provide a basis for action.
>
> W. Edwards Deming, 1938

Data is similar to mail. Some of it is junk: it has no value, but may clutter a large space in a database. Some is collected to provide a long-lasting record and is directed to an archive. Finally, some data is acquired so that the user may take action.

While the previous chapter dealt with how data can be transformed into long-term asset evaluation, this chapter considers the use of data acquisition for decision-making.

Field examples

The objective of collecting log data may be to answer questions by "Yes" or "No."

- Is the log data collection successful? Was all required information collected?
- Is the MWD tool still functioning correctly? Is the downhole battery dead?
- Is the cement quality satisfactory? Is a squeeze job required?
- Is the mud weight too high (with the risk of fracturing the formation)?
- Is the mud weight too low (with a risk of blowout)? Should weighting agents be added? Which ones and how much?
- Has the top of a specific formation been reached? Is it time to start coring or to set casing?

- Does the well need to be steered down? Up? Left? Right?
- Is the borehole staying in the high-porosity fractured zone?
- Has the well crossed the fault?
- Are these two wells hydraulically connected?

3.1 Requirements of decision-making

3.1.1 Timeliness

In decision-making, time is of the essence. For instance, the MWD information acquired in real time and made available to the user through a hydraulic or electromagnetic telemetry has value only in the time before the logging equipment comes back to surface and the memory data is accessed.

Timeliness requires that the **correct** amount of data is made available to the decision-maker. Too little data leads to poorly informed decisions, while too much leads to tardy decisions (especially if the data flow is limited by the telemetry). Compression and decimation schemes are critical to the delivery of the correct amount of data.[1]

3.1.2 Accuracy

Measurement accuracy may be sacrificed to some extent to improve timeliness. Collecting data that is not completely corrected because of limited telemetry bandwidth may be appropriate if this data is available early.

Still, accuracy cannot be completely sacrificed. Correctness becomes critical when the measurement is close to a decision threshold.

Example

A porosity measurement is assumed to have a constant uncertainty[2] of 3 pu over the entire measuring range. In a high-porosity sand, the tool reads 27 pu. From a different source, the oil company knows that the field can be developed if porosity is above 22 pu. With the largest uncertainty, the measurement still lies above the economic limit. Conversely, in a low-porosity carbonate of 3 pu, the measurement may read between 0 and 6 pu, from a respectable reservoir to nothing at all.

---

[1] This could be achieved by sacrificing resolution. The compression of a 256-kbytes picture to 2 kbytes eliminates some details but does not prevent the user from recognizing the silhouette of the model.

[2] Uncertainty is defined in Chapter 4.

3.1.3 Cost-effectiveness

Collecting a large amount of data may facilitate decision-making, but in a totally uneconomical way. Also, adding more information may not be a guarantee of increased knowledge [7]. A balance between completeness and complexity must be struck.

3.1.4 Planning

While measurements are available only when logging, large amounts of information are already available a long time before data collection. Critical thresholds and measurement specifications fall in this category. In addition, rough geometrical shapes are known through surface seismic and studies of basin geology. Information from nearby wells is available in some cases. From these sources, it is possible to estimate the values that will be measured. The user is able to predict if measured values are likely to be close to decision thresholds and hence select the measurements whose specifications enable successful decision-making.

Example

From the previous example about low-porosity carbonates, it is concluded that the 3-pu uncertainty is unacceptable. Alternatives are possible, but they need to be decided before the job is performed. They include

- changing the measurement type (e.g., using density-derived porosity instead of neutron-derived porosity)
- changing the tool type (e.g., spectral instead of global)
- changing the calibration procedure
- changing the logging speed
- changing the filtering
- changing borehole and mud conditions
- changing the measurement company!

3.2 Data quality objectives method

The previous section includes a few recommendations about decision-making. Some organizations, including the United States Environmental Protection Agency (EPA), have developed the data quality objectives (DQO) method [1].

The DQO methodology can be readily applied to the oilfield environment as shown in Sections 3.2.3, 3.2.4 and 3.3.

3.2.1 Description of the method

The DQO method attempts to answer the following questions:

- What is the problem and the desired result?
- Is the problem important?
- How is the collected information to be used to resolve the problem?

There are seven core steps:

- State the problem.
- Identify the decision.
- Identify inputs to the decision.
- Define the study boundaries.
- Develop a decision rule.
- Specify limits on decision errors.
- Optimize data collection.

Each step is strongly dependent on a rigorous completion of the previous steps. For instance, if the problem is poorly stated, there is little chance that a proper decision rule will be established, and hence, it will be impossible to manage the uncertainties adequately and optimize acquisition design.

3.2.2 Example of DQO in the field of environmental protection

State the problem: Residents near a given site are concerned about contamination of an aquifer by the chemical agent AA.
Identify the decision: Determine whether AA concentration is above a limit established as safe.
Identify the inputs to the decision: AA concentrations are sampled in two wells.
Define the boundaries of the study: Data from the two wells is considered to be relevant to assess the aquifer.
Develop a decision rule: If the lower concentration of the two wells is above a level c_1, then the site is assessed as contaminated. If the larger concentration of

the two wells is below a level c_2, smaller than c_1, then the site is said to be free of contamination. In other cases, additional measurements are needed. This situation is called the gray region.

Specify limits on decision errors: Measurement errors have an upper limit defined by the specifications of the measuring equipment and procedures. Decision levels c_1 and c_2 take into account the maximum value of the error.

Optimize the design: Select the best (cost-effective but sufficient) sampling and analysis. Document the design for further reference and certification.

Wireline logging techniques have been adapted to facilitate decision-making in potentially contaminated sites [2].

3.2.3 Hydraulic separation with a cement job

State the problem/Identify the decision: One simple and classical question raised from a cement job is: Is the quality of the cement good enough to ensure that zone A is separated hydraulically from zone B?

Identify the inputs to the decision: The question can be answered by running a cement bond log or similar measurement. But, how is it possible to relate log readings to the decision that the cement quality is adequate?

Two parameters are used:

- the amplitude of the acoustic log
- the length of the interval where amplitude reaches the desired value.

Define the boundaries of the study: The decision thresholds are very specific to the casing under consideration [6]. In an example, it could be a 7-in., 29-lbm/ft casing. For any other conditions, a new rule would apply.

Develop a decision rule/Specify limits on decision errors: From these two parameters, a decision rule can be developed: **If** the acoustic amplitude is less than 9.4 mV for over 10 ft (3 m), **then** cement quality is acceptable.

Optimize the design: Managing uncertainty will be achieved by verifying calibrations, ensuring adequate tool positioning, that the repeat section reasonably matches the main log and that logging speed conforms to the tool specifications.

3.2.4 Calculating the length of the producing drain

Optimization of well design can be guided by simple modeling, the collection of the right amount of data to support and check the reservoir model, and the establishment of clear decision rules.

State the problem: In a specific example [3], a North Sea company optimized well trajectories to comply with the following constraints:

- The producing zone has a 20-m standoff from the oil-water contact (OWC).

- The well capacity exceeds that of the electric submersible pumps used in the completion.

- The well eventually penetrates the OWC to serve as additional depth control, enabling future monitoring of OWC changes and later conversion to water injection.

Identify the decision: Once it is verified that sufficient well productivity will be attained, then the well angle is dropped to reach the OWC.

Identify the inputs to the decision: Real-time MWD density log and grain size from sieve analysis of cuttings are used to derive formation porosity (ϕ). Permeability k using permeability/porosity transforms that have been previously established from log-core comparison is then derived from ϕ. A vertical well productivity index is calculated with the standard reservoir parameters.

Develop a decision rule: **When** well productivity exceeds 30 b/d/psi, **then** the well angle is dropped.

Several wells have been drilled using the previously described approach. Further optimization of well design has resulted in significant cost savings and enhanced recovery.

3.3 Other decision-making examples

3.3.1 Quality of the data required for quantitative evaluation

A complete quantitative formation evaluation may take days or weeks. Still, assessing that the acquired data will be useful requires a much shorter time and needs to be performed much earlier than the evaluation itself. This decision should be taken before the measurement company has rigged down the equipment and, especially, before irreversible actions, such as setting casing, are completed.

3.3.2 Hydraulic continuity between two wells

State the problem: Two wells are drilled in the same field. Each has a pay zone at approximately the same depth. Are these pay zones in hydraulic communication?

Identify the decision: If at the same true vertical depth, the two wells display the same pressure, it can be reasonably established that the two wells are communicating. If the pressures are significantly different, they are not. The key issue is to define what **significantly different** means.

Identify the inputs to the decision: Reservoir pressures can be measured with formation testers. They provide two parameters:

- the true vertical depth computed from measured depth and a directional survey
- the formation pressures.

The data is plotted (*Fig. 3.1*) so that fluid (oil, water or gas) contacts are easily identified. Depth difference ΔD and pressure difference Δp are extracted from this plot.

Develop a decision rule: **If** ΔD is smaller than the reasonable depth discrepancy expected at the depth of the reservoir for two separate depth measurements [8] **and** Δp is smaller than the sum of the expected pressure gauge inaccuracies, **then** the two zones are in communication.

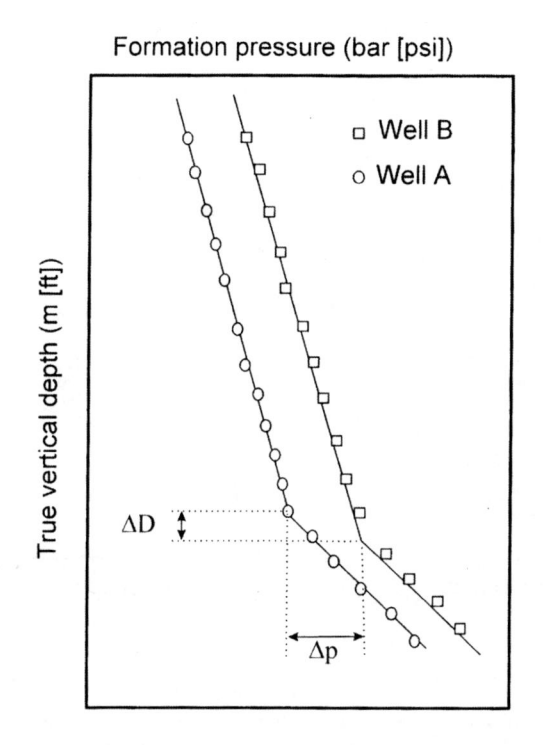

FIG. 3.1 : Test of hydraulic continuity between two wells.
Courtesy of SPWLA, adapted from [9]

Example [9]

$$\Delta D = 2.21 \text{ m } (7.25 \text{ ft}) \qquad \Delta p = 0.80 \text{ bar } (11.6 \text{ psi})$$

The difference of 2.21 m is within the expected combined absolute depth accuracy between the two wells. However, the combined expected inaccuracies on the pressure gauges are 0.35 bar (5.1 psi). The method then suggests that the two pay zones are not in communication. This was confirmed by PVT analysis.

3.3.3 Geosteering

Wells require increasingly complex trajectories in order to position longer sections of the drainhole in layers of highest producibility.

Originally, wells were drilled according to the well plan. Generally, little modification to the plan was carried out because limited downhole geological information was available except for cuttings and torque and drag measurements. The advent of MWD that provides real-time acquisition of downhole geological information has changed the perspective. It is now possible to deviate from the original plan as fresh information validates (or invalidates) the geological prognosis. *Figure 3.2* shows a simplified approach to geosteering, the technique that uses MWD to steer the well.

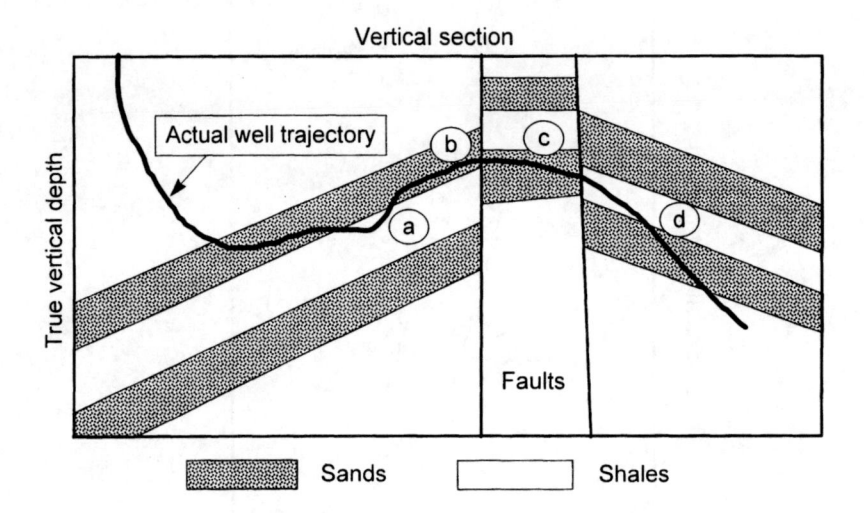

FIG. 3.2 : Geosteering example; the well path is the result of the decisions:
  a. The gamma ray curve indicates the well is in a shaly zone. Drilling proceeds by steering upward.
  b. According to the geological model, a fault is about to be crossed. The drilling is steered down.
  c. The previous assumption is confirmed as the gamma ray curve indicates the zone to be a sand zone. No change of direction is performed.
  d. The well has crossed a second fault and is now in the shaly zone. Drilling is steered downward until sand formations are recognized by the gamma ray curve.

Courtesy of W. Lesso, adapted from [4]

Geosteering has evolved and has been refined. It is based on the modeling of the formation log response followed by a comparison of acquired data with the model [5].

References

1 Blacker, S., "Decision-making process for streamlining environmental restoration: risk reduction using the data quality objective process," 19th annual national energy and environmental quality division conference, 1994.

2 Ellis, D. V., Perchonok, R. A., Scott, H. D., Stoller, C., "Adapting wireline logging tools for environmental logging applications," paper C, SPWLA 36th annual logging symposium, Paris, 1995.

3 Harrison, P. F., Mitchell, A. W., "Continuous improvement in well design optimizes development," SPE 30536, *Trans.* SPE 70th annual technical conference and exhibition, Dallas, 1995.

4 Lesso, W. G. Jr., Zoons, C. W., Sapijanskas, M. D., "Back to basics in extended reach medium to long radius horizontal drilling," 8th offshore South East Asia conference, Singapore, 1990.

5 Lesso, W. G., Jr., Kashikar, S. V., "The principles and procedures of geosteering," IADC/SPE 35051, *Trans.* IADC/SPE drilling conference, New Orleans, 1996.

6 Rouillac, D., *Cement evaluation logging handbook*, Éditions Technip, 1994.

7 Saleri, N. G.,"Reengineering simulation: managing complexity and complexification in reservoir objects ," SPE 36696, pp. 5-11, *SPE reservoir evaluation and engineering*, Vol. 1, No. 1, 2-1998.

8 Sollie, F., Rodgers, S., "Towards better measurements of logging depth," SPWLA 35th annual logging symposium, Tulsa, 1994.

9 Rodgers, S., Theys, P., "Le log," *The Log Analyst*, Vol. 39, No. 2, 3-1998.

4

Elements of metrology I: error analysis

> A significant problem that requires further investigation is accuracy and precision. One of the things I found surprising as a geochemist is the frequency with which numbers are presented without errors. This lack of attention to errors makes me very uncomfortable.
>
> *Claude J. Allègre*

The science and techniques of measuring are gathered in a field called metrology. Logging attempts to mimic measurements that are routinely performed in the laboratory environment. Because it is performed in less than comfortable conditions for the logging sensors **and** the logging engineer,[1] it is sometimes forgotten that logging is a segment of the measurement industry at large and, as such, should abide by its rules. Before proceeding to the study of log acquisition techniques, it is useful to review some definitions related to metrology [III, IV, 2 and 3].

4.1 True value and errors

4.1.1 True value of a physical characteristic

The objective of the experimentalist is to quantify the characteristics of a phenomenon. In the geosciences, these parameters could be formation density, dielectric

---

[1]Large hydrocarbon deposits tend to lie under deserts, jungle and rough seas.

constant and sound velocity. The system under investigation has intrinsic values for each of these categories, which we define as the **true values**. These values are abstract since they cannot be perfectly known through any physical process.

4.1.2 Error

A measurement assigns a number to a specific physical characteristic of a system through an experiment. This number could be derived from the deflection angle of a needle or the counting of pulses initiated by electric discharges. The number so obtained is different from the **true value** and the difference is defined as the **error**:

$$\text{error} = \text{measurement} - \text{true value}.$$

The error cannot be determined exactly. Otherwise, it would be used to indirectly obtain the true value. Nevertheless errors can be **estimated** and **managed**.

4.1.3 Types of error

The first step in the process of managing errors is to recognize that there are different types of error.

It is generally accepted that there are three types of errors:

(a) systematic errors
(b) random errors
(c) blunders.

A **systematic error** is a reproducible inaccuracy introduced by faulty design, failing equipment, inadequate calibration, inferior procedure or a change in the borehole environment. By opposition, a **random error** cannot be reproduced and is mostly imputable to the physics of measurement. **Blunders** are defined as "gross, stupid, or careless mistakes."[2]

To clarify the meaning of these types of errors, consider an analogy with target practice (*Fig. 4.1*). The distribution of the shots on a given target results from two sources: the imperfection of the gun sight puts the average position of the bullets away from the center; it represents a **systematic** error. Around this average position, the dispersion of the shots is due to random causes; for instance, the slight movement of the shooter caused by breathing. This represents a **random** error. The shot position associated with the **blunder** may originate from a distraction imposed on the shooter. While systematic and random errors may position the shots at a short distance away from the center of the target, the blunder could bring the shot far away from the target itself.

---

[2] From a Norwegian word related to the behavior of a person as if blind.

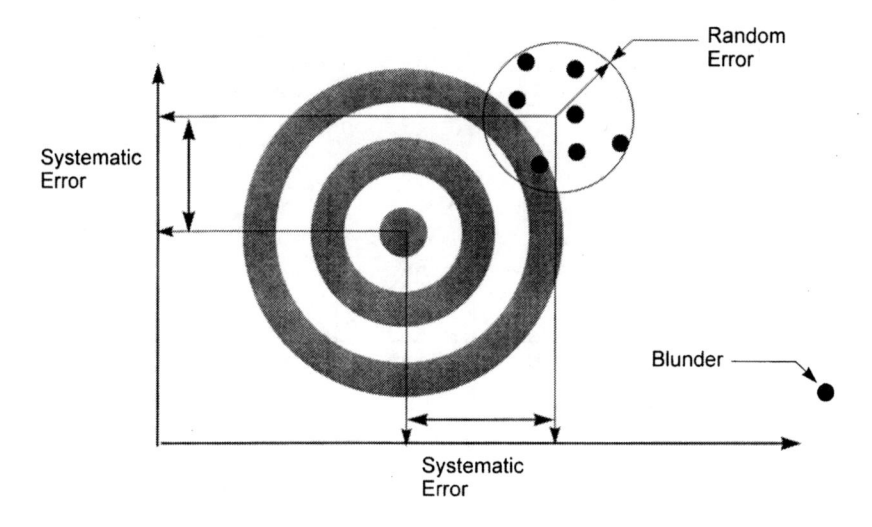

FIG. 4.1 : Target practice and types of error.

4.1.4 Error analysis

Definition

Error analysis is the detection, recognition and quantification of the contributing errors. It is required for reducing error because different causes of error justify different remedies. Half the work of the metrologist is dedicated to derivation of a measurement while the other half is devoted to estimation of the measurement error.

While random error can be reduced by multiplying the number of measurements (reduced logging speed, logging the same interval several times and use of multiple detectors), systematic error is not modified by such steps. Systematic error needs to be recognized and corrected. Blunders are eliminated by a stringent quality assurance program. Quality control of log data is the means for verifying that standard operating practice has been followed and is indispensable in the process of removing blunders. This is reviewed in Chapter 21.

Once the blunders are totally eradicated, any error can be viewed as the superposition of systematic and random errors (*Fig. 4.2*). As the number of measurements increases, random error decreases while systematic error stays constant. Random and total errors are largest for a single measurement.

Importance of error analysis: example

A petrophysicist, for example, uses natural gamma ray logging to estimate clay volume. To obtain the desired result, the total number of gamma rays must be known

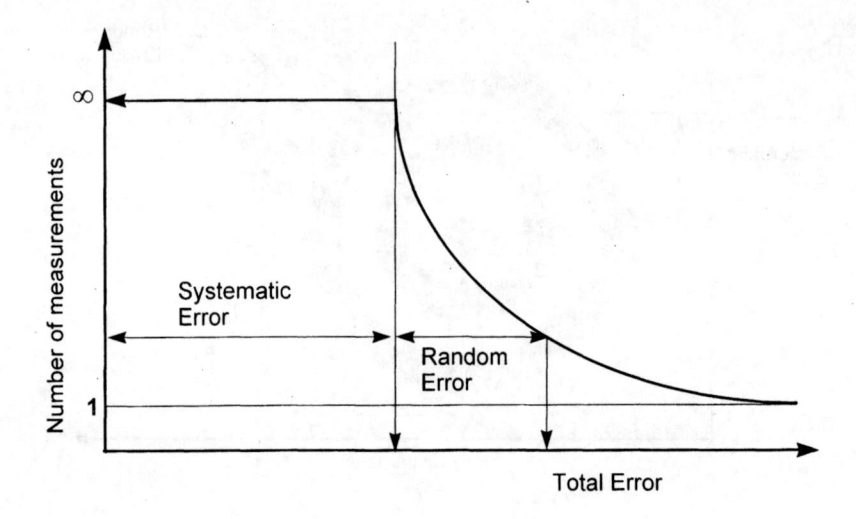

FIG. 4.2 : Variation of total error with the number of measurements. Total error is largest for a single measurement and smallest for an infinity of measurements.

within ±5 API of the "true value." The total error can be reduced by the following steps:

(a) Reduce logging speed to increase the counting rate received from one unit of volume of the formation. This reduces random error.
(b) Increase counting rates by using a larger or more efficient detector. This again affects random error.
(c) Improve calibration. In the current operating mode, radioactive sources are not removed from the calibration area and this parasitic signal introduces a systematic error in the measurement.
(d) Ensure that the logging procedures have been well defined and are rigorously adhered to in order to eliminate blunders.
(e) Use a caliper measurement as input to the environmental correction. This reduces systematic error.

4.2 Evaluating systematic error

Systematic error is a troublemaker in the game of error reduction. It cannot be tracked by repeating the measurement because it will appear with the same magnitude on repeat runs, **if the same operating procedures are followed**. It will be detected only by a thorough analysis and control of the conditions prevailing during the measurement[3] or by finding other ways of deriving the same parameter, then checking the consistency. Once recognized, systematic error can be easily characterized. Again, **if operating procedures have not changed**, the systematic error can be removed from previous measurements.

Example 1

The response of a density tool has been established with an artificial limestone block improperly characterized by a third party. The response algorithm forces the density to be 2.69 g/cm^3, the erroneous value attributed to this block, when the counting rates are similar to the ones read in the artificial formation. The logs recorded with this algorithm are all systematically shifted. Later, the artificial formation is correctly characterized and a new density, 0.02 g/cm^3 higher, is assigned to it. As the culprit, the response algorithm, has been used for all the faulty logs, they can all be corrected.

Example 2

In the thermal neutron calibration procedure, there is no provision for temperature effects. However, the logging engineers using their college reminiscences, think that a nuclear reaction involving neutrons should be temperature dependent. For this reason, they carefully record the calibrating fluid temperature. Later, the dependence of the calibration to fluid temperature is characterized and a correction algorithm designed. All previous logs can be corrected.

These examples show that stability of the procedure, even **if the procedure is not perfect**, is essential. **Logging procedures must be carefully described and deviations from standard operations reported**. If this is achieved, analysis of the operation can correct for systematic errors, even **after** the operation has been completed.

4.3 Estimating random error

4.3.1 General

While systematic error can be quantified and removed, random error can only be estimated. To evaluate a physical property, whose true value is X, a common procedure is to perform n independent measurements, yielding values x_1 to x_n. After all systematic errors have been removed and if n is large enough, a reasonable estimate of X is the average $\bar{x}$ of the n measurements.

---

[3]Including tool response function and calibration.

$$\bar{x} = \frac{\sum_{i=1}^{n} x_i}{n}.$$

It cannot be affirmed that X is strictly equal to $\bar{x}$, but it is possible to affirm there is a probability p that X belongs in a **confidence interval**, such that

$$\bar{x} - q(p) \times \sigma < X < \bar{x} + q(p) \times \sigma,$$

in which σ is the **standard deviation** defined as

$$\sigma = \sqrt{\frac{\sum_{i=1}^{n} (x_i - \bar{x})^2}{(n-1)}}$$

and $q(p)$ is a number function of p, the probability [1]. Depending on the number of measurements available, $q(p)$ takes different values as listed in *Table 4.1*.

TABLE 4.1
$q(p)$ VALUES

n	2	3	4	5	6	7	8	9	10
$q(p = 95\%)$	6.4	1.30	0.72	0.51	0.40	0.33	0.29	0.26	0.23
$q(p = 99\%)$	31.8	3.01	1.32	0.84	0.63	0.51	0.43	0.37	0.33

In logging, it is neither practical nor economical to repeat a survey several times. An alternative approach is therefore used to estimate the confidence interval. Once a standard deviation is estimated, the inequality above still applies to limit the confidence interval where the true value of the desired characteristic can be expected.

The alternative approach consists of designing a statistical model from the analysis of the physical process on which the measurement is based. Since measurements cannot be repeated at will, it is more practical to define the distribution of the results of hypothetical repeated measurements.

For instance, it is possible to predict that a tossed coin will come up heads 50% of the time when the number of trials becomes large. In a similar way, it can be anticipated by how much the true value under investigation differs from one, two or three measurements.

To handle the prediction, it is necessary to elaborate the statistical model that best matches the physical process controlling the experiment that leads to the measurement [VIII]. In these models, the measurement can be defined as the counting of the successes resulting from a given number of trials. Each trial can have only one of two outcomes. It is a success or it is not. The probability of success is assumed to be constant for all trials. In the next paragraphs, three types of statistical models, following the binomial, Poisson and Gauss distributions [III] and [1], are described.

4.3.2 Binomial distribution

If n is the number of trials for which each trial has a constant probability of success p, then the predicted probability of counting x successes is

$$p(x) = \frac{n!}{(n-x)!x!} p^x \times (1-p)^{n-x}.$$

This distribution is defined only for integer values of n and x. The mean of this distribution is

$$\bar{x} = \sum_{x=0}^{n} x \times p(x) = np.$$

The standard deviation is σ, such that

$$\sigma^2 = \sum_{x=0}^{n} (x-\bar{x})^2 p(x) = np \times (1-p).$$

4.3.3 Poisson distribution

The binomial distribution can be considerably simplified when the number of trials n is large and the probability of success p is low. With the previous notations, it can be written

$$p(x) = \frac{(pn)^x \times e^{-pn}}{x!}.$$

The mean has a value identical to the mean of the binomial distribution:

$$\bar{x} = np.$$

The value of the probability can be rewritten:

$$p(x) = \frac{(\bar{x})^x \times e^{-\bar{x}}}{x!}.$$

The standard deviation σ is such that

$$\sigma^2 = \sum_{x=0}^{n} (x-\bar{x})^2 p(x) = pn = \bar{x}.$$

The Poisson distribution applies well to most nuclear counting experiments in which a large number of particles participates in a large number of trials, associated with a low number of successes.

4.3.4 Gaussian distribution

If, in addition to the constraints mentioned (large number of trials and low probability of success), the mean value is large, then the mathematical expression of the probability $p(x)$ supports further mathematical simplification:

$$p(x) = \frac{1}{\sqrt{2\pi x}} \times e^{-\frac{(x-\bar{x})^2}{2x}}.$$

This distribution is only valid pointwise and for integer values of x. By definition, the mean is $\bar{x}$. The standard deviation is equal to the square root of the mean, as could be anticipated.

When the population of data is large, then a continuous function with similar characteristics is of more practical interest. The probability $\mathrm{d}P(x)$ for observing a value within the interval $[x, x + \mathrm{d}x]$ is given by

$$\mathrm{d}P(x) = \frac{1}{\bar{x} \times \sqrt{2\pi}} e^{-\frac{1}{2}(\frac{x-\bar{x}}{\bar{x}})^2}.$$

A diagram of the probability Gaussian distribution is shown in *Fig. 4.3*.

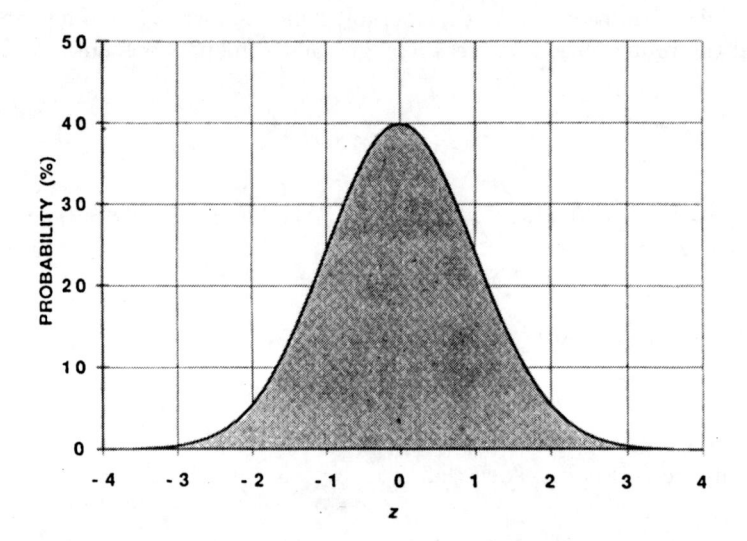

FIG. 4.3 : Gaussian distribution.
$z = (x - \bar{x})/\sigma$.

It is particularly interesting to evaluate the cumulative probability that the values of x are different from $\bar{x}$ by less than a given value, $\Delta x = x - \bar{x}$. To do so, it is necessary to integrate $\mathrm{d}P(x)$ between $\bar{x} - \Delta x$ and $\bar{x} + \Delta x$:

$$P(\Delta x) = \int_{\bar{x}-\Delta x}^{\bar{x}+\Delta x} \mathrm{d}P(x)\mathrm{d}x.$$

The values of this integral cannot be easily evaluated because there is no analytical form, but it can be derived by approximation [III]. If $z = (x-\bar{x})/\sigma$, then the cumulative probability that z is below a given value z_0 is[4]

$$P(z_0) = \frac{2}{\sqrt{\pi}} e^{-z_0^2/2} \sum_{k=0}^{k=\infty} \frac{(z_0/\sqrt{2})^{2k+1} 2^{2k} k!}{(2k+1)!}.$$

The values of $P(z_0)$ as a function of z_0 are listed in *Appendix 3*.

Exercise 1[*,5]

For a population of measurements following a Gaussian distribution, we want to know, in percentage of the standard deviation, the range of values $[x_{\min}, x_{\max}]$ so that 98% of the points belong to this interval.

4.3.5 One-σ, two-σ, three-σ

It is rare that an error bar appears on the table of results of a petrophysical interpretation because every input or output is generally assumed to be error free. When the error bar is shown, it could be put under the following format:

Density $= 2.267 \pm 0.017$ (g/cm^3).

What is then the meaning of ± 0.017g/cm^3? That, with 100% certainty, the true value of the density is between 2.250 and 2.284 g/cm^3? Generally, no. Assume that the distribution of all possible measurements follows a Gaussian curve with an average of 2.267 g/cm^3 and a standard deviation of 0.017 g/cm^3. Then, there is a 68.3% chance that the true value is between $[2.267 - 0.017]$ g/cm^3 and $[2.267 + 0.017]$ g/cm^3.

If 68.3% seems to be too loose a control on the measurement, then it is possible to widen the interval of confidence and bring the probability that the true value belongs to this interval to a higher value. Tripling the length of the interval, to 2×0.051 g/cm^3, would increase the probability to about 99%.

The relation between the range of the confidence interval and the probability that the true value belongs to this confidence interval is shown for some specific values (*Table 4.2*).

---

[4]The terms of the sum with $k > 10$ are smaller than 10^{-4} if z is smaller than 2. In practice, only Σ_0^{10} is computed.
[5]Solutions to the exercises marked with an asterisk are found in Appendix 4.

TABLE 4.2
RELATION BETWEEN CONFIDENCE
LEVEL AND LENGTH OF THE
CONFIDENCE INTERVAL

	Probability (%)	Interval (g/cm^3)
0.67-σ	50.0	[2.267 ± 0.011]
1.00-σ	68.3	[2.267 ± 0.017]
1.64-σ	90.0	[2.267 ± 0.028]
1.96-σ	95.0	[2.267 ± 0.033]
2.58-σ	99.0	[2.267 ± 0.044]
3.00-σ	99.7	[2.267 ± 0.051]

When an error bar appears beside a measured value, it is necessary to qualify what it means. Laymen tend to think in terms of three-σ probability while physicists and experimentalists are used to one-σ probability [X]. In the rest of this book, one-σ probability will be used unless indicated otherwise.

4.4 Precision and accuracy

4.4.1 Definitions

Although most dictionaries consider precision and accuracy synonyms, metrologists and physicists assign different meanings to these words [III and 2]:

(a) **Accuracy** of an experiment is how close the result comes to the true value. It is a measure of the correctness of the result. It is generally dependent on how well the systematic error is compensated for and controlled.

(b) **Precision** of an experiment is how exactly the result is determined, regardless of its proximity to the true value. Precision can also be understood as the closeness of several measurements made with similar methods. It is then formulated as the standard deviation of these multiple measurements:

$$\sigma = \sqrt{\frac{\sum_{i=1}^{n}(x_i - \overline{x})^2}{(n-1)}}.$$

The precision of an experiment is mostly dependent on how well random error is analyzed and overcome.

An accurate measurement strives for results with a high probability p even with a wide interval of confidence. Attempting to make a precise measurement is similar to trying to make the interval of confidence $[\bar{x} - q(p)\sigma, \bar{x} + q(p)\sigma]$ as small as possible, but maybe with a low probability value p.

4.4.2 Importance of precision and accuracy

(a) **Accuracy** and **precision** must be considered simultaneously. There is not much point in being sure that the true value X is in the interval of confidence if this interval is large. Measuring a density of 2.2 g/cm^3 with an interval of confidence from 1.7 g/cm^3 to 2.7 g/cm^3 is accurate but not precise. Conversely, measuring a density with a true value of 2.221 g/cm^3 with a range 2.525 g/cm^3 to 2.527 g/cm^3 is precise but not accurate.

(b) An analysis of the behavior of a sensor often includes terms related to accuracy and precision, which means that a "good" measurement depends on "good" accuracy and "good" precision. For instance, a pressure gauge includes four terms, mean quadratic deviation (MQD),[6] hysteresis,[7] temperature dependence—all impacting accuracy—and precision.

(c) In the logging industry, precision can be obtained quantitatively by the repeat analysis, which compares the readings recorded during different passes over the same interval.

(d) Accuracy cannot be quantified easily. "Good" accuracy is the result of

1. rigorous control of the tool response
2. minimization of calibration errors
3. consistent operating procedures
4. stable behavior of the downhole equipment under high pressure and temperature
5. tight control of the environment.

Figure 4.4 represents accuracy and precision for a true value X and two measurements x_1 and x_2. This diagram shows that precision can be easily evaluated as x_1 and x_2 can be known, while it is not possible to quantify accuracy as X, by definition, cannot be perfectly known [2].

Figure 4.5 represents the different possibilities offered to measurements in terms of accuracy and precision. In real conditions, the important attribute of similar diagrams is the scale on the x axis.

---

[6] Estimation of the quality of the fit between the theoretical model and the actual transducer response.
[7] Defined in Chapter 6.

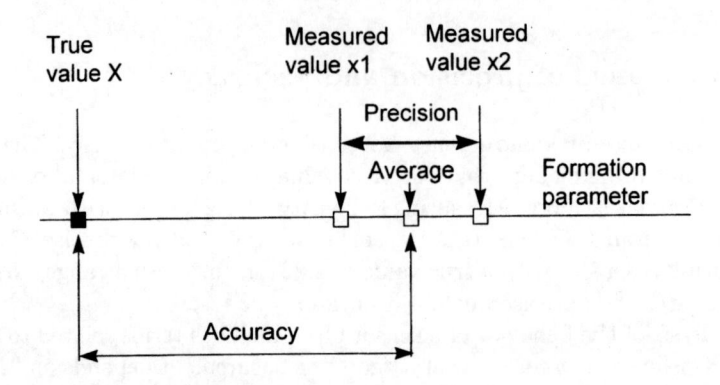

FIG. 4.4 : Accuracy and precision.
X is the true value to be measured. x_1 and x_2 are two measurements.

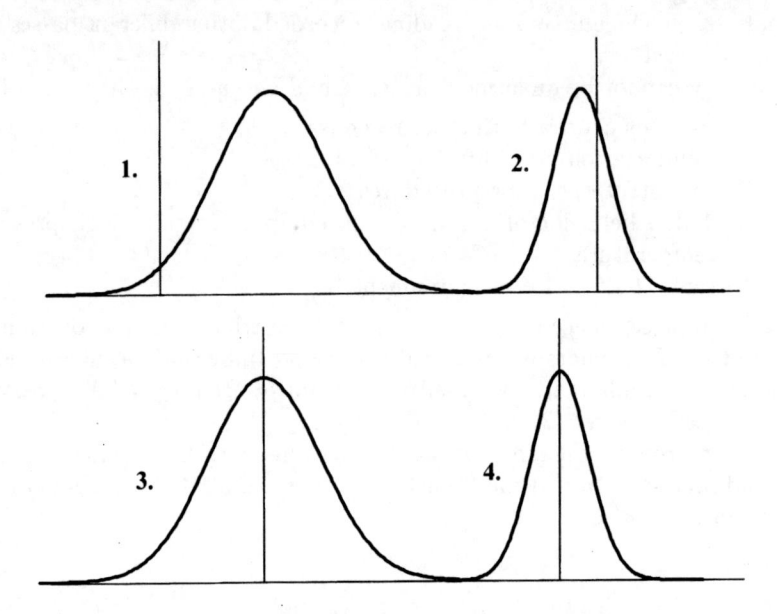

FIG. 4.5 : Different types of measurements.
The vertical line represents the true value X. The Gaussian curve in each case is the locus where measurements plot.

1. Inaccurate and imprecise measurement,
2. Inaccurate and precise measurement,
3. Accurate and imprecise measurement,
4. Accurate and precise.

Courtesy of SPWLA [4]

4.5 Repeatability and reproducibility

4.5.1 Repeatability

Repeatability is an in-situ estimation of precision. It is the difference in the magnitude of two measurements made under the same conditions, with the same equipment, same engineer, same environment. The repeatability test is not able to detect most systematic error.

Stability is used to quantify how a sensor repeats at the same location and in similar conditions, but for extended periods. Stability is a desired characteristic of the pressure sensors used in long-duration tests.

4.5.2 Reproducibility

Reproducibility[8] is the difference in the magnitude of two measurements made with the same method but possibly with different equipment and different personnel. Good reproducibility is a crucial attribute for a logging measurement, especially in fieldwide and multiwell evaluations.

The absolute value of a key petrophysical parameter derived from a single measurement may not be that important. Indeed, this value will be crosschecked and eventually corrected through the use of other measurements. What may be more important is that the measurements are consistent or reproducible, in other words that they are reading values that may or may not be shifted, but with a shift that is **constant**.

This point is highlighted by an example. Consider a reservoir with true density of 2.2 g/cm^3 and 2.3 g/cm^3 at two wells. The density difference represents a trend that may have important economic implications on the reserves. The two wells are logged with two tools. The first tool reads the correct value, 2.2 g/cm^3. The second tool reads systematically lower by -0.1 g/cm^3 as a result of inferior manufacturing procedures. As a consequence, the reading in the second well is also 2.2 g/cm^3, a value accepted by the analyst because it matches the first well value. Generally, on development wells, limited logging and coring programs are undertaken, and this erroneous value may not be detected. The porosity variation in the field will pass temporarily undetected, with considerable economic implications.

Reproducibility can be checked by operating different tools in artificial formations or in the same well.

Table 4.3 lists repeatability and reproducibility characteristics for a thermal neutron tool. The instrument-to-instrument checks were performed with 11 instruments run in calibration pits and calibrated with six different calibrators.

---

[8]Reproducibility is sometimes called intertool repeatability as opposed to intratool repeatability, which considers several measurements performed by the same instrument.

TABLE 4.3
REPEATABILITY AND REPRODUCIBILITY OF A THERMAL NEUTRON POROSITY TOOL (IN pu)

Response of 11 series 2435 CN instruments in Western Atlas test pit.

Apparent mean ϕ_N	Single instrument three-pass standard deviation	Instrument-to-instrument standard deviation
25	0.45	0.62
14	0.14	0.30
4	0.07	0.17

Courtesy of SPWLA [VII]

4.6 Uncertainty

Definition

The definitions of uncertainty extracted from dictionaries are clouded by a lack of certainty. The whole book referred to as [3] attempts to quantify this word. The formal definition proposed in the introductory chapter is as follows:

Uncertainty (of measurement): parameter, associated with the result of a measurement, that characterizes the dispersion of values that could reasonably be attributed to the measurand.[9]

An actionable definition of uncertainty is proposed in [4]:

$$\text{Uncertainty} = \sqrt{\text{Precision}^2 + \text{Accuracy}^2}.$$

References

1 Degurse, A. M., Gozard, F., Rosenfeld-Gipch, L., Soulié, L., *Physique*, Hatier, Paris, 1987.

2 International Organization for Standardization, *Quality assurance requirements for measuring equipment*, ISO/DIS 10012-1, Geneva, Switzerland, 1992.

3 International Organization for Standardization, *Guide to the expression of uncertainty in measurement*, ISBN 92-67-10188-9, Geneva, Switzerland, 1995.

4 Kimminau, S., "Traceability—making decisions with uncertain data," *The Log Analyst*, Vol. 35, No. 5, September-October 1994.

---

[9]The measurand is the physical quantity whose measurement is attempted.

5

Elements of metrology II: volume considerations

A measurement can be both accurate and precise so that the measured value is close to the true value. This is necessary for a **good** measurement, but it is not sufficient. The measurement may fail to be useful if the volume of investigation that is scanned is not related to the formation or area of interest.

Examples

1. The transmitter-to-receiver spacing of the induction tool is several feet. The resulting measurement integrates large volumes of formation and is unable to detect fractures or thin beds.

2. Conversely, a sensor has an excellent vertical resolution, but it does not investigate the formation deeper than a few inches. The sensor describes the borehole and the invaded zone, but it is unable to gather information on the virgin zone, the volume that is of most interest to the log user.

3. A logging device may deliver a global measurement while an azimuthal evaluation is desired. The CBT*[1] Cement Bond Tool performs the former while the CET* Cement Evaluation Tool and other scanning devices are able to do the latter. The CBT tool is unable to detect cement chanelling.

Figure 5.1, next page, shows two measurements, one having geometrical relevance and one without any useful information.

---

[1] An asterisk (∗) represents a trade mark of Schlumberger.

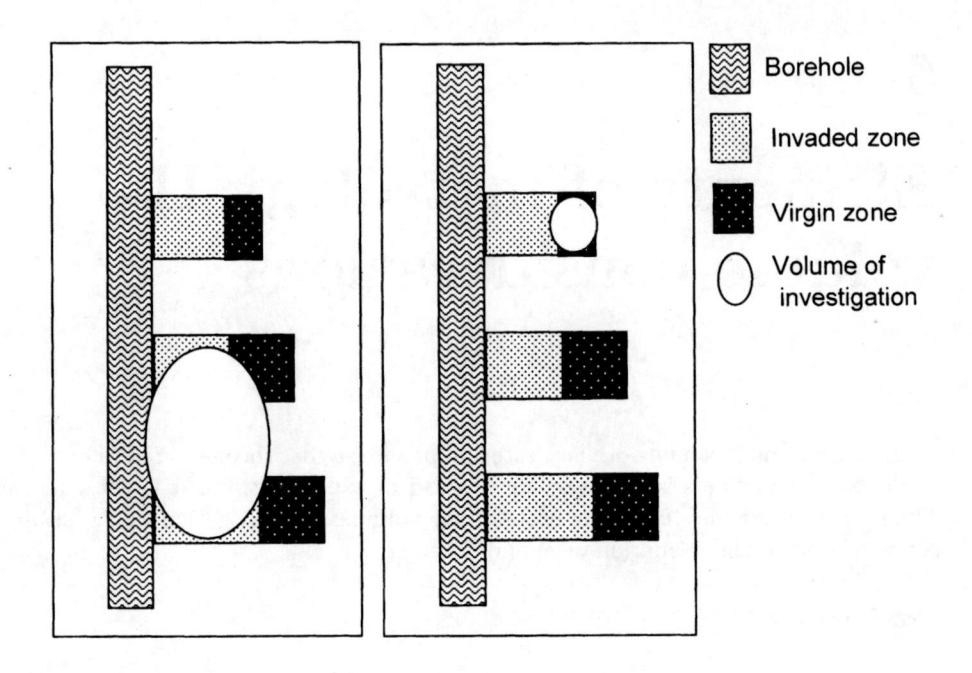

FIG. 5.1 : Importance of geometrical relevance.
On the left, the measurement is unable to distinguished thinly laminated beds. In addition, the reading is too shallow. On the right, the measurement is reading deep—assessing the virgin zone. In addition, it is focused and separates thin beds. Indeed, this second measurement is the dream of the log user.

5.1 Geometrical factor

In a logging operation, the response of the formation can be spontaneous, as in natural gamma ray emission or the spontaneous potential, or induced, when the formation is excited by radiation or a field created by the logging tool. In both cases, it is important to know the relative contributions of the different volumes of formation as a function of their position relative to the logging tool.

The volume characteristics of a logging measurements cannot be clearly defined without the concept of geometrical factor.

5.1.1 Definition

The total signal originating from the formation is S. This signal is contributed by n cells indexed by i from 1 to n. Signal s_i comes from cell i. The **geometrical factor**[2] of cell i is g_i, such that

$$g_i = \frac{s_i}{S}.$$

It follows that

$$\sum_1^n g_i = 1.$$

In other words, g_i is the weighting factor associated with cell i. It varies as a function of the location of the cell relative to the logging detector and, indeed, with the configuration of the tool. In some cases, this weighting factor becomes nil, which means the logging device is blind to some volume of formation. For instance, the borehole volume has a relatively small contribution to the overall signal of the deep induction. Also, the proximity to the logging tool does not directly imply a large weighting factor.

A simple application of the geometrical factor concept concerns induction logging. In this case, the whole space can be divided into only three volumes: the borehole of conductivity C_m, the invaded zone of conductivity C_{xo}, and the virgin zone of conductivity C_t. g_m, g_{xo} and g_t are the respective geometrical factors for these zones. The total signal observed at the sonde is

$$C_{ind} = (g_m \times C_m) + (g_{xo} \times C_{xo}) + (g_t \times C_t).$$

Exercise 2*

This exercise relates to a field example where the induction log exhibited unexplained readings. The geometrical factor approach sheds some light on the problem and validates the log readings.

1. Assuming $R_w = 0.1 \, \Omega/\text{m}$, $m = n = 2$ and $\phi = 30$ pu, compute R_t with Archie's formula for $S_w = 5$ su and $S_w = 7.5$ su.

2. Using the information in *Table 5.1*, with $R_m = 0.1 \, \Omega/\text{m}$, $R_{mf} = 0.1 \, \Omega/\text{m}$, $S_{xo} = 74.5$ su, compute the corrections to the medium and deep induction and shallow focused resistivity logs for the borehole mud conductivity and the invaded zone. Assume a 9-in. borehole and a 1.5-in. standoff.

3. Compute the expected deep and medium induction resistivity readings. What kind of anomaly can be expected?

---

[2] This definition assumes linear signals.

TABLE 5.1
GEOMETRICAL FACTORS

	Shallow focused resistivity	Medium induction	Deep induction
Borehole →	-	0.0004	−0.0002
Invasion diameter ↓			
15	0.18	0.0075	−0.0024
20	0.33	0.035	−0.0035
25	0.43	0.080	−0.0030
30	0.51	0.15	0.0000

Courtesy of C. Clavier

5.1.2 Elemental geometrical factor

The relation tying C_{ind} to C_m, C_{xo} and C_t assumes that conductivities are constant over large volumes. In the general case and for any logging tool, only a small cell as described in *Fig. 5.2*, of dimensions dr, $rd\theta$, dz and whose location is defined by r, θ and z, would have characteristics that are constant.

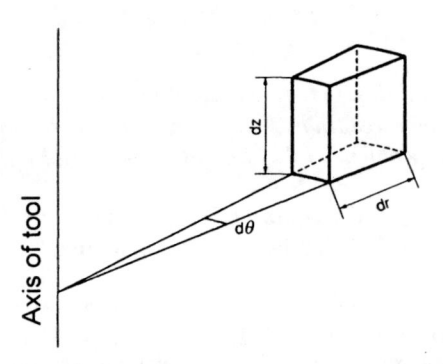

FIG. 5.2 : Elemental formation cell.

A signal $s(r, \theta, z)$ originates from this cell and the corresponding geometrical factor is $g(r, \theta, z)$. The total signal S is such that

$$S = \int_0^\infty \int_0^{2\pi} \int_{-\infty}^\infty g(r, \theta, z) \times s(r, \theta, z) \times \mathrm{d}r \mathrm{d}\theta \mathrm{d}z.$$

5.1.3 Successive integrations of the geometrical factor

5.1.3.1 2D geometrical factor

If the formation and the sensor configuration present a cylindrical symmetry, the geometrical factor can be assumed to be constant regardless of the azimuth. For any two azimuths θ_1 and θ_2, $g(r, \theta_1, z) = g(r, \theta_2, z)$. $\gamma(r, z)$, the 2D geometrical factor, can be described as a function of two variables only, r and z (*Fig. 5.3* for several induction tools):

$$\gamma(r, z) = \int_0^{2\pi} g(r, \theta, z) \mathrm{d}\theta.$$

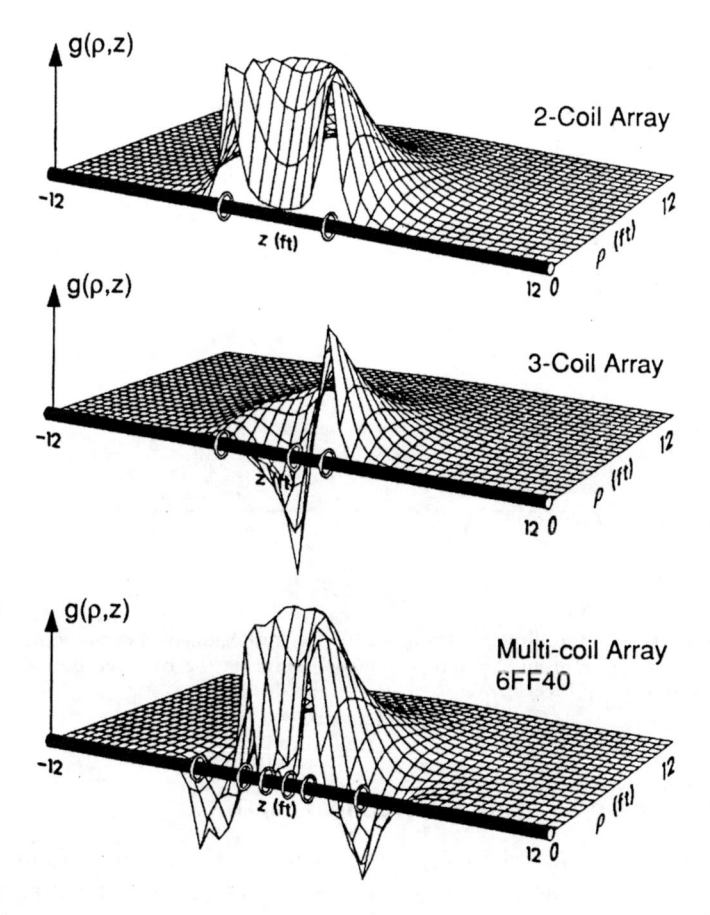

FIG. 5.3 : 2D geometrical factor contour map for the induction tools.

Courtesy of Schlumberger

5.1.3.2 Radial geometrical factor

If, in addition, shoulder-bed effect is negligible, the resulting $\gamma(r, z)$ is independent of z because, for whichever two values of z, z_1 and z_2,

$$\gamma(r, z_1) = \gamma(r, z_2).$$

This yields the function of a single variable r, called the **radial geometrical factor** $G(r)$ (*Fig. 5.4* for the *6FF40* induction tool):

$$G(r) = \int_{-\infty}^{\infty} \gamma(r, z) \mathrm{d}z.$$

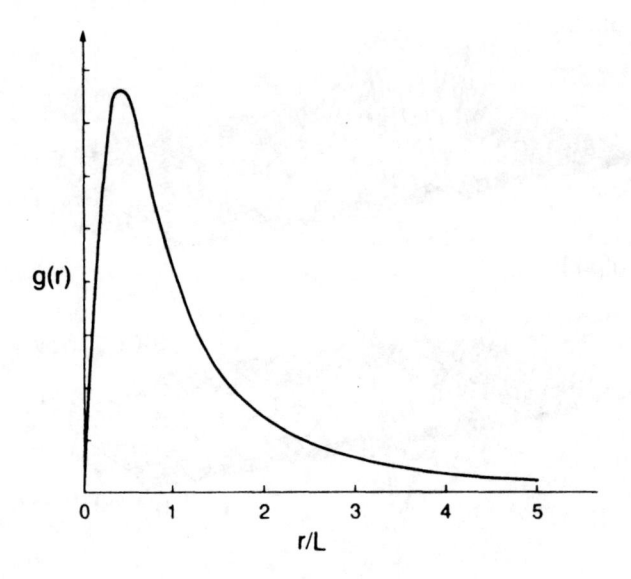

FIG. 5.4 : Radial geometrical factor for induction measurements. r is the radius from the axis of the tool and L is the distance between transmitter and receiver.
Courtesy of Academic Press [XIX]

5.1.3.3 Integrated radial geometrical factor

Another integration provides the percentage of signal originating from the volume limited by the logging sonde and a cylinder located at a distance r from the axis of the sonde:

$$\Gamma(r) = \int_0^r G(r) \mathrm{d}r.$$

5. Elements of metrology II: volume considerations

By definition, this integrated geometrical factor tends towards 1 when r increases. From $\Gamma(r)$, it is possible to derive the percentage of signal coming from an annulus defined by radii r_1 and r_2:

$$\Gamma_{r_1 \to r_2} = \Gamma(r_2) - \Gamma(r_1).$$

The integrated geometrical factor provides a way to consider how a logging device "illuminates" the zone of interest.

Figure 5.5 is an example of the integrated geometrical factor of the induction tool. For the medium induction (ILM) information, $\Gamma(50$ in.$)$ is 50% and $\Gamma(100$ in.$)$ is 71%. It can be deduced that, for a homogeneous formation, 21% of the signal originates from a zone limited by two cylinders of radii 25 in. and 50 in.

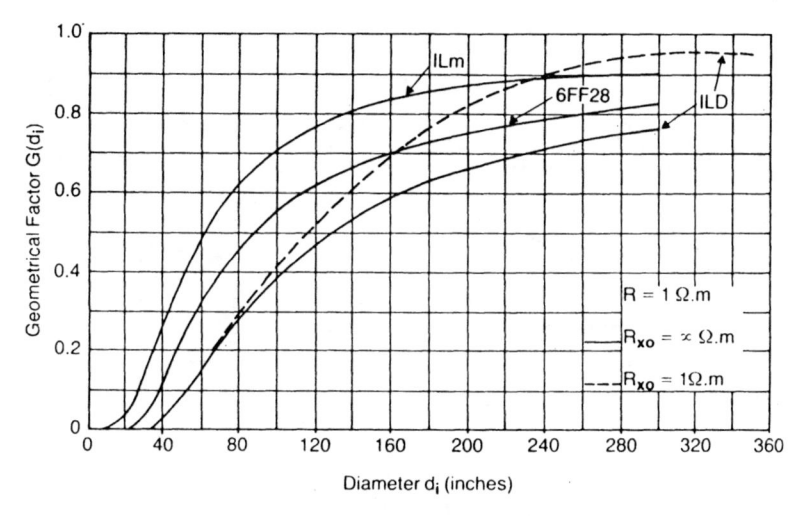

FIG. 5.5 : Integrated geometrical factors of induction tools.

Courtesy of Schlumberger

Example of radial geometrical factor

From a distance 0 to $d_h/2$ from the axis of the induction tool, the volume is filled with mud of conductivity C_m. From $d_h/2$ to $d_i/2$, the volume corresponds to the invaded zone:

$$g_m = \Gamma_{0 \to d_h/2} \qquad g_{xo} = \Gamma_{d_h/2 \to d_i/2} \qquad g_t = \Gamma_{d_i/2 \to +\infty}.$$

5.1.4 Pseudogeometric factor

For other logging tools, a concept similar to the geometrical factor is introduced. Because it is not possible to define a true geometrical factor, since such a parameter would vary considerably with the characteristics of the formation, the concept of pseudogeometric factor is introduced.

For the laterolog, the pseudogeometric factor J is:

$$R_{LLD} = J \times R_t + (1 - J) \times R_{xo}.$$

J is plotted as a function of twice the distance from the borehole axis (*Fig. 5.6*).

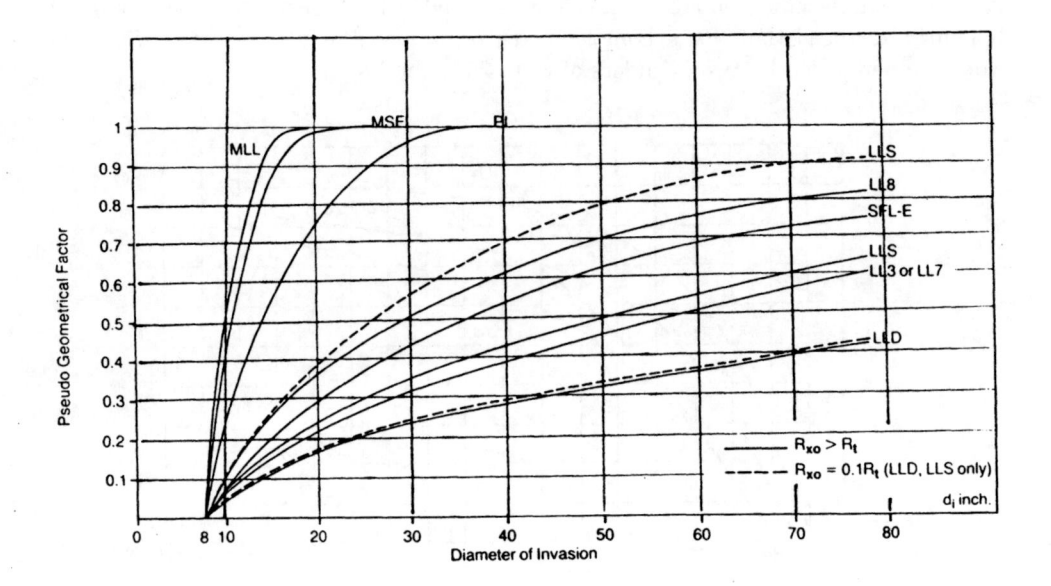

FIG. 5.6 : Pseudogeometric factor of the resistivity logs.

Courtesy of Schlumberger

Pseudogeometric factors are similarly defined for nuclear tools. It could apply to single-detector counting rates (yielding the percentage of counting rates originating from a volume fraction of the formation) and also to global parameters such as formation density and neutron porosity. *Figure 5.7* represents the pseudogeometric factor of a thermal neutron porosity tool.

Pseudogeometric factors generally depicted in the logging literature correspond to narrow conditions of application. The pseudogeometric factor of the dual laterolog is drastically modified by the thickness of the investigated bed and the resistivity contrast of shoulder beds (*Fig. 5.8*).

Figure 5.9 shows the pseudogeometric factors of the nuclear tools at 22 pu and 35 pu.

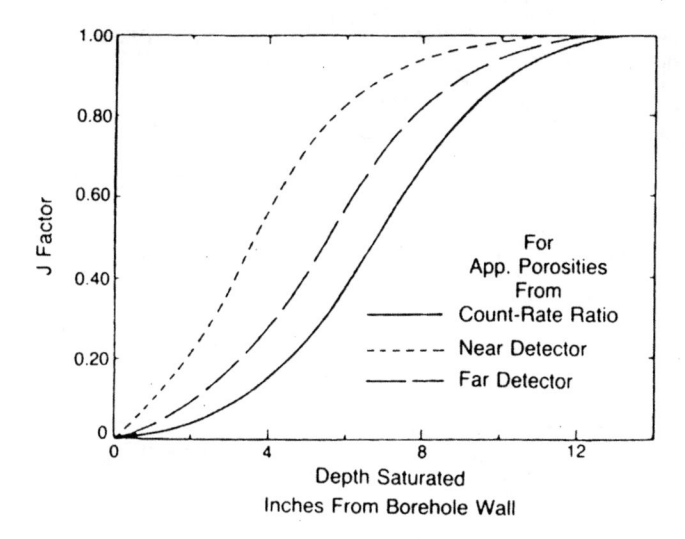

5. Elements of metrology II: volume considerations   55

FIG. 5.7 : Pseudogeometric factor of *Schlumberger* CNL* compensated neutron logging tool.

Courtesy of Academic Press [XIX]

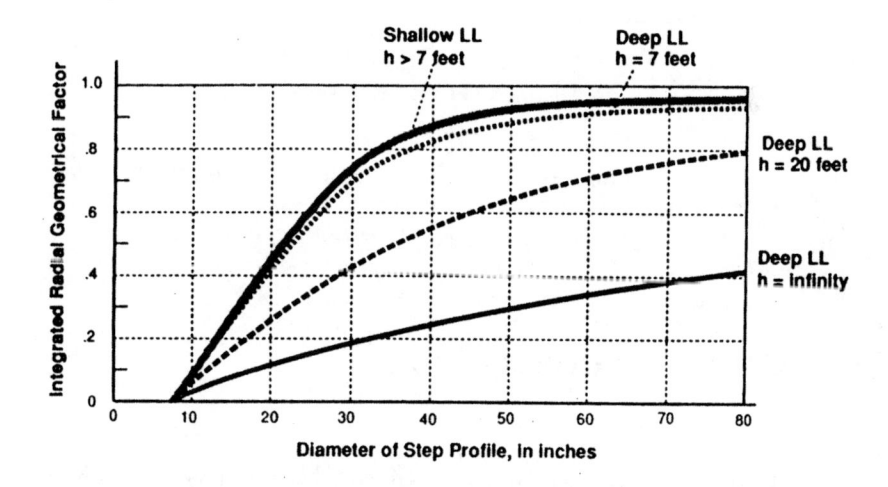

FIG. 5.8 : Change of the pseudogeometric factor of the laterolog with formation thickness.

Courtesy of Schlumberger

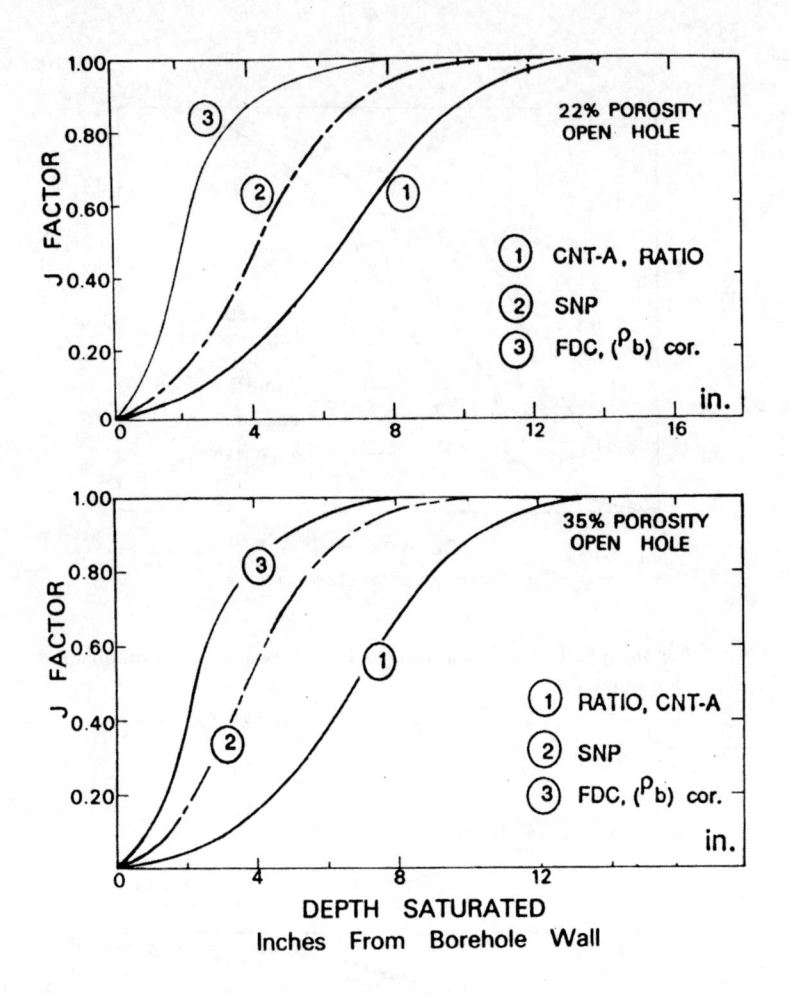

FIG. 5.9 : Comparison of the pseudogeometric factors at 22 pu and 35 pu for the nuclear tools used in the 1970s.
Courtesy of Schlumberger [7]

5.2 Depth of investigation

5.2.1 Conventional definition of depth of investigation

Before the nineties, the definition of depth of investigation depended on tool type, and had a different meaning for resistivity and nuclear logging tools.

5.2.1.1 Resistivity tools

Depth of investigation for resistivity tools was directly derived from the integrated radial geometrical or pseudogeometric factor curve. It was defined as the radius r from the borehole axis such that this factor equalled 50%. For instance, for the deep laterolog tool, it was 1.25 m (50 in.). As there was no standard in the industry, some definitions used the radius and some the diameter.

5.2.1.2 Nuclear tools

For nuclear tools, a similar definition was selected, but the formation under consideration, located within a distance r from the borehole axis was such that the pseudogeometric factor was 90%. While the depth of investigation and integrated geometrical factor curves of the resistivity and conductivity devices were generally derived by analytic methods (mathematical modeling), similar information for nuclear devices was obtained from experiments. The setup used for the determination of the depth of investigation of the neutron tool is shown in *Fig. 5.10*.

FIG. 5.10 : Experimental setup for the determination of the depth of investigation of nuclear tools.

Courtesy of Schlumberger [7]

Exercise 3

From the previous definitions of the depth of investigation and from *Figs 5.6* and *5.7*, derive the conventional depth of investigation of the deep laterolog, shallow laterolog and compensated neutron porosity.

5.2.2 Ambiguity of the conventional definition of the depth of investigation

The conventional definition of depth of investigation gives a less complete picture than the full description of the cumulative geometrical factor. To illustrate this point, consider the geometrical factors of three imaginary devices of exactly the same depth of investigation (defined as the radius for which the integrated radial geometrical factor is 50%) of 50 in. The geometrical factors for these devices are illustrated in *Fig. 5.11*. Device 1 has a flat response with the geometrical factor independent of the radial distance up to a maximum, and zero response beyond. Device 2 is highly sensitive to the shallow invaded zone but still has significant response in deeper parts of the formation. Device 3 is sensitive to an "intermediate" zone and is not sensitive to the shallow or deep regions.

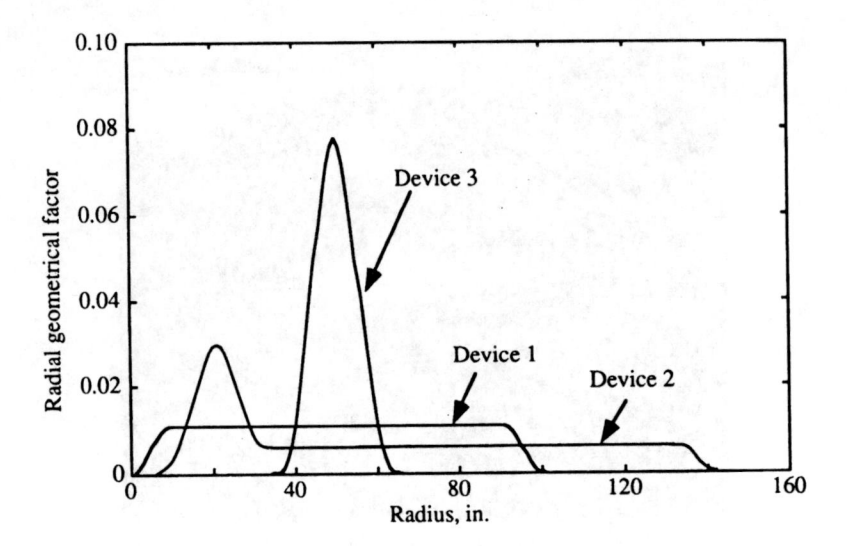

FIG. 5.11 : Radial geometrical factors of three hypothetical logging devices.

Courtesy of SPWLA [6]

Figure 5.12 shows the integrated radial geometrical factors for these devices. Although all three devices have exactly the same conventional depth of investigation, they do not investigate the same part of the formation.

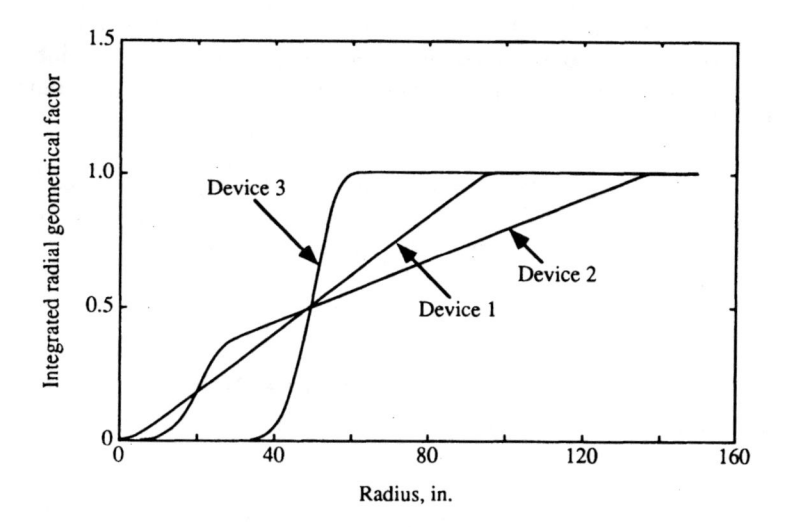

FIG. 5.12 : Tools with the same depth of investigation but with different geometrical characteristics.
Courtesy of SPWLA [6]

5.2.3 Modern definition of depth of investigation

It is difficult to account for all the problems introduced by the conventional definition of depth of investigation without introducing tedious and complicated sets of specifications. A workable compromise between detailing the real behavior and an overly simplistic parameterization is still possible.

Since the geometry can depend on conditions, the specific conditions for which the numbers are quoted must be part of the specification. These conditions must be chosen within the range of most common applications of the device. For devices where the dependence on conditions is very great, two or more sets of numbers should be specified at the extremes of the reasonable range of applicable conditions.

Describing depth of investigation with a single parameter, even in cases where geometrical behavior is independent of formation characteristics, is inadequate. One must be able to predict the importance of different (radial) regions of the formation by looking at the tool specifications.

One way to parameterize a complex radial response function is to present three numbers corresponding to 10%, 50% and 90% levels of the integrated radial geometrical factor. This is a scale-independent method to describe the shape of the integrated radial response.

Turning back to the example of the three hypothetical devices described in *Figs 5.11* and *5.12*, this approach clearly differentiates their radial behavior. Even though the 50% levels are the same (50 in.) for all three devices, the 10% levels are 13, 17 and 43 in., respectively, while the 90% levels are 85, 120 and 56 in., respectively.

These specifications can be used to choose the optimum device for a particular application. If the perturbed (invaded) formation region extends 30 in. into the formation, device 3 is the obvious choice. On the other hand, if this region extends 60 in., device 3 would be of little use, and device 2, in spite of the "peak" at 20 in., is the preferred choice.

Finally, since radial and vertical behaviors may be interdependent, the above specifications can only apply to "infinitely thick" beds (i.e., beds much thicker than the vertical resolution of the device). Such a statement should also be an integral part of the specification.

5.2.4 Variation of depth of investigation with formation and environment

Regardless of the selected definition, depth of investigation, like the geometrical factor, varies with formation and environment characteristics (*Table 5.2*).

TABLE 5.2
THERMAL DECAY TIME DETECTORS DEPTH OF INVESTIGATION

Environment		Depth of investigation (in.)	
Borehole	Mud	Near detector	Far detector
5.5-in. casing 1-in. cement annulus	Freshwater invasion Saltwater invasion	8.0 10.6	8.7 12.4
Uncased	Freshwater invasion Saltwater invasion	8.8 11.7	8.8 13.1

Courtesy of Schlumberger [2]

5.2.5 Electric diameter

The concept of electric diameter was introduced when dipmeter tools were developed. The computation of the dip is linked to trigonometry and geometry considerations. Because dipmeter tools generate currents that penetrate the formation, the spatial position of the bed boundaries used to compute the dip is not at the intersection with the borehole wall but somewhat inside the formation [5]. To compensate for this effect, a distance, called incremental electric diameter, is added to the physical borehole diameter to derive correct dips.

For the first dipmeter devices, the value of the incremental electric diameter was established and validated on field data. For recent tools, full 3D modeling is performed to derive accurate values of the electric diameter. The incremental electric diameter (in the order of a few inches) bears no direct relation to the depth of investigation (in the order of many inches) because the tool response at bed boundaries is quite shallow.

5.3 Vertical resolution

5.3.1 Conventional definitions

There are many definitions of **vertical resolution**, many of which are clouded by conflicting reports. The literature reports vertical resolution of the induction measurement, for example, as 1.5 to 2.4 m (5 to 8 ft). Mathematical modeling shows that a 1.8-m (6-ft) bed with an actual 25-Ω/m resistivity and shouldered by beds at 1 Ω/m is depicted as a peak reaching a maximum of 11 Ω/m (*Fig. 5.13*). Vertical resolution is also too readily equated to the physical distance between transmitters/sources and receivers/detectors. This may yield erroneous values. An example of such a mistake is the photoelectric effect measurement, which identifies beds 10-cm (4 in.) to 20-cm (8 in.) thick even though the source-detector spacing is approximately 36 cm (14 in.).[3]

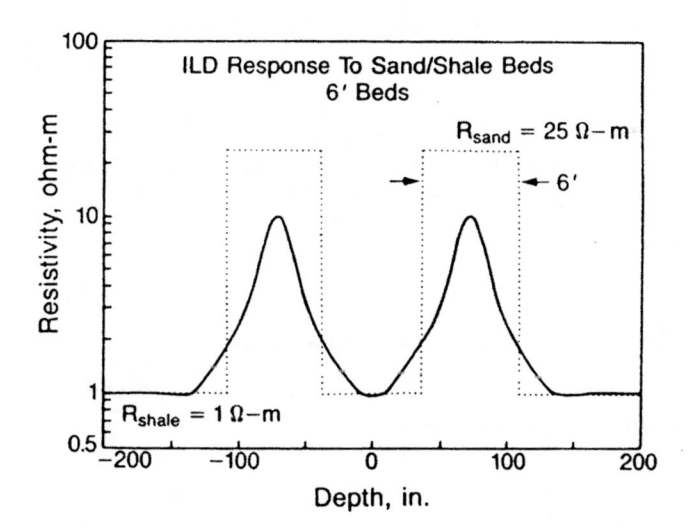

FIG. 5.13 : Readings of the induction tool in a 1.8-m (6-ft) bed.

Courtesy of Schlumberger [1]

---

[3] For the *Schlumberger* Litho-Density tool.

5.3.1.1 Theoretical definition

The theoretical definition of vertical resolution is as follows: *The full width at half maximum of the response of the measurement to an infinitesimally short event (Fig. 5.14)*. Generally, this definition has limited appeal to the user.[4]

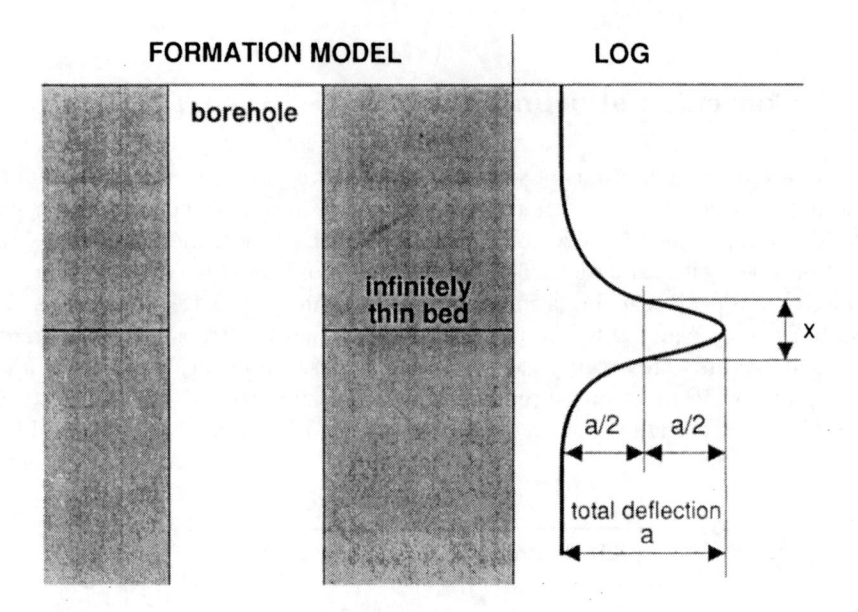

FIG. 5.14 : Theoretical definition of vertical resolution.
Vertical resolution is x. For log deflections whose shapes are Gaussian with a standard deviation σ, the vertical resolution, or full width at half maximum, is 2.35σ.

5.3.1.2 50% height and 50% area estimates

One definition of vertical resolution is derived from the vertical geometrical factor. The width at half of maximum is selected as vertical resolution. A more representative value is the width for a 50% area (which means that the area above is equal to the area below). As observed on *Fig. 5.15*, the values derived from these two definitions could be drastically different.

50% height estimate: 20 in.   50% area estimate: 200 in.

---

[4]Dirac-shaped infinitesimally thin formations are seldom found in nature. In any case, they are of little economic interest.

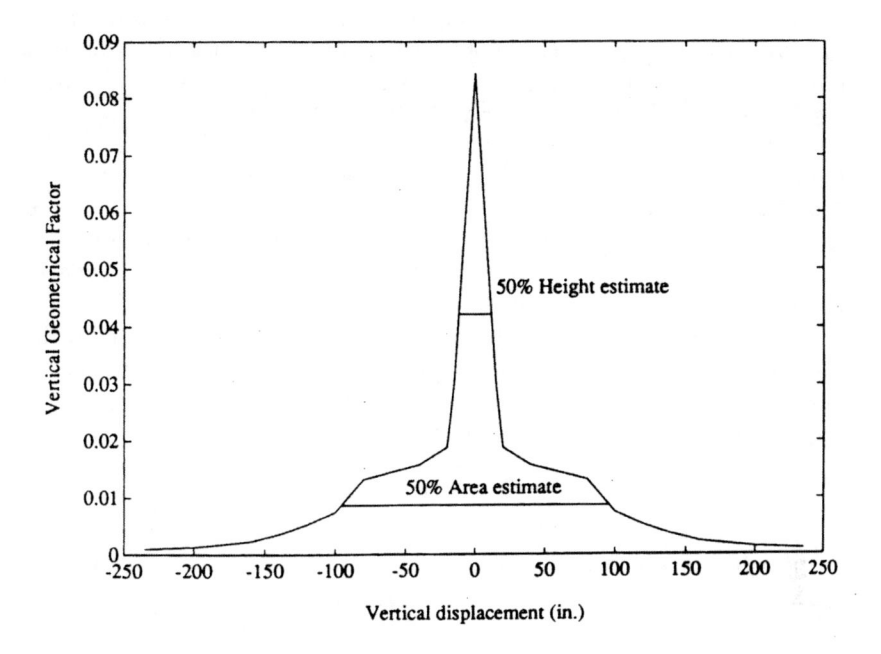

FIG. 5.15 : 50% height and 50% area estimates for a hypothetical device.

Courtesy of SPWLA [6]

5.3.1.3 Other definitions

Other definitions are also found in the literature:

(a) **Qualitative vertical resolution**: Vertical resolution is the minimum distance x such that the logging tool is able to resolve distinct events separated by this distance .

(b) **Quantitative vertical resolution**: Vertical resolution is the minimum bed thickness[5] for which the sensor measures, possibly on a limited portion of the bed, a parameter related to the real value of the formation.

Still, these definitions leave some vagueness:

(a) Most of the time, the measured value could be quite different from the true value of the formation, since this measured value needs to be corrected for shoulder bed or other environmental effects. For the 1.8-m (6-ft) bed mentioned, the correction chart brings the original value of 11 Ω/m up to the correct value of 25 Ω/m.

(b) Vertical resolution depends on the contrast between the reference bed and the neighboring bed. The induction tool would do a better job on a transition between 1 and 2 Ω/m than on a transition between 1 and 25 Ω/m.

---

[5] In this context, the bed is assumed to be perpendicular to the borehole axis.

(c) Vertical resolution may also depend on whether the reference bed corresponds to a positive or negative change of the measured parameter in comparison with the neighboring beds. The resolution is different for a transition from low to high values than a transition from high to low values. The **contrast** aspect is illustrated on the examples given by the dielectric MWD sensors. Starting from a reference resistivity of 1 Ω/m, thin conductive beds are well identified. *Fig. 5.16* shows that a 1.2-m (4-ft) bed is well characterized by the two curves R_{ps} and R_{ad}. The R_{ps} curve even does a good job appraising a 0.3-m (1-ft) bed at the correct value. Conversely, still starting from a 1-Ω/m reference line, but now investigating resistive beds, the vertical resolution is altered. Neither of the two measurements is able to quantify a 10-Ω/m–1.2-m (4-ft) bed.

(d) Vertical resolution also relates to the overall range of the measured parameter. For example, the vertical resolution of the induction tool is likely to be more difficult to assess around 200 Ω/m than around 1 Ω/m.

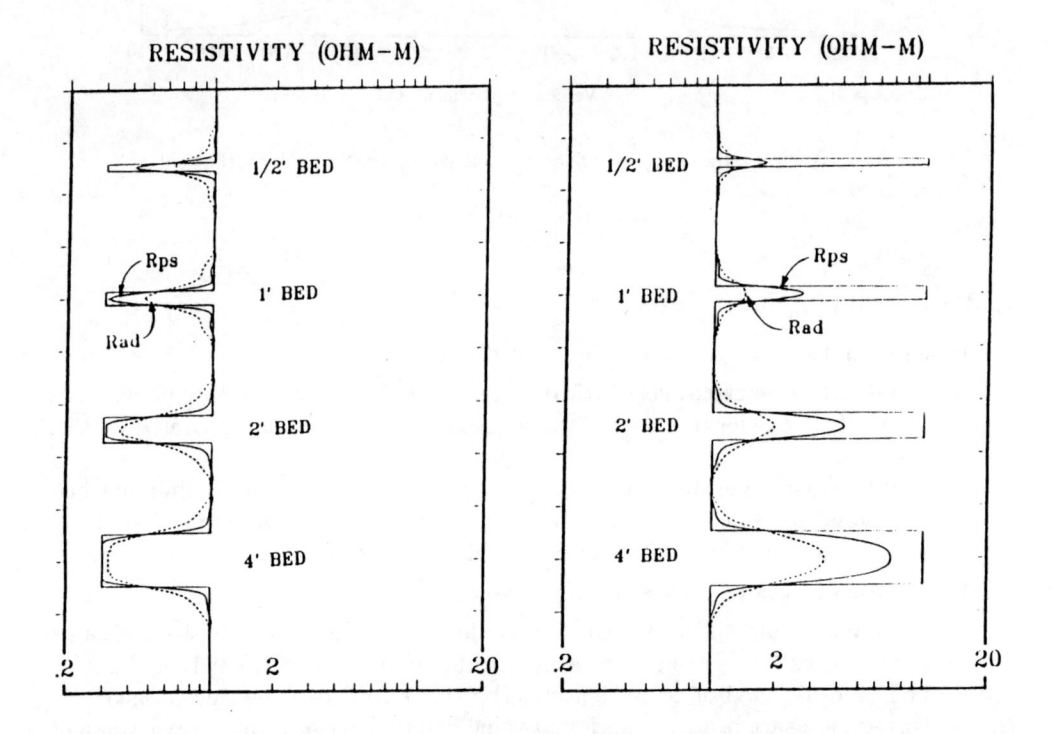

FIG. 5.16 : Effect of resistive and conductive shoulder beds on vertical resolution.
The readings of the *Schlumberger* MWD resistivities, R_{ps} and R_{ad} are related to the contrast. In one case, the thinner beds are more conductive (left). In the other, they are more resistive (right).

Courtesy of SPWLA [4]

A practical way to evaluate vertical resolution in an area is to run the logging tool over an interval of laminations of different thicknesses, if available, and controlled by core information. The thickness of the thinnest bed identified corresponds to the **local vertical resolution**.

Table 5.3 gathers the estimated vertical resolutions of the most common logging tools supplied by *Schlumberger*. All these values are subject to a number of limiting conditions which are explained in [6].

TABLE 5.3
VERTICAL RESOLUTION OF LOGGING TOOLS

Measurement	Vertical resolution	
	(in.)	(mm)
Formation MicroScanner*	0.2	5
Stratigraphic High-Resolution Dual Dipmeter	0.4	10
High-Resolution Dipmeter	0.5	13
EPT* Electromagnetic Propagation: Attenuation	2	50
Transit time	2	50
Microlog	2 to 4	50 to 100
Micro Spherical Focused Log	2 to 3	50 to 76
Phasor*-Induction SFL: Deep	84 to 96	2000
Medium	60 to 72	1500
Enhanced Resolution Phasor Induction: Deep/Medium	36	910
Spherically Focused Resistivity	30	760
Laterolog	24	610
Litho – Density*: Density	15	380
Enhanced resolution density	4	100
P_e	2	50
Compensated Neutron: Porosity	15	380
Resolution matched porosity	24	610
Enhanced processing porosity	12	310
Gamma Ray	8 to 12	200 to 310
Natural Gamma Ray	8 to 12	200 to 310
Array – Sonic*: Standard mode	48	1200
6-in. Δt mode	6	150
Borehole compensated sonic	24	610

From The Technical Review, courtesy of Schlumberger [1]

5.3.2 Modern definition of vertical resolution

The following procedure, extracted from [6], yields a specification that reflects more closely the actual geometrical behavior of a logging tool.

1. Give the set of conditions of application.

2. Derive two vertical response functions, one for total response and the other for the deeper 50% of the formation.

3. Specify three widths such that the areas above them and bound by the vertical response function correspond to 10%, 50% and 90% of the total area under the response function curve.

4. Repeat the previous steps for the two defined vertical response functions.

Example: a deep induction measurement

The integrated deep and total responses of a hypothetical deep induction tool are shown *Fig. 5.17*.

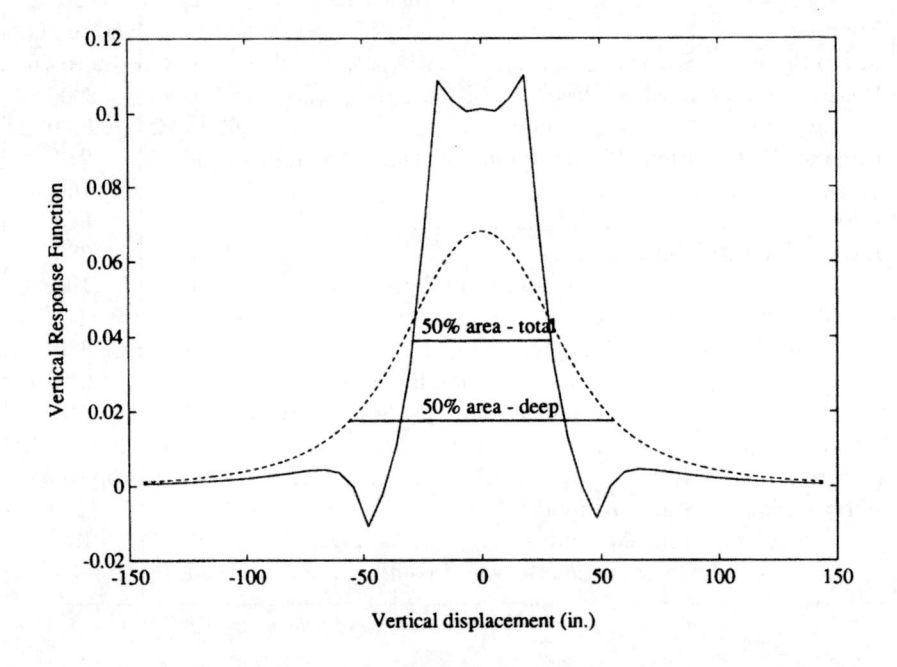

FIG. 5.17 : Vertical geometrical factor of a deep induction measurement.
Courtesy of SPWLA [6]

The following geometrical specifications are obtained:

TABLE 5.4
VERTICAL RESOLUTION OF AN INDUCTION TOOL

Measurement characteristic	in.		
Integrated radial geometrical factor	24	63	300
Total vertical width	42	57	151
Deep vertical width (beyond 63 in.)	50	109	243

Courtesy of SPWLA [6]

5.3.3 Reduction or enhancement of vertical resolution

Effects of filtering and signal processing on vertical resolution are described in Chapter 11. Processing techniques that allow the enhancement of vertical resolution are reviewed in Chapter 12.

5.3.4 Combination of measurements with different vertical resolutions

Most logging curves are derived through algorithms combining several raw measurements having different vertical resolutions. It is necessary to understand how these measurements are combined before making any inference on the vertical resolution of the usable data.

Example: Schlumberger thermal neutron porosity

The raw measurements are the instantaneous near and far counting rates. They are used to derive two parameters, ϕ_1 and ϕ_2. ϕ_1 is derived from the direct ratio of the counting rates. ϕ_2 is derived from the raw counting rates, but before being integrated in an algorithm, the counting rates are depth matched and resolution matched: the near-spacing detector counting rate resolution is **degraded** to match the long-spacing detector counting rate resolution. For this reason, ϕ_2 has a vertical resolution of 61 cm (24 in.).

This resolution matching allows the combination of two measurements relating to the same volume of formation, which improves the relevance of the resulting curve. *Fig. 5.18* illustrates the resolution matching concept.

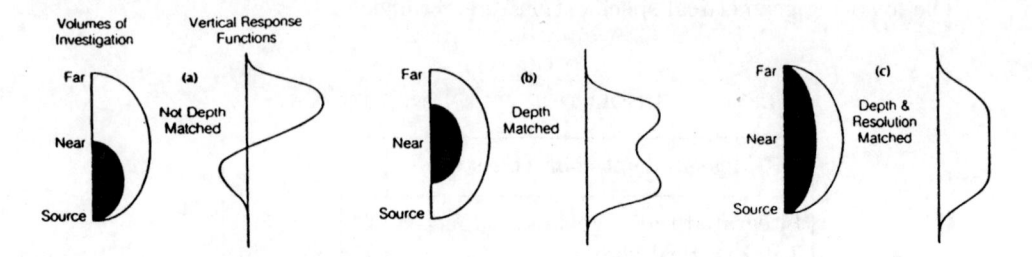

FIG. 5.18 : Methods of combining measurements from two sensors having different vertical resolutions. In each case, the volumes of investigation and the corresponding vertical response functions are shown:
  a. Traditional method for combining thermal neutron measurements (no depth matching).
  b. Traditional method for combining density measurements (depth matching only).
  c. Ideal case for both (depth and resolution matching).

Courtesy of Schlumberger [1]

5.3.5 Consonant sensors

Inaccuracies develop when measurements originating from sensors scanning different volumes of formation are combined to derive the petrophysical characteristics. The concept of consonance [3] quantifies the ability of two sensors to observe the same formation.

The ratio of the volumes investigated by two sensors is called consonant ratio, a value of one being computed when the volumes are identical. The consonant ratio is derived from the azimuthal coverage, the depth of investigation and the vertical resolution. These values are gathered in *Table 5.5* for some selected sensors. The resulting consonant ratios are listed in *Table 5.6*.

TABLE 5.5
VOLUMETRIC CHARACTERISTICS OF SCHLUMBERGER LOGGING TOOLS

The high resolution measurements are averaged over 1 ft. The depth of investigation of nuclear tool is taken at 90% of geometrical factor. The azimuth angle is estimated from the tool axis.

	CMR	EPT	R_{xo}	P_e	ϕ_{epith}	ρ_b	Σ	ϕ_{th}	Sonic
Azimuth....................	23	23	23	23	45	45	45	360	360
Depth of investigation (in.)	1.5	1.5	3	3	7	4	6	9	9
Vertical resolution (in.)....	12	12	12	12	12	12	12	24	30

Courtesy of SPWLA [3]

TABLE 5.6
CONSONANCE OF TOOLS

	ϕ_{epith}	ρ_b	Σ	ϕ_{th}	Δt
CMR/EPT	44	14	32	1152	1440
R_{x0}/P_e	11	3.5	4	288	360
ρ_b	3	?	2,25	81	101
U	1	1	1.36	81	101
ϕ_{epith}	?	3	1.36	27	34

Courtesy of SPWLA [3]

5.4 Relation between depth of investigation and vertical resolution

With the development of high resolution devices, some claims are made regarding the coupling of centimeter-scale (inch-scale) vertical resolution and meter-scale (foot-scale) depth of investigation. Considering that most physical fields are spherical, it can be intuitively deducted that depth of investigation and vertical resolution should be roughly of the same magnitude. However, in some cases, such as that of the laterolog, focusing changes the picture.

References

1 Allen, D., Anderson, B., Barber, T., Everett, B., Flaum, C., Hemingway, J., des Ligneris, S., Morriss, C., "Advances in high resolution logging," pp. 4-15, *The Technical Review*, Vol. 36, No. 2, 4-1988.

2 Antkiw, S., "Depth of investigation of the dual-spacing thermal neutron decay time logging tool," paper CC, *Trans.* SPWLA 17th annual logging symposium, 1976.

3 Casu, P. A., Andreani, M., Klopf, W., "Using consonant-measurement sensors for a more accurate log interpretation," paper NN, *Trans.* SPWLA 39th annual logging symposium, Keystone, 1998.

4 Clark, B., Lüling, M. G., Jundt, J., Ross, M., Best, D., "A dual depth resistivity measurement for FEWD," paper A, *Trans.* SPWLA 29th annual logging symposium, San Antonio, 1988.

5 Faivre, O., Catala, G., "Dip estimation from azimuthal laterolog tools," paper CC, *Trans.* SPWLA 36th annual logging symposium, Paris, 1995.

6 Flaum, C., Theys, P. P., "Geometrical specifications of logging tools: A need for new definitions," paper ZZ, *Trans.* SPWLA 32nd annual logging symposium, Midland, 1991. Also presented at the SPWLA European symposium, London, 1991.

7 Sherman, H., Locke, S., "Depth of investigation of neutron and density sondes for 35% porosity sand," paper Q, *Trans.* SPWLA 16th annual logging symposium, 1975.

6

Elements of metrology III: other attributes

6.1 Sensitivity to the environment

6.1.1 Effect of temperature

In addition to influencing electronic circuitry, high temperature directly affects the sensors; in particular, the nuclear detectors. A thallium-activated sodium iodide detector, described in Section 9.2.2.1 and commonly used in the oil field, suffers a light output loss of 30% at 140°C (284°F), while a bismuth germanate detector would withstand a drop of efficiency by an order of magnitude at 110°C (230°F) [3 and 4]. In addition, the optical coupling between the detector and the photomultiplier, generally provided by silicon compounds, tends to degrade above 175°C (350°F). It is possible to reduce temperature effects by using Dewar flasks, but this can be achieved only to the detriment of detector size with, as a consequence, a reduction of counting rates. *Fig. 6.1* shows the behavior of different detectors under increased temperatures.

6.1.2 Effect of pressure

High pressure may affect logging sonde geometry and thus may change the distance between transmitters and receivers. Induction sondes should be characterized at downhole pressure in order to have proper sonde error compensation. This issue is further discussed in Section 17.6.3.1.

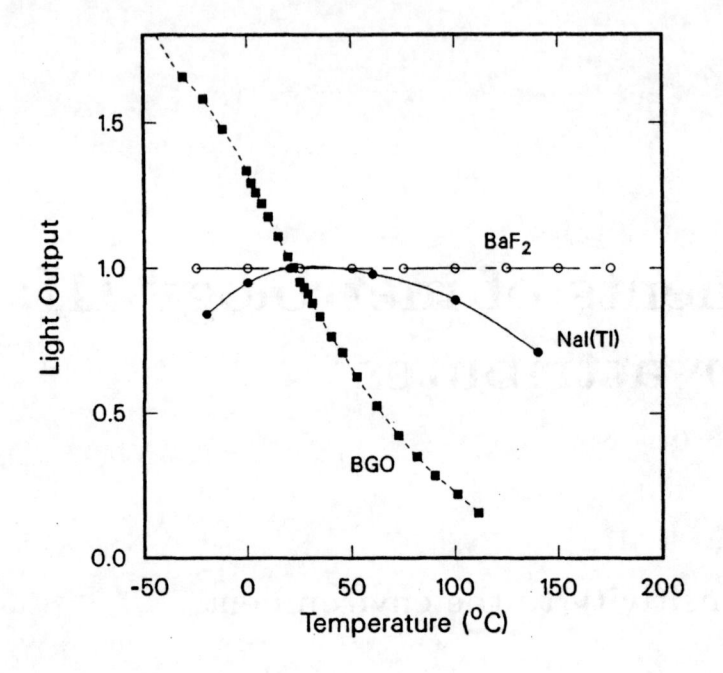

FIG. 6.1 : Changes in characteristics of a nuclear detector with temperature.
The temperature dependence of three types of scintillation detectors [NaI(Tl), BGO and BaF$_2$] is shown with the light output normalized to unity at 22°C.
Courtesy of J. Schweitzer [4]

6.1.3 Effects related to magnetism

Magnetic fields affect all scintillator crystal/photomultiplier systems by disturbing the path of the photons as they pass from one dynode to the next. This can be critical in the case of a spectrometric (energy-dependent) detection [1].

Temperature and magnetism effects in nuclear detectors are generally compensated for by the use of a reference source with a constant energy. Any anomaly may be detected with the quality control curves further detailed in Chapter 18.

Magnetic interferences induced by the tool, tubulars or the drillstring affect directional surveys (Chapter 20).

6.1.4 Effect of flux of other particles

Nuclear detectors often operate in a mixed flux of neutrons and gamma rays. These particles can create parasitic signals that may affect the intended measurement. For instance, natural gamma rays cannot be differentiated from the gamma rays induced

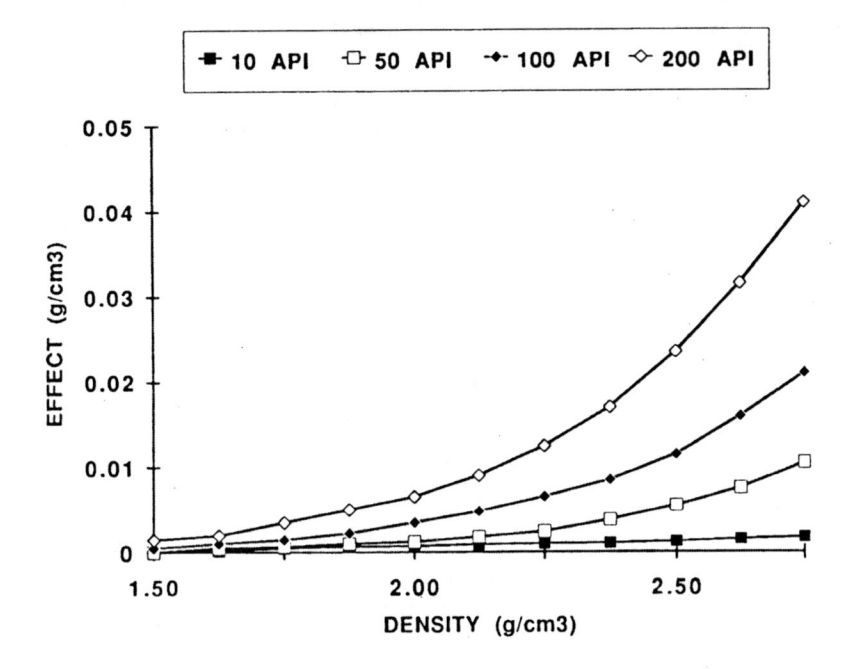

FIG. 6.2 : Effect of natural gamma rays on density readings. Nonspectrometric detectors are unable to differentiate induced and natural radiation. In a high natural gamma ray environment, formations appear lighter.
Courtesy of Schlumberger

by a controlled source. For some density detectors, natural gamma rays would affect the density measurements. This effect is quantified in *Fig. 6.2*.

Another unwanted interference is when nuclear detectors are exposed to visible light during maintenance or repair. The photon flux can cause noticeable deterioration of detector performance. Special handling procedures preclude this occurrence.

Finally, the formation may be activated by a neutron source, the consequence being an increase of gamma rays unrelated to the natural radioactivity of the formation.

6.2 Time response

6.2.1 Hysteresis

Hysteresis corresponds to the discrepancy observed on a measurement made in two different conditions—after an increase of the output signal, and after a decrease.

Hysteresis is particularly critical for pressure gauges. Special operating procedures should be set so that the measurement is always performed only by successive increases or decreases. A hysteresis cycle is depicted in *Fig. 6.3*.

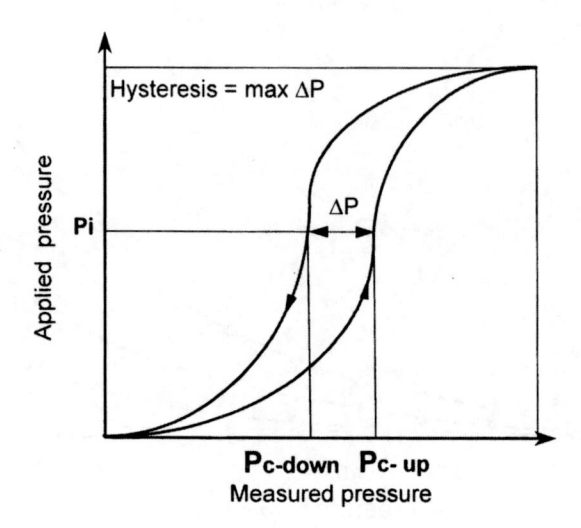

FIG. 6.3 : Effect of hysteresis.

Courtesy of SPWLA [5]

6.2.2 Dead time

Dead time is the delay required before a measurement can be repeated. The notion of dead time is illustrated on a device that suffers from it, the Geiger-Müller tube (Section 9.2.1).

Immediately after the creation of a pulse, the Geiger-Müller tube is unable to react to any incoming radiation. The tube is inoperative and is described as dead. The dead time is defined as the period of time between the reference pulse detection and a second Geiger-Müller discharge. In practical terms, the electronics located downstream of the tube would not be able to detect any pulse below a certain threshold. By extension, dead time may also describe the time necessary for the electronics to be capable of detecting a second pulse. Corrections for dead time can be designed through theoretical modeling or can be measured explicitly.

Recovery time defines the time necessary for the tube to return to its original state. Dead time and recovery times are depicted on *Fig. 6.4*.

A phenomenon similar to dead time affects the pressure gauges. **Relaxation time** is the time needed after a high temperature and pressure exposure for the gauge to recover a stabilized signal.

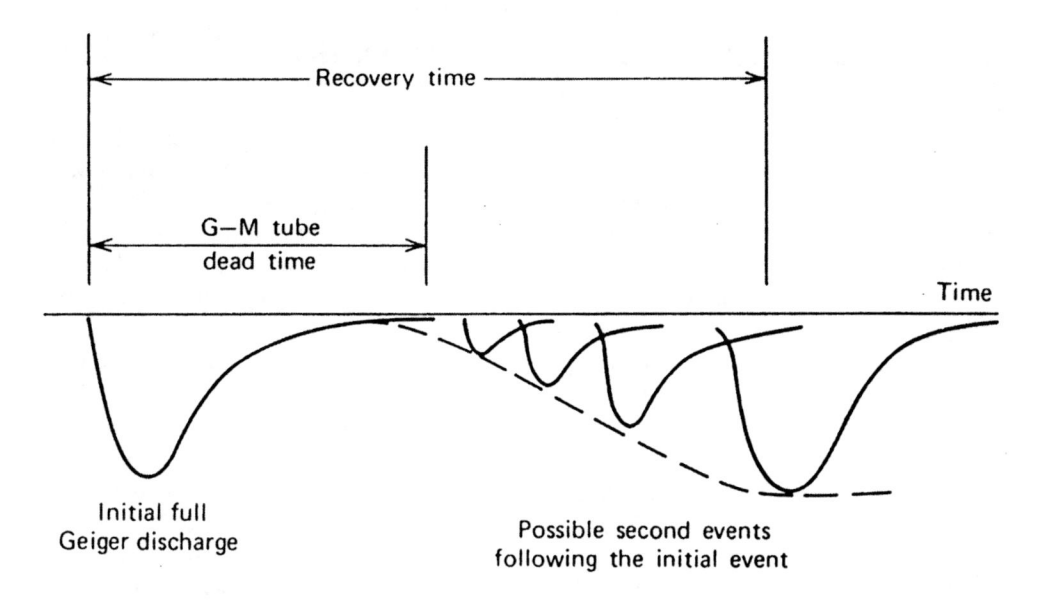

FIG. 6.4 : Dead time and recovery time of a Geiger-Müller counter.
Courtesy of Wiley & Sons [VIII]

6.3 Effect of the measuring apparatus on the measurement

Numerous logging devices affect the measurement they attempt to perform. For instance, a conductive accessory could affect the induction measurement by creating an additional signal. A fluid density measurement can be affected by the metallic masses of the tool. Also, a protective sleeve could affect the log accuracy during drillpipe conveyed operations. The dual laterolog K factor, which controls the final value of the resistivity, varies with the length and nature of the metallic accessories located above and below the sonde. Shifts as high as 10% can be observed.

Exercise 4

After review of paragraphs 6.1 to 6.3, establish for each effect the impact on the measurement error in terms of random and systematic components.

6.4 Sensitivity to shocks and vibrations

At the onset of the technique, logging tools were very inefficient spudding devices. Repeated usage of the tools to negotiate a formation bridge or ledge was often followed by failure or damage of the equipment. The addition of logging sensors to the drillstring has changed this status. While wireline logging tools are built to sustain tens of g accelerations, the MWD tools can bear accelerations of several hundred g.[1] In addition, the latter are tested for repeated levels of acceleration to ensure resistance to fatigue. They often undergo several thousand torture cycles.

An analysis of shocks versus time shows that wireline tools encounter the highest accelerations during transportation, typically at the docks or on rough roads. This is why a verification before survey (Chapter 18) is a recommended practice. Conversely, MWD tools experience downhole shocks 10 times larger than during transportation. These tools are not likely to be affected by tough transportation conditions and wellsite prejob checks are less critical.

MWD building techniques and knowhow are currently being transferred to wireline logging tools. PLATFORM EXPRESS* has been designed with MWD design methods.

6.5 Downhole access

The ability of a logging tool to reach the zones to be evaluated is an important feature. Sensor or instrumentation size is often critical. For instance, it is almost impossible to build a small-diameter inertial platform. Size has also been a major impediment to particle accelerometers. Other metrological characteristics such as accuracy and precision may sometimes be sacrificed to size and the customer needs to balance the different characteristics to optimize drilling and logging programs.

Tool size is often reduced by the size of the tubulars that need to be accessed before reaching the logging interval (which could have a much larger borehole or diameter size). This is the case of through-tubing and pumpdown operations. In the latter, logging tools are pushed down inside drillpipes before reaching an openhole interval.

Tool diameter is not the only constraint that controls access to downhole formations. Hole curvature is also of importance. This is generally expressed in the industry as dogleg severity. *Figure 6.5* shows the relation between maximum dogleg severity and hole size for a set of tools designed for slim and deviated holes.

Lesser sensitivity to severe doglegs is obtained through the use of special accessories, such as knuckle joints, which give flexibility to the entire tool string. These joints have the drawback of lowering the reliability of the tool set by adding connec-

---

[1] 250 g is approximately equivalent to the force felt by a driver when stopping a car travelling 240 km/h (150 miles/h) over a distance of 1 m (3 ft).

tions and removing the sensors farther away from the bottom of the hole. Their use needs to be carefully planned.

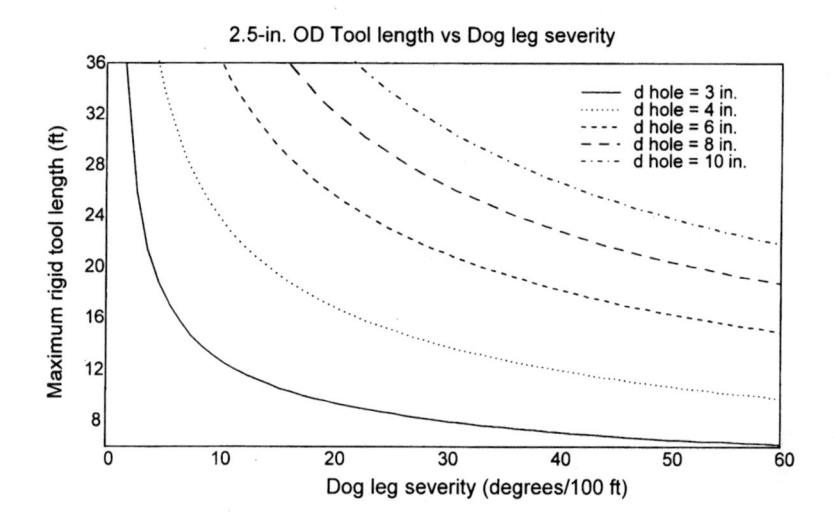

FIG. 6.5 : Dogleg severity as a function of hole size for the SlimAccess* logging set.

Courtesy of Schlumberger

6.6 Reliability

Reliability is not usually considered an attribute of a measurement. However, it is to be stressed that in case of failure, the measurement may be nonexistent. Reliability cannot be directly observed, but it can be derived from the failure rate λ. Reliability, R, can then be expressed as

$$R = e^{-\lambda t},$$

in which t is time and e is the base of natural logarithms (2.718).

Failure rate is usually not constant throughout the life of the equipment and displays the behavior shown on *Fig. 6.6*.

The drilling and logging industry uses two complementary indicators to monitor reliability:

(a) mean time between failure (MTBF) (equal to the inverse of the failure rate)
(b) jobs between jobs with lost time failure (JbJwLTF).

Evaluation of reliability requires knowing which elements of a system are indispensable for its operation and identifying subsystems that function in parallel and

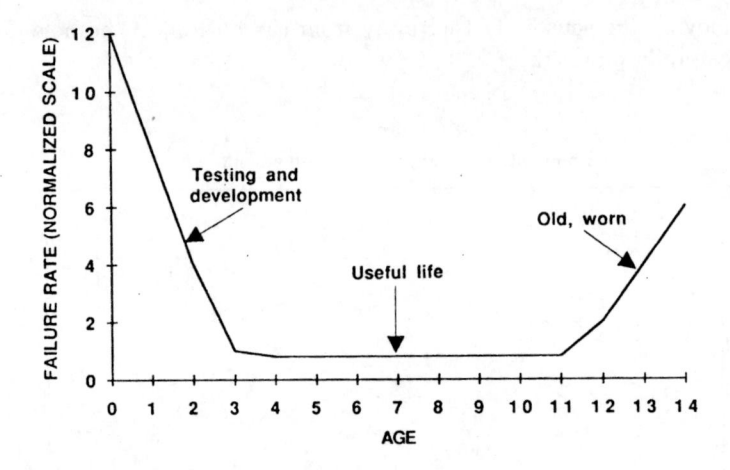

FIG. 6.6 : Reliability versus age of the tool.

provide some redundancy [2]. The reliability (or MTBF) of each subsystem is then evaluated. The total system reliability is then equal to one divided by the sum of the failure rates.

Exercise 5*

Analyze a MWD Triple Combo system. Assuming that each measurement sub has a MTBF of 1200 h and that the telemetry sub has a MTBF of 800 h, compute the system MTBF.

Reliability can be improved by the use of redundancy and by preventive maintenance (replacement or check of equipment at a frequency slightly higher than the frequency of failures) [IV].

References

1 Aitken, D., personal communication.

2 Martin, C., Philo, R. M., Decker, D. P., Burgess, T. M., "Innovative advances in MWD," *IADC/SPE 27516*, presented at the IADC/SPE drilling conference, Dallas, 2-1994.

3 Melcher, C. L., Schweitzer, J. S., "Gamma ray detector properties for hostile environments," pp. 876-878, *IEEE Trans. on Nuclear Sciences*, Vol. NS-35, No. 1, 2-1988.

4 Melcher, C. L. , Schweitzer, J. S., Liberman, A., Simonetti, J., "Temperature dependence of fluorescence decay time and emission spectrum of bismuth germanate," *IEEE Trans. on Nuclear Sciences*, Vol. NS-32, No. 1, 2-1985.

5 Veneruso, A., Machtalère, V., "Pressure gauge specifications for well testing," *The Log Analyst*," Vol. 35, No. 5, 9-1994.

7

Mathematical preliminary: propagation of errors

> The man who closes his door to errors also closes it to Truth.
>
> *Rabindranath Tagore*

7.1 Derivatives

7.1.1 Function of one variable

For a function depending on one variable, it is simple to define the derivative relative to this variable. For instance,

$$y = x^2$$

$$\frac{dy}{dx} = 2x.$$

Derivatives of higher orders are similarly defined:

$$\frac{d^2y}{dx^2} = \frac{d}{dx}\left(\frac{dy}{dx}\right) = 2.$$

Petrophysical example

The porosity of a rock, ϕ_D, with known matrix and fluid[1] densities ρ_{ma} and ρ_{mf}, can be expressed as a function of the log-derived density, ρ_b:

$$\phi_D = \frac{\rho_{ma} - \rho_b}{\rho_{ma} - \rho_{mf}}.$$

Selecting $\rho_{ma} = 2.65 \text{ g/cm}^3$ and $\rho_{mf} = 1.0 \text{ g/cm}^3$,

$$\phi_D = \frac{2.65 - \rho_b}{1.65} = 1.61 - 0.6\rho_b.$$

The derivative of ϕ_D, relative to ρ_b is

$$\frac{d\phi_D}{d\rho_b} = -\frac{1}{\rho_{ma} - \rho_{mf}} = -0.6.$$

This derivative quantifies the sensitivity of ϕ_D to the density ρ_b. When ρ_b increases by as much as 1.0 g/cm^3, ϕ_D decreases by 0.6, equivalent to 60 pu.

7.1.2 Function of several variables

Most relations in physics and petrophysics associate one quantity with several variables. It is therefore necessary to introduce partial derivatives. A function y of two variables u and v is assumed:

$$y = f(u, v).$$

Two partial derivatives $\frac{\partial f}{\partial u}$ and $\frac{\partial f}{\partial v}$ can be defined for y. The practical method to obtain the partial derivation of y relative to variable u, noted as $\frac{\partial f}{\partial u}$, is to assume that v is constant. Similarly, $\frac{\partial f}{\partial v}$ is computed by finding the derivative of f as a function of variable v while assuming u constant.

Petrophysical example

It is likely that Archie's formula is the one most used by petrophysicists. If R_w, a, m, and n are assumed to be constant, then R_t is a function of only two variables, ϕ and S_w:

$$R_t = \frac{aR_w}{\phi^2 S_w^2} = aR_w \phi^{-2} S_w^{-2}.$$

To derive the partial derivative in relation to ϕ, S_w is assumed to be constant. Then Archie's formula is similar to $y = \alpha x^{-2}$ with $x = \phi$ and α is a constant:

$$\frac{\partial R_t}{\partial \phi} = -2\frac{aR_w}{\phi^3 S_w^2}.$$

---

[1] The fluid contained by the rock at the vicinity of the borehole is mostly mud filtrate.

In the same way,
$$\frac{\partial R_t}{\partial S_w} = -2\frac{aR_w}{\phi^2 S_w^3}.$$

Exercises 6* and 7*

6.* Given the function y, such that
$$y = f(x,u) = 3x^2 + 4xu^3,$$
determine all nonzero partial derivatives.

7.* Perform the same exercise with y:
$$y = 5\log u + 8\log v^2.$$

7.2 Setting the problem of error propagation

It often happens that a desired characteristic of the formation cannot be directly measured. However, it is generally possible to estimate this parameter as a function of measurable inputs. For instance, no physical process allows direct observation of the formation density in a well, but this quantity can be derived from the process of counting gamma rays with two detectors. The counting rates are n_1 and n_2, and

$$\rho_b = f(n_1, n_2).$$

Defining the function relating ρ_b to n_1 and n_2 is essential, but it is also important to evaluate how the uncertainties of the measurable inputs n_1 and n_2 are propagated to the final result.

Earth scientists face a cascade of propagating errors. Taking the example mentioned above, we can consider that ρ_b is used to derive ϕ_D in conjunction with two additional inputs, ρ_{ma} and ρ_{mf}. The density-derived porosity, ϕ_D, is used with other inputs—sonic, neutron and core porosities (respectively ϕ_S, ϕ_N and ϕ_C)—to evaluate the final porosity, ϕ. ϕ can then be used to compute water saturation, S_w, and finally hydrocarbon pore volume, HCPV (*Fig. 7.1*).

Also, depending on the scale considered, the type of error may change when the integration from the well scale to the field scale is attempted. Systematic error can be reclassified as random. In other words, a systematic error, when neither detected nor quantified, can be limited by an estimated maximum error. Systematic error can then be treated the same way as random error. *Table 7.1* gives a qualitative evaluation of the errors in the process of the field integration.

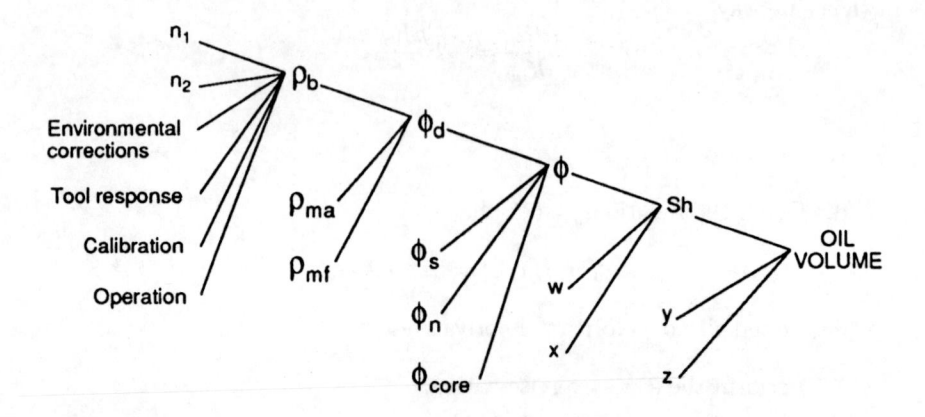

FIG. 7.1 : Tree of input parameters to hydrocarbon pore volume computation; each input is associated to an error.

TABLE 7.1
EVOLUTION OF ERROR TYPE DURING FIELD INTEGRATION

Origin of uncertainty	Summation	
	Over the reservoir	Over the field
Acquisition		
Nuclear	Random	Random
Faulty calibration	Systematic	Random
Environmental effects	Systematic	Systematic
Processing		
Depth mismatch	Systematic/ Random	Random
Interpretation model	Systematic	Systematic
Interpretation parameters	Systematic	Systematic
Estimation for missing logs	Systematic	Systematic
Field mapping		Random

Courtesy of C. Clavier

The quantification of the propagated error can be performed two ways, by analysis [III] or by number-crunching [1].

7.3 Analytical quantification of propagated error

The propagation of errors is analyzed when it is possible to describe a parameter y as an analytical function[2] of two inputs u and v:

$$y = f(u, v).$$

We assume that observations u_1 to u_n and v_1 to v_n are made and that

$$\overline{u} = \frac{1}{n} \sum_{i=1}^{n} u_i$$

is the mean value of the set u_1 to u_n. $\overline{v}$ is the mean value of the set v_1 to v_n.
It is generally assumed[3] that the most probable value of x is given by the formula

$$\overline{y} = f(\overline{u}, \overline{v}).$$

The variance of y, σ_y^2, is defined as

$$\sigma_y^2 = \lim_{n \to \infty} \frac{1}{n} \sum_{i=1}^{n} (y_i - \overline{y})^2,$$

in which $y_i = f(u_i, v_i)$. σ_y is the standard deviation.

In most cases, there are few observations (in logging, it is generally limited to two observations, the main and repeat passes run over short intervals). Nevertheless, as the conditions under which the measurements are performed are known, estimation of variance of u and v is generally available. It is then possible to derive a relation linking the variance of y to the variances of u and v.

The difference $y_i - \overline{y}$ can be developed in a Taylor series:

$$y_i - \overline{y} \approx (u_i - \overline{u}) \frac{\partial y}{\partial u} + (v_i - \overline{v}) \frac{\partial y}{\partial v}.$$

This limited series assumes that the errors are **small**. Otherwise, it would be necessary to include higher order partial derivatives. Combining the expressions of σ_y^2 and of $y_i - \overline{y}$, we obtain the following

$$\sigma_y^2 = \lim_{n \to \infty} \frac{1}{n} \sum_{i=1}^{n} [(u_i - \overline{u}) \frac{\partial y}{\partial u} + (v_i - \overline{v}) \frac{\partial y}{\partial v}]^2$$

$$\sigma_y^2 = \lim_{n \to \infty} \frac{1}{n} \sum_{i=1}^{n} [(u_i - \overline{u})^2 (\frac{\partial y}{\partial u})^2 + (v_i - \overline{v})^2 (\frac{\partial y}{\partial v})^2 + 2(u_i - \overline{u})(\frac{\partial y}{\partial u})(v_i - \overline{v})(\frac{\partial y}{\partial v})].$$

---
[2] In addition, the function f must have a proper behavior: successive derivatives of f can be defined and f is continuous. These conditions are not studied any further and are assumed to be met.
[3] This is reasonable if u and v are not covariant.

We recognize

$$\sigma_u^2 = \lim_{n\to\infty} \frac{1}{n} \sum_{i=1}^{n} (u_i - \bar{u})^2,$$

$$\sigma_v^2 = \lim_{n\to\infty} \frac{1}{n} \sum_{i=1}^{n} (v_i - \bar{v})^2,$$

$$\sigma_{uv}^2 = \lim_{n\to\infty} \frac{1}{n} \sum_{i=1}^{n} (u_i - \bar{u})(v_i - \bar{v}).$$

The equation above can be written as

$$\sigma_y^2 \approx \sigma_u^2 \left(\frac{\partial y}{\partial u}\right)^2 + \sigma_{uv}^2 \left(\frac{\partial y}{\partial u}\right)\left(\frac{\partial y}{\partial v}\right) + \sigma_v^2 \left(\frac{\partial y}{\partial v}\right)^2.$$

Consider the middle term:

$$\sigma_{uv}^2 \left(\frac{\partial y}{\partial u}\right)\left(\frac{\partial y}{\partial v}\right).$$

Only if fluctuations of u and v are uncorrelated, should we expect, on the average, to find approximately as many equal negative values for this term as positive values. We would expect this term to vanish in the limit of a large selection of observations. In that case, the equation above reduces to

$$\sigma_y^2 \approx \sigma_u^2 \left(\frac{\partial y}{\partial u}\right)^2 + \sigma_v^2 \left(\frac{\partial y}{\partial v}\right)^2.$$

If y depends on a larger number of variables,

$$y = f(u_1, u_2, \ldots, u_p),$$

then the formula is generalized to

$$\boxed{\sigma_y^2 \approx \sum_{j=1}^{p} \sigma_{uj}^2 \left(\frac{\partial y}{\partial u_j}\right)^2.}$$

7.4 Applications

7.4.1 Sums or differences

Assume

$$y = u - v,$$

$$\frac{\partial y}{\partial u} = 1 \qquad \frac{\partial y}{\partial v} = -1,$$
$$\sigma_y^2 = \sigma_u^2 + \sigma_v^2.$$

A common application of this formula to a nuclear measurement is the subtraction of the background [VIII]:

$$\text{net counts} = \text{total counts} - \text{background counts}.$$

Numerical application

Total counts are $u = 877$ and background counts are $v = 123$. Then, the net counts y are 754. For any radioactive decay measurement, $\sigma^2 = $ counts (Section 10.1.3 for details). Then you have the following:

$$\sigma_u^2 = 877 \qquad \sigma_v^2 = 123$$
$$\sigma_y^2 = 877 + 123 = 1000$$
$$\sigma_y = 31.6$$
$$\text{net counts} = 754 \pm 31.6.$$

In addition to the practical application to background evaluation, this simple example teaches an important lesson. Even though input v is subtracted from input u, the corresponding errors, and hence the squares of the standard deviations, add up.

7.4.2 Multiplication by a constant

$$u = \frac{x}{a}$$
$$\frac{\partial u}{\partial x} = \frac{1}{a}$$
$$\sigma_u = \frac{\sigma_x}{a}.$$

If x counts are recorded over a time τ, then the counting rate u is

$$u = \frac{x}{\tau}.$$

In most practical cases, time is measured with a high certainty and therefore only the uncertainty on the counts needs to be considered ($\sigma_\tau = 0$).

Example

$x = 1120$ counts and $\tau = 5$ s. The counting rate is $1120/5 = 224$. The associated standard deviation is

$$\sigma_u = \sigma_x/\tau.$$

$$\sigma_u = \frac{\sqrt{1120}}{5} = 6.7$$
$$u = 224 \pm 6.7 \text{ cps}.$$

7.4.3 Combination of independent measurements

y is defined as a linear function of p independent measurements y_i, each weighted by a coefficient a_i:

$$y = \frac{a_1 x_1 + a_2 x_2 + \cdots + a_p x_p}{a_1 + a_2 + \cdots + a_p}$$

$$\sigma_y^2 = \sum_{i=1}^{p} \sigma_{x_i}^2 \left(\frac{\partial y}{\partial x_i}\right)^2 = \frac{N}{D},$$

where $N = \sum_{i=1}^{p} a_i^2 \sigma_{x_i}^2$ and $D = (\sum_{i=1}^{p} a_i)^2$.

It is possible to find a set of factors a_i with i varying from 1 to p so that the resulting uncertainty σ_y is minimum. In order to have σ_y^2 minimum, the partial derivative of σ_y^2 with respect to the weighting factors a_i must be zero:

$$\frac{\partial \sigma_y^2}{\partial a_i} = 0.$$

With N and D defined as above, the p equations are equivalent to

$$\frac{\partial N}{\partial a_i} \times D = \frac{\partial D}{\partial a_i} \times N$$

or

$$a_i \sigma_{x_i}^2 \left[\sum_{i=1}^{p} a_i\right]^2 = \left[\sum_{i=1}^{p} a_i\right] \left[\sum_{i=1}^{p} a_i^2 \sigma_{x_i}^2\right].$$

This relation is simplified when the weighting factors are normalized, which corresponds to $\sum_{i=1}^{p} a_i = 1$. Then,

$$a_i = \frac{1}{\sigma_{x_i}^2} \times \left[\sum_{i=1}^{p} a_i^2 \sigma_{x_i}^2\right].$$

Each contributor x_i should be weighted inversely as the square of its own error. This result is often intuitively applied. Data with high uncertainty is often disregarded. Data with small uncertainty is given more weight.

For instance, assume two measurements with equal weights:

$$y = 0.5u + 0.5v.$$

$$\sigma_y = \sqrt{0.5^2\sigma_u^2 + 0.5^2\sigma_v^2} = 0.5\sqrt{\sigma_u^2 + \sigma_v^2}.$$

σ_y is most reduced when u and v have uncertainties of the same magnitude.[4] Conversely, if $\sigma_v > 3\sigma_u$, then the resulting uncertainty is larger than the one obtained by using only u.

Application to porosity

The effective porosity, ϕ_e, is assumed to be derived only from the thermal neutron logging instrument, which yields ϕ_N, and from the density tool, which gives ϕ_D. Then:

$$\phi_e = \alpha\phi_D + (1-\alpha)\phi_N,$$

in which α, the weighting factor, can be optimized so that σ_{ϕ_e} is minimal:

$$\alpha = \frac{1}{\sigma_{\phi_D}^2}[\frac{1}{\sigma_{\phi_D}^2} + \frac{1}{\sigma_{\phi_N}^2}]^{-1}.$$

$[1/\sigma_{\phi_D}^2 + 1/\sigma_{\phi_N}^2]^{-1}$ is somewhat constant—at low porosities, σ_{ϕ_D} is large while σ_{ϕ_N} is small; inversely, at high porosities, σ_{ϕ_D} is small and σ_{ϕ_N} is large. So α is essentially the inverse of σ_{ϕ_D}. α changes with the range of porosities under study.[5]

7.5 Examples of propagation of errors

Mathematical example

$$y = u^2 + v^2$$

$$\frac{\partial y}{\partial u} = 2u \qquad \frac{\partial y}{\partial v} = 2v$$

$$\sigma_y^2 = 4u^2\sigma_u^2 + 4v^2\sigma_v^2.$$

Petrophysical example: density

The density tool response, the relation between ρ_b and the counting rates collected by the downhole detectors, is:[6,7]

$$\rho_b = A + B\ln N_{ls} - C\ln N_{ss}.$$

---

[4] If $\sigma_u \simeq \sigma_v$, then $\sigma_y \simeq \sqrt{2}/2\sigma_u$. If $\sigma_u \simeq 3\sigma_v$, then $\sigma_y \simeq \sigma_u$.

[5] The weighting coefficients introduced here are different from the empirical coefficients on ϕ_N and ϕ_D used in well known relations such as $\phi = 0.5\,(\phi_N + \phi_D)$ and which are applied regardless of the porosity range.

[6] See Chapter 16 for a detailed explanation of this equation and Section 10.1.3, for derivation of the value of the standard deviation.

[7] The effect of background counts is neglected.

Therefore,

$$\frac{\partial \rho_b}{\partial N_{ls}} = B/N_{ls} \qquad \frac{\partial \rho_b}{\partial N_{ss}} = -C/N_{ss},$$

$$\sigma_{N_{ls}}^2 = N_{ls} \qquad \sigma_{N_{ss}}^2 = N_{ss},$$

$$\sigma_{\rho_b}^2 = N_{ls}(B/N_{ls})^2 + N_{ss}(C/N_{ss})^2,$$

$$\sigma_{\rho_b} = \sqrt{(B^2/N_{ls}) + (C^2/N_{ss})}.$$

Typical values for B and C are respectively 0.55 and 0.48 for type I tool and 0.63 and 0.70 for type II tool. In an aluminum block with an apparent density of about 2.59 g/cm^3, the following counting rates are read:[8] for type I tool, $N_{ls} = 500$ cps and $N_{ss} = 900$ cps; for type II tool, $N_{ls} = 2730$ cps and $N_{ss} = 3160$ cps. As a result, $\sigma_{\rho_b} = 0.029$ g/cm^3 for type I tool and $\sigma_{\rho_b} = 0.017$ g/cm^3 for type II tool.

For a given tool, a lower density gives higher counting rates and consequently a lower standard deviation σ_{ρ_b}. The uncertainty of the density-derived porosity is

$$\sigma_{\phi_D} = 0.6 \sigma_{\rho_b}.$$

For type I tool, $\sigma_{\phi_D} = 1.7$ pu and for type II tool, $\sigma_{\phi_D} = 1.0$ pu. Type I tool is equipped with detectors counting gamma rays regardless of their energy. Discrimination of unwanted low-energy gamma rays is performed with a cadmium shield. Type II tool has spectral detectors.

Exercises 8, 9*, 10*, 11* and 12*

8. If a table is round and its diameter is determined to within 1%, how well is its area known? Would it be better to determine its radius to within 1%?

9*. Derive the uncertainty on the porosity, ϕ_D.

1. Assume that ρ_{ma} and ρ_{mf} are not perfectly known ($\sigma_{\rho_{ma}} \neq 0$ and $\sigma_{\rho_{mf}} \neq 0$). Derive the general formula.

2. Compute σ_{ϕ_D} for $\sigma_{\rho_b} = 0.017$ g/cm^3, $\sigma_{\rho_{ma}} = 0.015$ g/cm^3, $\sigma_{\rho_{mf}} = 0.010$ g/cm^3, $\rho_b = 2.6$ g/cm^3, $\rho_{ma} = 2.65$ g/cm^3 and $\rho_{mf} = 1.1$ g/cm^3.

---

[8] Much smaller uncertainty levels are obtained with longer counting sequences. A 1-s measurement is equivalent to a logging speed of 9 m/min (1800 ft/h).

10*. Derive the uncertainty on the saturation.

1. Assume Archie equation with $m = n = 2$. R_w, m and n are known with no uncertainty.

2. Derive the general formula for the relative error σ_{S_w}/S_w.

3. Which input parameter is more important, ϕ or R_t?

4. Compute the relative error $\frac{\sigma_{S_w}}{S_w}$ assuming $R_w = 0.06\ \Omega/\text{m}$, $\phi_d = 30$ pu, $R_t = 10\ \Omega/\text{m}$ for technologies A and B. Assume $\frac{\sigma_{C_t}}{C_t} = 3\%$ for technology A and 2% for technology B. Assume that C_t and ϕ_D are independent variables and take $\sigma_{\phi_D} = 1.7$ pu for technology A and 1.0 pu for technology B.

11*. Find the analytical expression of σ_{S_w} derived from σ_ϕ, σ_{C_t}, σ_{R_w} and σ_m with Archie's formula. Assume $n = m$.

12*. Find the analytical expression of σ_{S_w} derived from σ_ϕ, σ_{R_t}, σ_{R_w}, $\sigma_{R_{sh}}$ and $\sigma_{V_{sh}}$ with Simandoux's equation:

$$\frac{1}{R_t} = \frac{1}{\beta} S_w^2 + \gamma S_w$$

with $\quad \dfrac{1}{\beta} = \dfrac{\phi^2}{aR_w} \quad\quad \gamma = \dfrac{V_{sh}}{R_{sh}}.$

Numerical application

1. $V_{sh} = 0$.

2. $V_{sh} = 0.2$ (or 20%); $R_{sh} = 1\ \Omega/\text{m}$; $R_w = 0.1\ \Omega/\text{m}$;

   $R_t = 10\ \Omega/\text{m}$; $\phi = 0.2$ (or 20 pu); $a = 0.8$.

   Take successively $\dfrac{\sigma_{V_{sh}}}{V_{sh}}$, $\dfrac{\sigma_{R_{sh}}}{R_{sh}}$, $\dfrac{\sigma_\phi}{\phi}$, $\dfrac{\sigma_{R_t}}{R_t}$ and $\dfrac{\sigma_{R_w}}{R_w}$ equal to 1 (or 100%) with all other relative uncertainties equal to 0. What can be concluded from the results?

7.6 Propagation of errors derived from a numerical method

The analytical approach to the propagation of errors assumes well-behaving data distributions, represented essentially by Gaussian or at least continuous functions. In practice, the information on the probability distribution of a parameter may be limited to few discrete values. In this case, a numerical method is more suitable to derive the error transmitted.

7.6.1 Theory

A numerical simulation takes the probability distribution of several input data and generates the distribution of the function derived from these inputs. The advantage of this method is that it handles poorly behaving distributions established with only few discrete values. Asymmetrical distributions, the nightmare of the analytical statistician, are smoothly digested.

7.6.2 Examples

Example 1: Rectangular distribution

The values selected are gathered in *Table 7.2*. The distributions of the values of resistivity, porosity and m are represented in *Figs 7.2a, b* and *c*.

TABLE 7.2
INPUT PARAMETERS FOR EXAMPLE 1

	m	R_w	R_t	ϕ
		(Ω/m)	(Ω/m)	(pu)
1	1.800	0.0200	19.50	12.00
2	1.825	0.0206	19.62	12.12
3	1.850	0.0212	19.75	12.25
4	1.875	0.0218	19.87	12.37
5	1.900	0.0225	20.00	12.50
6	1.925	0.0231	20.12	12.62
7	1.950	0.0237	20.25	12.75
8	1.975	0.0243	20.37	12.87
9	2.000	0.0250	20.50	13.00

Figure 7.2d shows the distribution of the frequency of the saturation values as well as the cumulative distribution of the saturation values. In this case, the saturation has a mean value of 22.5 su and a standard deviation of 2.9 su.

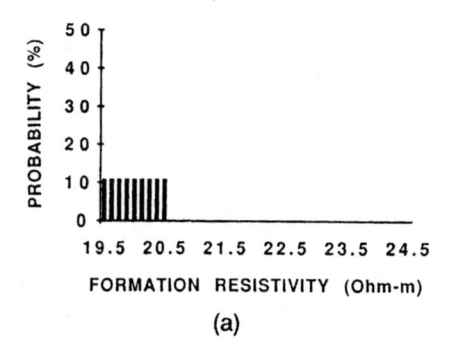

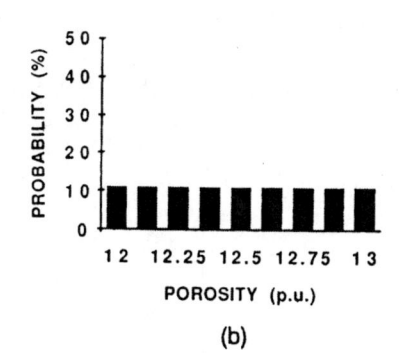

a. Input resistivity for S_w evaluation.

b. Input porosity for S_w evaluation.

c. Input cementation factor for S_w evaluation.

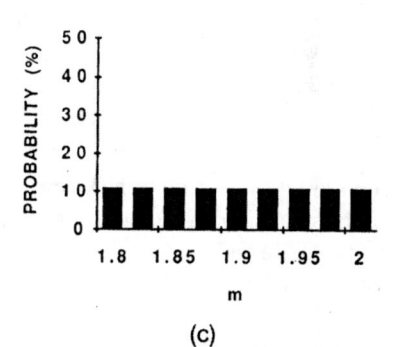

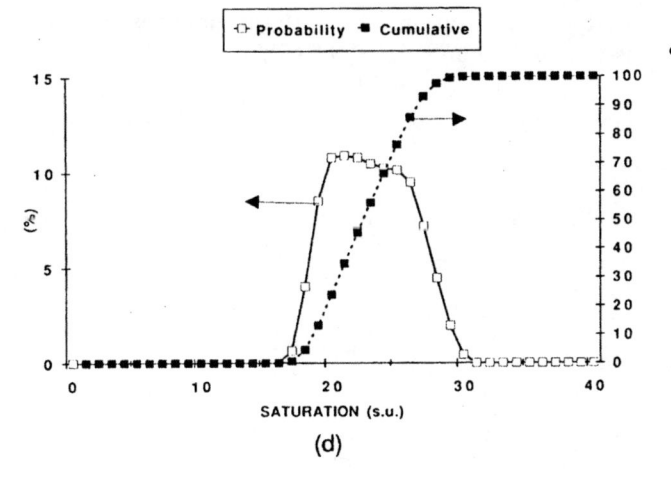

d. Probability distribution of the resulting saturation. The solid curve is the probability on the y-axis (scale 0 to 15% on the left, with the empty squares) that a given saturation (on the x-axis) results from the input distribution. The dashed curve with the black square is the cumulative probability (scale on the right from 0 to 100%) that the saturation is less than a given saturation value (x-axis).

FIG. 7.2 : Numerical simulation: example 1.

Example 2: Nonlinear distribution

Assume that the R_w and ϕ distributions are unchanged, but that the m distribution is weighted heavily towards 2.0. The spread of R_t values indicates that larger uncertainties are anticipated at high resistivity. The new input selection is shown in *Table 7.3*.

TABLE 7.3
INPUT PARAMETERS FOR EXAMPLE 2

	m	R_w (Ω/m)	R_t (Ω/m)	ϕ (pu)
1	1.80	0.0200	19.50	12.00
2	1.85	0.0206	19.62	12.12
3	1.90	0.0212	19.75	12.25
4	1.95	0.0218	19.87	12.37
5	2.00	0.0225	20.00	12.50
6	2.00	0.0231	21.00	12.62
7	2.00	0.0237	22.00	12.75
8	2.00	0.0243	23.00	12.87
9	2.00	0.0250	24.00	13.00

The new distributions of R_t and m are displayed in *Figs 7.3a* and *c*. *Figure 7.3d* shows the resulting saturation distribution. Notice the lack of symmetry and resemblance to a Gaussian curve. The saturation has a mean of 24 su and a standard deviation of 3.4 su.

Exercise 13*

1. Compute the most likely saturation value with the inputs listed in *Table 7.4* (assume $m = 2$ and $R_w = 0.03$ Ω/m).
2. How many values should be computed?
3. Make the histogram of the probability distribution of the saturation.

TABLE 7.4
DISTRIBUTION OF R_t AND ϕ

	R_t (Ω/m)	ϕ (pu)
1	25	18
2	30	19
3	35	23

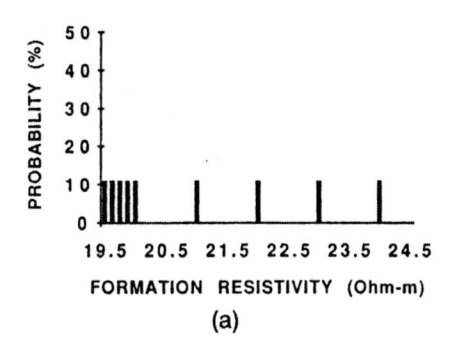

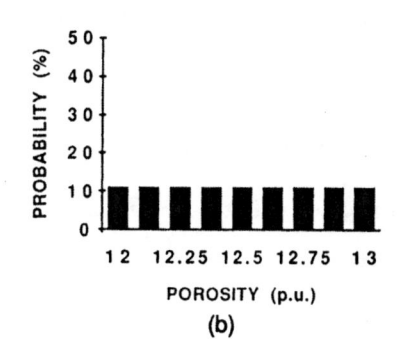

a. Input resistivity for S_w evaluation.

b. Input porosity for S_w evaluation.

c. Input cementation factor for S_w evaluation.

d. Probability distribution of the resulting saturation; same scales and codes as *Fig. 7.2*.

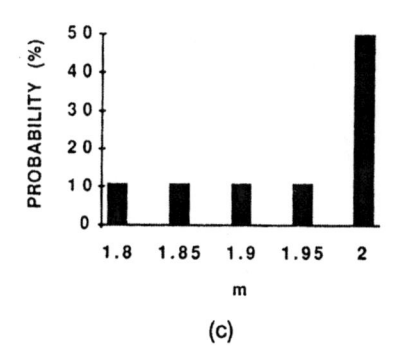

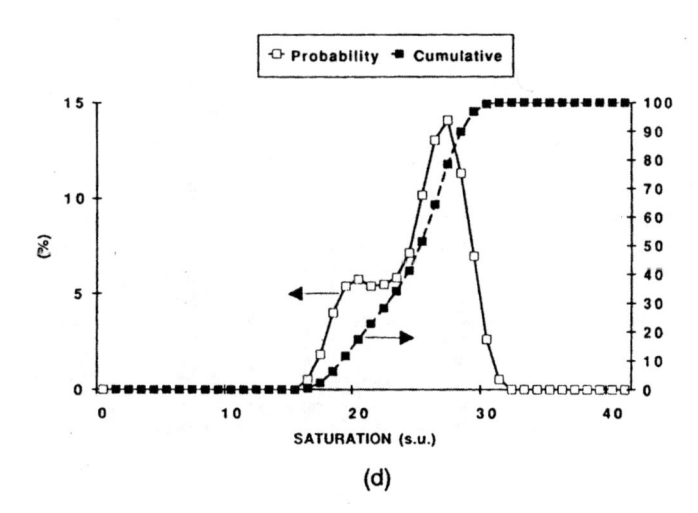

FIG. 7.3 : Numerical simulation: example 2.

7.6.3 Practical use of numerical methods

The exercise helps evaluate the time that would be taken if many different values are attributed to each input. The real application of numerical methods calls for dedicated computer software. These programs are classified under the category of risk analysis. They are currently used by several oil companies to predict the range of the uncertainties of porosity and hydrocarbon pore volume.

Example: Realistic computation of water saturation

Instead of using a single value that is unrealistic for the inputs to Archie's equation, the analyst can consider a range of equally probable values. In physical terms, it represents the one-σ interval of confidence around a mean value. *Table 7.5* is an example of such a spread [3].

TABLE 7.5
DUAL INPUT VALUES FOR ARCHIE'S EQUATION

Input	R_w (Ω/m)	ϕ (pu)	m	R_t (Ω/m)	n
Low	0.0350	29.80	1.780	1.84	1.81
High	0.0380	31.50	1.790	2.10	1.85
Average	0.0365	30.65	1.785	1.97	1.83

Figure 7.4 represents the probability distribution of the resulting water saturation.

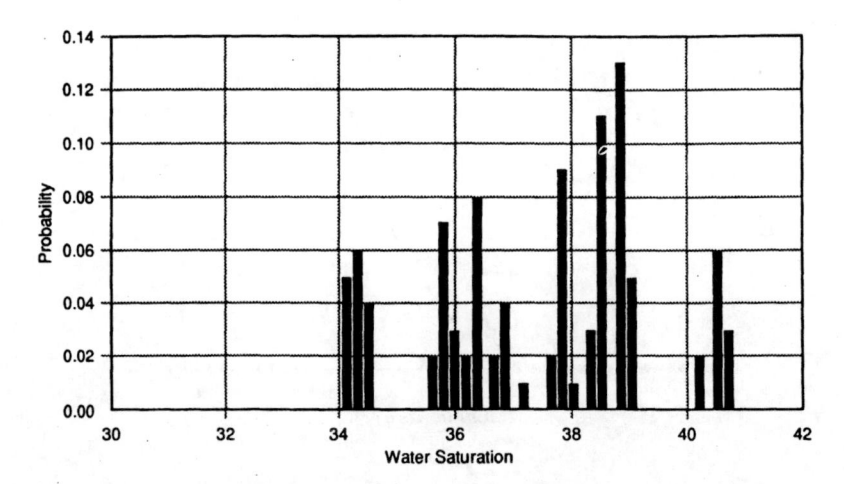

FIG. 7.4 : Risk simulation results.

Courtesy of SPWLA [3]

Table 7.6 lists the characteristics of the statistical distribution of water saturation. As expected, a range of values is derived.

Depending on the decision to be taken, the minimum, the maximum or a given percentile value can be selected.

TABLE 7.6
PROBABILITY DISTRIBUTION OF THE RESULTS

Characteristics		Saturation (su)
Minimum		34.03
Maximum		40.79
Mean		37.40
Standard deviation		1.92
Skewness		−0.16
Kurtosis		2.10
Percentile values	20%	35.84
	40%	36.84
	50%	37.82
	60%	35.56
	80%	38.83

Courtesy of SPWLA [3]

7.7 Applications of error propagation

7.7.1 Ranking of critical parameters

An important benefit of performing the analytical or numerical methods is evaluating the sensitivity of the output function to the different input parameters and, consequently, identifying those parameters that require the greatest precision and accuracy.

This approach directly applies to reserve evaluation [2]. Depending on reservoir characteristics, the uncertainty on the volume of hydrocarbons in place is not related to the input parameters in a unique way.

Table 7.7 shows that reserve evaluation is more sensitive to porosity, cementation and saturation exponents in a southern North Sea Rotliegende gas field than in a northen North Sea Brent oil field.

TABLE 7.7
SENSITIVITY ANALYSIS RESULTS

Each input parameter is allowed to vary by ±10%. The resulting change in hydrocarbon volume is computed and expressed as a percentage. Notice the lack of symmetry shown by negative and positive changes.

a. Rotliegende gas field: In this case ϕ and m are highly critical.

b. Brent oil field: ϕ is critical but m is less important.

	Correct value	±10% error	S_h $(1-S_{wi})$	GIIP (BCF)	Change (%)
			Rotliegende gas field		
R_w	0.018	0.0162	0.575	1041.52	+4.15
				1000.00	
		0.0198	0.531	960.51	−3.96
ϕ	0.15	0.135	0.503	819.1	−18.09
				1000.00	
		0.165	0.593	1180.90	+18.09
R_t	4.0	3.6	0.528	956.20	−4.38
				1000.00	
		4.4	0.573	1037.65	+3.77
m	2.0	1.8	0.630	1139.80	+13.98
				1000.00	
		2.2	0.459	831.00	−16.90
n	2.0	1.8	0.591	1069.20	+6.92
				1000.00	
		2.2	0.518	938.60	−6.14
			Brent oil field		
R_w	0.11	0.099	0.770	101.64	+1.64
				100.00	
		0.121	0.746	98.44	−1.56
ϕ	0.25	0.225	0.730	86.80	−13.20
				100.00	
		0.275	0.779	113.20	+13.20
R_t	30.0	27.0	0.744	98.27	−1.73
				100.00	
		33.0	0.769	101.49	+1.49
m	2.0	1.8	0.789	104.14	+4.14
				100.00	
		2.2	0.721	95.25	−4.75
n	2.0	1.8	0.793	104.66	+4.66
				100.00	
		2.2	0.724	95.60	−4.40

Courtesy of Gordon & Breach Science Publishers [2]

7.7.2 Worth of a porosity unit

Before acquiring or interpreting log data, it is useful to compute the worth of one porosity unit in the reservoir under investigation. Two reservoir configurations are depicted in *Fig. 7.5*.

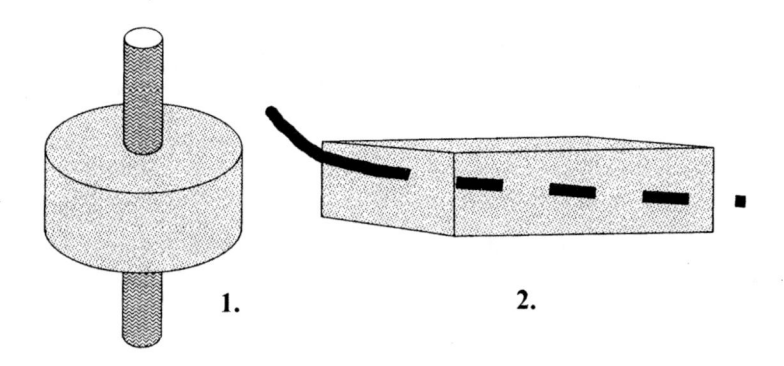

FIG. 7.5 : Reservoir configurations.
1. Vertical well.
2. Horizontal well.

Example 1: vertical well

Reservoir thickness	100 ft
Drainage radius	1500 ft
Water saturation	20 su
Oil in place equivalent to one porosity unit	1,000,000 bbl

Example 2: horizontal well

Reservoir thickness	50 ft
Length of pay	500 ft
Drainage radius	1500 ft
Water saturation	30 su
Oil in place equivalent to one porosity unit	200,000 bbl

Through this computation it is possible to propagate potential errors on one porosity unit to the size of the reservoir.

References

1 Aguilera, R., "Uncertainty in log calculation can be measured," pp. 126-128, *Oil & Gas Journal*, Vol. 31, No. 9, 9-1979.

2 Owens, J., Cockcroft, P., "Sensitivity analysis of errors in reserve evaluations due to core and log measurement inaccuracies," pp. 381-394, in *Advances in core evaluation accuracy and precision in reserves estimation*, edited by P. Worthington, Gordon & Breach Science Publishers, 1980.

3 Theys, P., "A serious look at repeat sections," paper F, *Trans.* 35th SPWLA annual logging symposium, Tulsa, 1994.

Part II

Data acquisition

8
Data acquisition

8.1 The data cycle

Each measurement is associated with an uncertainty. To quantify these uncertainties, it is necessary to identify and quantify potential sources of error.

To achieve this, it is useful to assess the potential error associated with each step of data acquisition. The full data cycle is depicted in *Fig. 8.1*.

In the first step, information representing the formation reaches the detectors or the sensors of the logging device, having been distorted to some degree by the downhole environment. Concurrently, auxiliary sensors are collecting data describing this surrounding environment. Downhole detection is more or less effective depending on the design of the logging tool and the technology used. For some measurements, this performance is affected by logging speed (or rate of penetration), tool acceleration, tool configuration, tool centralization and tool operation.

Data acquired downhole is then transmitted to the surface through a cable or through a mud pulse telemetry. The transmission mode, whether it is digital or analog, affects the original data. At this point, data is generally indexed to time. It is then subjected to a time-to-depth conversion.

This raw data is then transformed through algorithms defined at the time the tool response is established. In addition, a calibration transform is applied. This establishes a link between the general response of a device and the behavior of a specific tool. The definition of the tool response cannot be completed for every possible measurement range and therefore introduces some error.

At this point, environmental corrections are performed. The overall environmental corrections have been defined beforehand, at the time of the determination of the tool response. Inputs to the environmental corrections programs are also of importance. The complexity of the environment precludes a totally accurate description and sizable errors originate from the need to assume a number of environmental parameters.

Also, some error results from the inaccuracy of the measured input to the correction algorithm.

Most data is finally submitted to a certain degree of signal processing, usually filtering, before being displayed on a screen, printed on a film or paper copy and recorded on tape, floppy disc or CD-ROM. This last step represents the end of the data acquisition process and is before the interpretation and archiving processes in the cycle.

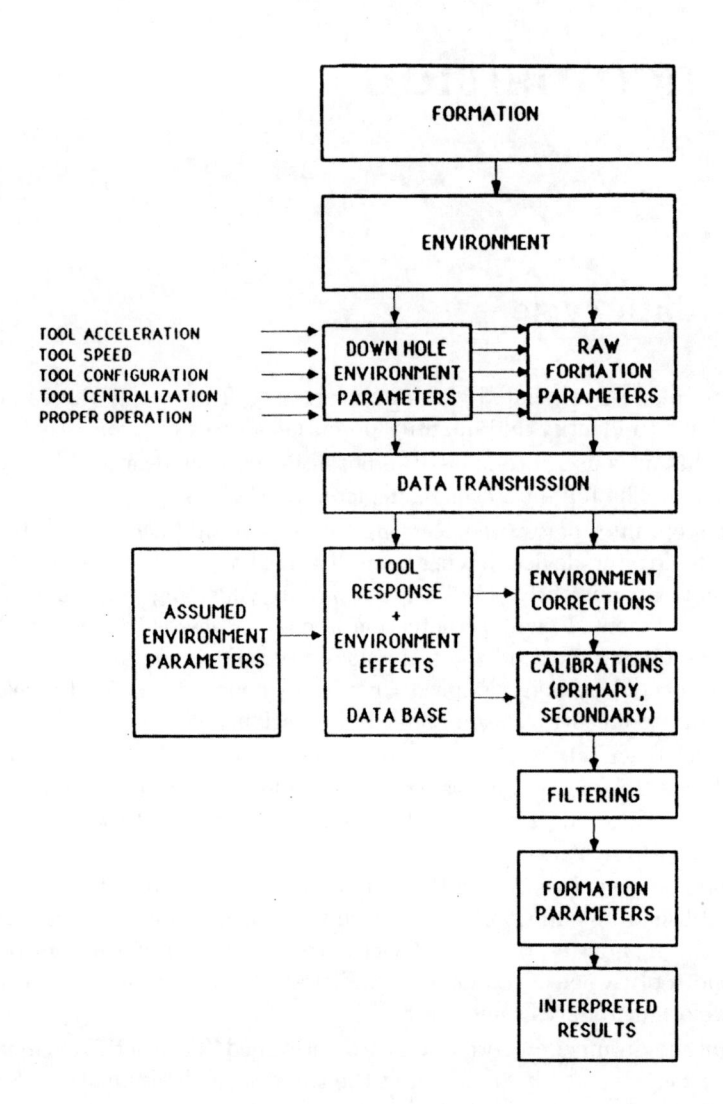

FIG. 8.1 : The data cycle.

[XVIII]

8.2 Causes of error

At each step of the data acquisition cycle, there are potential sources for error. These errors are listed in three groups:[1]

1. Random errors affecting the precision of the measurement

    The magnitude of random errors is affected by the following factors:

    (a) sensor technology
    (b) conversion from time to depth
    (c) logging speed or rate of penetration; sampling rate; telemetry limitations
    (d) signal processing and filtering
    (e) methods of enhancement of vertical resolution.

2. Systematic errors affecting the accuracy of the measurement

    Systematic errors are introduced in the measurement chain as follows:

    (a) definition of tool response and of environmental corrections
    (b) reporting of the inputs to the environmental corrections; propagation of the error attached to these inputs
    (c) poor representation of the actual environment
    (d) calibration procedures
    (e) tool stability under downhole conditions
    (f) failure of logging equipment.

3. Error on the location of the sensors

    (a) absolute measured depth errors
    (b) depth-matching errors
    (c) errors on true vertical depth and on the horizontal position of the sensor at the time of the measurement.

---

[1] The dreaded blunders may occur in any part of the data cycle and do not follow predictable patterns. The way to manage them is reviewed in Chapter 21.

Figure 8.2 shows the measurement chain considering the steps that affect precision [1]. They are reviewed in Chapters 10 to 12.

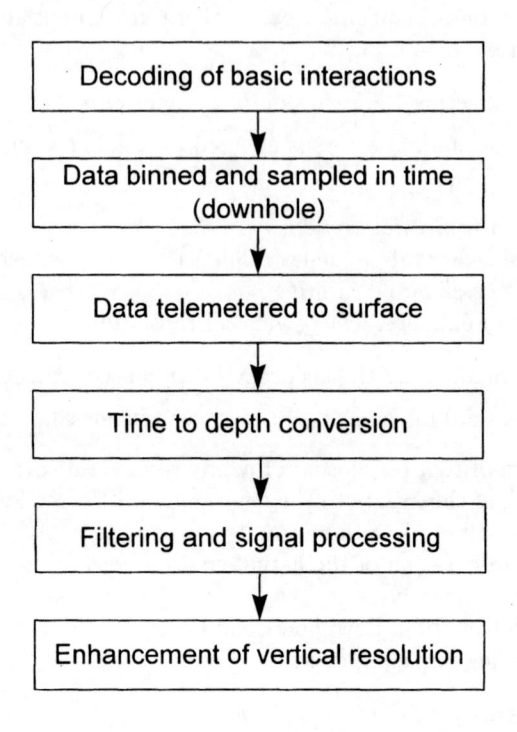

```
┌─────────────────────────────────────┐
│   Decoding of basic interactions    │
└─────────────────────────────────────┘
                  ↓
┌─────────────────────────────────────┐
│   Data binned and sampled in time   │
│             (downhole)              │
└─────────────────────────────────────┘
                  ↓
┌─────────────────────────────────────┐
│      Data telemetered to surface    │
└─────────────────────────────────────┘
                  ↓
┌─────────────────────────────────────┐
│       Time to depth conversion      │
└─────────────────────────────────────┘
                  ↓
┌─────────────────────────────────────┐
│     Filtering and signal processing │
└─────────────────────────────────────┘
                  ↓
┌─────────────────────────────────────┐
│   Enhancement of vertical resolution│
└─────────────────────────────────────┘
```

FIG. 8.2 : The measurement chain. Steps that affect precision.

Figure 8.3 represents the complete chain, now considered under the aspects of accuracy. This can be viewed as the succession of links data is going though if the measurements were infinitely long.

There is a first link between the measurements made on a single prototype tool (or the tools of a pilot series) and international standards. The tool T_0 is characterized at temperature t_0, pressure P_0 in conditions S_0. The tool response is further refined through mathematical modeling, extending use at temperatures t_m, pressures P_m and environments S_m.

The behavior of different tools T_m can also be characterized. These "ideal" tools are then linked to real tools T_c through calibration (t_c, P_c, S_c). The mild calibration conditions are linked to the real pressure, P, and temperature, T, encountered in the well. A selection of environmental conditions S_e are compensated for in the cycle. This selection is different from the exact conditions that are prevailing downhole.

The steps that affect accuracy are reviewed in Chapters 13 to 18. Measurements of depth and location in the well are found in Chapters 19 and 20.

International standards (m, kg, s, etc,)		
↕		
Data collected by reference tool in reference conditions	To Po	to So
↕		
Data extrapolated through mathematical modeling	Tm Pm	tm Sm
↕		
Data collected by tool t calibrated to reference tool to	Tc Pc	t Sc
↕		
Data at T and P but not environmentally corrected	T P	t Sr
↕		
Data corrected with selected environmental corrections	T P	t Se
↕		
Data collected downhole	T P	t S

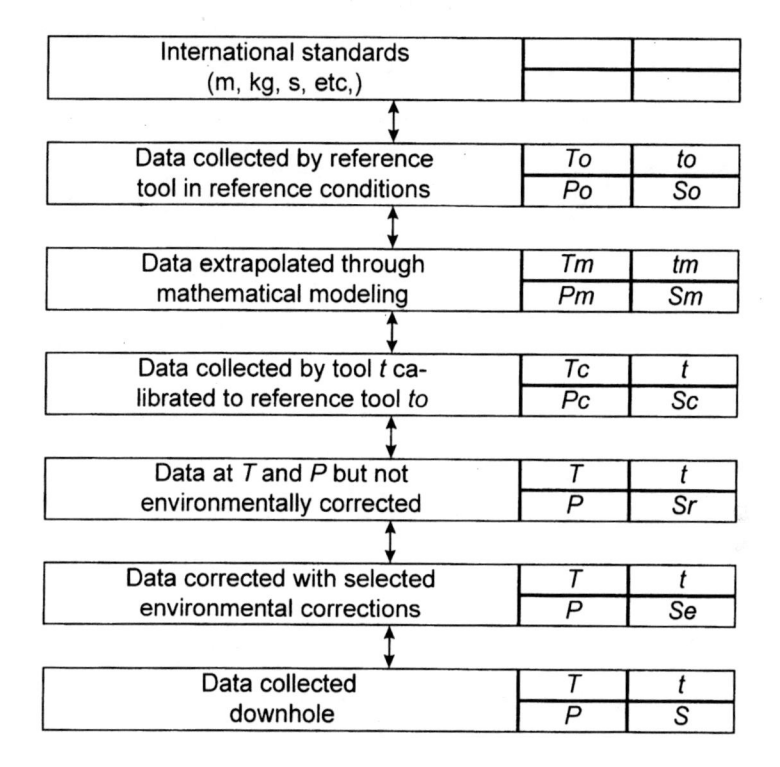

FIG. 8.3 : The measurement chain.
Steps that affect accuracy. For each step, a temperature, pressure and set of environmental conditions are defined.

Courtesy of SPWLA [2]

References

1 Theys, P., "A serious look at repeat sections," paper F, *Trans.* SPWLA 35th annual logging symposium, Tulsa, 1994.

2 Theys, P., "Accuracy—Essential information for a log measurement," paper XX, *Trans.* SPWLA 38th annual logging symposium, Houston, 1997.

9

Sensor and source technology

The primary components of the logging system involved in the generation of information are the logging sensors and sources. While the logging sources (transmitters, nuclear sources, electrodes) generate particles, fields, energy, etc., the sensors convert them into electric signals that can then be processed. A nonexhaustive list of sensors includes electrodes, nuclear detectors/photomultipliers, acoustic transducers and pressure gauges. Source and sensor technology and the underlying physics play an important part in the error of measurement.

9.1 Logging sources

9.1.1 Nuclear sources

Nuclear sources come in two categories—chemical and particle accelerators. The yield of emitted particles directly impacts precision as the number of particles counted at the detector is proportional to the particles emitted by the source. This yield is expressed in Bq.[1] 10^8 particles per second correspond approximately to 2000 GBq. Though high source yields are beneficial to the precision of the measurements, they cannot be used because they constitute considerable health hazards.

Not all particles emitted are "useful." A source may be characterized by a high Bq number but the emitted particles could be of the α type that are easily stopped by a sheet of paper. The emission of $4 \; 10^4$ αs is required to produce one neutron for a AmBe source.

---

[1] The becquerel (Bq) is the activity of a radionuclide decaying at the rate of one spontaneous nuclear transition per second. Another unit, the curie (Ci) is often used. 1 Ci = $3.7 \; 10^{10}$ Bq.

Chemical sources generate neutrons or α, β and γ rays. Typically, cesium (^{137}Cs) sources are used for density logging. They emit 662-keV γ rays that result from the reaction:

$$\mathrm{Cs}_{55}^{137} \longrightarrow \mathrm{Ba}_{56}^{137} + \mathrm{e}_{-1}^{0} + \gamma \ (662 \text{ keV}).$$

Americium beryllium (^{241}Am^{7}Be) sources[2] generate α particles, then 4.5-MeV neutrons. Californium (^{252}Cf) sources have been routinely used for the determination of aluminum (^{27}Al) concentrations in the formation.

Particle generators include linear accelerometers and minitrons. The development of the former has been slowed [3] because of the shear size of the device. Minitrons, on the contrary, have followed a continuous set of improvements. They are used as the starting point of induced gamma ray spectrometry. 14.06-MeV neutrons are generated by accelerating deuterium ions into a tritium target.

$$\mathrm{H}_1^2 + \mathrm{H}_1^3 \longrightarrow \mathrm{He}_2^4 + \mathrm{n}_0^1.$$

The challenge of this type of generator is to deliver higher yield rate (number of particles per second, in the order of 10^8) in a reliable and safe manner.

9.1.2 Sound sources

Sound sources come in two types, monopole and dipole. **Monopole sources** generate sound propagating in a spherical mode. They are built with materials that can be extensively deformed under the influence of an electric field (piezoelectric), or in a varying magnetic field (magnetostrictive).

Dipole sources are made with materials moving along one direction within a solenoid. They behave similarly to loudspeakers.

Downhole seismic sources, in parallel, have been developed to generate enough energy for surface detection without damaging the casing and the borehole.

9.1.3 Electromagnetic transmitters

Induction coils or electromagnetic antennae generate electromagnetic fields. Their configurations are optimized to minimize attenuation losses and reflected electromagnetic waves.

Current or voltage sources are metallic electrodes, the rustier, the more effective.

---

[2]They have replaced the plutonium-beryllium (^{238}Pu^{7}Be) and radium-beryllium (^{226}Ra^{7}Be) sources.

9.1.4 Natural sources

In many instances, downhole measurements can be performed without inducing fields, particles or mechanical vibrations because they are naturally produced by the Earth.

The **Earth's magnetic field** is extensively used to measure the orientation of the hole or the precession of spins in nuclear magnetic resonance methods. The measurement of the variations of the **Earth's gravity field** is the basis of gravimetry, which allows an evaluation of the density of rocks. The measurement of this gravity field is also used in surveying (Chapter 20).

Natural gamma rays are generated by minerals containing potassium, thorium or uranium. Because the energy levels of the rays emitted by these elements are different, it is possible to evaluate the elemental composition of these minerals.

Spontaneous potentials have two origins [2]. Most commonly, a difference of potential would be generated at a sand-shale-borehole interface, as ions can be displaced. The potential depends on the salinity of the mud and of the mud filtrate. Streaming spontaneous potentials are observed when, the hydrostatic pressure of the mud being larger than the formation pressure, some mud fluid filters through the mudcake into a permeable bed. The electromotive force (emf) is caused by filtration.

9.2 Logging sensors

9.2.1 Geiger-Müller detector

The Geiger-Müller counter, introduced in 1928, is still being used today in the oil industry because of its ruggedness, low cost and simplicity of its associated electronics.

This detector [VIII] reacts to all incoming radiation in exactly the same way, which precludes its use in spectrometric applications. In other words, low- or high-energy radiation, mostly unwanted, is detected as medium-energy radiation, the only one giving usable information in density logging. Geiger-Müller counters have a significant dead time, greatly exceeding the values observed for other commonly used detectors. Multiple pulsing (emission of several pulses for a single incoming radiation) is also a common occurrence. Therefore, a recommended procedure is quenching, reduction of the tube high voltage supply for a fixed time after a pulse to a value too low to support subsequent pulse creation.

9.2.2 Scintillation detectors

The scintillation process is available for the detection and spectrometry of a wide variety of radiation types. By this process, an incident radiation is converted into light, which is converted into an electric pulse by a photomultiplier. The photomultiplier adds only a small amount of noise.

The most widely used scintillators are inorganic alkali halide crystals, such as sodium iodide and the organic-based liquids and plastics. The inorganics have the best light output but have a slow response time. The organics are much faster, but yield less light.

9.2.2.1 NaI(Tl) detector

The thallium activated sodium iodide material was discovered in the early fifties. About 10^{-3} mole fraction of thallium is added to pure sodium iodide. Because of its hygroscopic nature, the material needs to be confined to a moisture-proof container. Another drawback of this detector is that iodine may be activated by incident neutrons, an occurrence that may result from the proximity of a neutron source. Still, this crystal has the highest light yield of any known scintillator. Its response to incident particles, electrons and gamma rays is close to linear over a wide range of energy.

9.2.2.2 BGO detector

Bismuth germanate ($Bi_4Ge_3O_{12}$) detectors are a popular replacement for NaI(Tl) detectors because of their higher detection efficiency.[3] However, BGO scintillation efficiency is strongly temperature dependent [4 and 5]. It decreases by a factor of 12 from $-50°C$ to $110°C$ ($-58°F$ to $230°F$). The light output also decreases following exposure to ultraviolet, gamma ray or neutron radiation. This decrease is difficult to model and varies from crystal to crystal.

9.2.3 Semiconductor detectors

A limitation of scintillation detectors is the energy resolution.[4] Radiation detectors based on semiconductors became available in the early sixties [6]. In addition to higher resolution, semiconductors make compact devices and have fast response times. Silicon- and germanium-based detectors are the most popular.

The performance of germanium detectors for detection of penetrating particles (for instance, gamma rays) can be enhanced by adding lithium to the original lattice by the so-called drifting process. These detectors must be stored and operated at reduced

---

[3]Scintillation efficiency is the percent of incident particle energy converted into visible light. It is also called light output.

[4]Energy resolution is related to the ability of the detector to distinguish between two radiations whose energies lie near each other. For more details, pp. 89-90, [VIII].

temperature to preserve the stability of lithium. This is why Ge(Li) detectors are located in vacuum cryostats connected to a constant supply of cooling agent. The significant size of this tank creates a technological challenge to logging tool design.

9.2.4 Other detectors

GSO are cerium-activated GD_2SiO_5 crystals with high density and atomic number. They enable the manufacturing of smaller diameter detectors suitable for through-tubing operations.

LSO are cerium-doped lutetium oxyorthosilicates that combine high counting efficiency with high light-flash intensity.

Table 9.1 summarizes the characteristics of the scintillation detectors.

TABLE 9.1

COMPARISON OF DETECTOR CHARACTERISTICS

Crystal type	Density (g/cm^3)	Atomic number Z	Relative light flash intensity	Sensitivity to shocks	Sensitivity to moisture
Na(Tl)	3.67	51	100	High	High
BGO	7.13	75	15	Low	Low
GSO	6.71	59	20	High	Low
LSO	7.40	66	75	Low	Low

Courtesy of Oilfield Review [1]

9.2.5 Helium detectors

Helium is often used as the detection material for neutrons. Neutrons interact with He in the following fashion:

$$He_2^3 + n_0^1 \longrightarrow H_1^3 + p_1^1.$$

The Q value of the reaction, corresponding to the energy liberated in the reaction following the neutron capture, is 0.765 MeV.

Detection efficiency can be improved by increasing the pressure in the detector.[5]

9.2.6 Sonic transducers

Sonic transducers can be sorted in two groups, magnetostrictive and piezoelectric transducers. For the second type of transducers, the piezoelectric effect converts

---
[5]Typical pressures are 10 atm.

mechanical energy into electric signals and vice versa. When subjected to a stress system, they develop a measurable electrostatic surface charge. Ceramic materials, such as $BaTiO_2$, have the proper conversion efficiency. Quartz crystals, another class of dielectrics, are also used.

9.2.7 Pressure gauges

Pressure data is collected by pressure gauges. These gauges are built around a pressure sensing element. In one type of pressure sensing device, the pressure is balanced by an opposing force, whose coupling to a transducer converts it into an electrical output. In another type, the sensor is exposed directly to the pressure. Quartz crystal gauges belong to this group. A vibrating quartz crystal is exposed to pressure that mechanically distorts the crystal. This deformation modifies the frequency of the signal emitted by the quartz. The change of frequency can be related to the surrounding pressure.

The critical specifications of pressure sensors are their static parameters, which include accuracy, resolution (minimum pressure change which can be detected), stability and sensitivity, and their dynamic characteristics, which describe transducer performance in transient conditions. Of particular interest is the behavior of the sensor submitted to temperature and pressure variations.

Figure 9.1 shows the transient response of a typical sensor submitted to a fixed temperature step. The stabilization time required to come back into the vicinity of the original pressure value often exceeds several minutes.

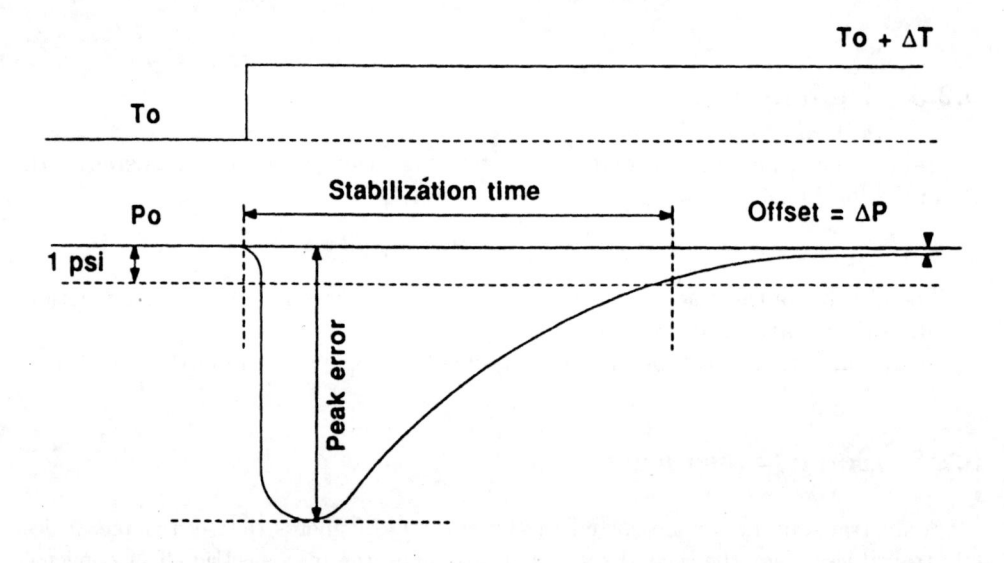

FIG. 9.1 : Pressure gauge time response.

Courtesy of Schlumberger

9.2.8 Magnetometers and accelerometers

Magnetic fields are measured with fluxgate magnetometers operating close to saturation. Accelerometers are servo-pendulums.

9.3 Optimization of sensor design

9.3.1 Compensation of effects influencing the measurement

Chapter 6 showed that a sensor could be affected by the borehole environment or by other pieces of the logging system, or even by itself. Sensors are designed to minimize sensitivity to these effects. These effects, however, cannot be completely eradicated. In order to compensate for them, these effects are analyzed and described in the most accurate and complete way.

Example 1: Dead time correction

While dead time reduces the number of interactions counted by a nuclear detector, it happens in a well-defined way for a given system. The dead time behavior of a detector can be modeled and corrected. Correction methods are developed in [VIII].

Dead time correction for the thermal neutron tool corresponds to the so-called nonparalyzable model. The fraction of time the detector is dead is given by $m\tau$, m being the measured counting rate and τ the dead time.

The true interaction rate, n, is then

$$n = \frac{m}{1 - m\tau}.$$

The value of τ is determined experimentally. A typical τ value for a neutron He detector (thermal detection) is 10^{-5} s.

Example 2: Detector resolution

An important property of a nuclear detector is its response to a source of radiation with a narrow energy band. The ideal response is a sharp spike. In practice, the response is a curve similar to a Gaussian distribution (*Fig. 4.3*). The larger the width, the worse the detector resolution. This resolution is expressed as a percentage, with smallest values corresponding to the best resolution.

Temperature affects the resolution of nuclear detectors. An increase of 100°C (210°F) would add a few percentage points to the resolution factor. This resolution loss creates some difficulties in induced gamma ray spectroscopy. Elemental standards established under perfect laboratory conditions are not useful for computing elemental yields from spectra measured with a crystal coping with downhole temperatures. Elemental standards are recomputed so that they correspond to a resolution that matches the one of the detectors working at downhole conditions.

9.3.2 Detector geometry

Big sensors mean heavy logging tools that put logging crews and cables to the test. Large sensors are still required for the following reasons:

(a) Detector efficiency increases with detector size.

(b) Logging tools and their accessories have well-known characteristics that can be fully assessed at surface during tool development. Conversely, the borehole environment is poorly known. The logging sonde should therefore occupy the largest portion of the hole possible. The use of fluid excluders on the *Schlumberger* GST* Induced Gamma Ray Spectrometry Tool corresponds to this requirement.

9.3.3 Optimization of detector design for the density tool

The following conditions are gathered to obtain the most efficient nuclear detectors:

(a) **maximum total counts**

This requirement directly relates to the formula linking uncertainty and counts. The higher the counts, the smaller the uncertainty. This is demonstrated in Section 7.5 with the comparison of the counts of density tools of different type.

(b) **maximum long-spacing density sensitivity**

The sensitivity is defined as the derivative of the natural logarithm of the counting rates with respect to the density. A large sensitivity means that the influence of the environment (mudcake) has minimum impact.

(c) **maximum angle between spine and rib**

The uncertainty on the mudcake correction can be described as the projection of the counting rates uncertainty along the rib onto the spine. Its magnitude is inversely proportional to the sine of the angle between the spine and the rib. The larger this angle, the smaller the corresponding error.

References

1 Adolph, B., Stoller, C., Brady, J., Flaum, C., Melcher, C., Roscoe, B., Vittachi, A., Schnorr, D., "Saturation monitoring with the RST Reservoir Saturation Tool," *Oilfield Review*, 1-1994.

2 Doll, H. G., "The S. P. log: theoretical analysis and principles of interpretation," *Petroleum Technology*, TP2463, 9-1948.

3 King, G. III, Becker, A. J., Corris, G. W., Boyce, J. R., Bramblett, R. L., "Density logging using an electron linear accelerator as the x-ray source," pp. 990-994, *Nuclear Instrumental Methods*, Vol. B24/25, 1987.

4 Melcher, C. L., "Correlation between thermoluminescence and radiation damage in bismuth germanate," pp. 465-467, *Nature*, Vol. 313, No. 6002, 1985.

5 Melcher, C. L., Schweitzer, J. S., "Gamma ray detector properties for hostile environments," *IEEE Trans. on Nuclear Sciences*, Vol. NS-35, No. 1, 1988.

6 Schweitzer, J. S., "Nuclear techniques in the oil industry," *Nuclear Geophysics*, Vol. 5, No. 1/2, 1991.

10

Effect of measurement duration on precision

10.1 Duration of the basic interactions experienced in logging

10.1.1 Induction, laterolog, microresistivity, electromagnetic propagation

The measurements are linked to electromagnetic fields. They propagate at the speed of light. Each measurement takes at most 1 ms.

10.1.2 Sonic and ultrasonic

The physical phenomenon supporting the measurement takes at the most 660 μs/m (200 μs/ft), corresponding to the transmission of sound through fluids. For the longest spacings used in the industry—3.65 m (12 ft)—the transit time from the transmitter to the receiver is 2.4 ms.

Each sonic transmitter emits from 5 to 20 sound bursts every second. A complete borehole compensated transit time measurement takes four bursts. In some cases, it is useful to stack more waveforms, up to a maximum of 60 for each measurement sequence. The complete measurement would then take several seconds.

Ultrasonic measurements are also performed at the speed of sound. Each radial measurement (in one direction) corresponds to the journey of the sound wave over a few inches and takes a few dozens microseconds.

10.1.3 Nuclear

Elementary interactions involve gamma rays and neutrons. Each interaction takes around 10 μs. Five to ten interactions may occur before the detector recognizes the incidence of a particle, which is then counted (Chapter 9). One count does not make a measurement, though. Precise nuclear measurements are only obtained after the accumulation of numerous counts. In fact, precision is directly related to the number of counts, which is strongly dependent on measurement duration. The next paragraphs are dedicated to the quantification of the precision of nuclear measurements.

Precision of nuclear measurements

We have seen in Chapter 4 that, considering several repeated experiments yielding as many measurements, the best estimate of the true value is the mean of these measurements. In addition, it is possible to estimate the precision of the experiments by the standard deviation of the measured values. Because cost constraints preclude repeated logging over the same interval, it is necessary to estimate the precision even though only one measurement is available. This is all the more important with nuclear tools, which may be affected by large random errors.

A single measurement x is available. This measurement has been drawn from a parent population whose statistical distribution is either Poissonian or Gaussian. Because x is the only available information, there is no better choice than taking the mean of the parent distribution as $\bar{x} = x$. Having selected $\bar{x}$ and making an educated guess that the distribution is Poissonian or Gaussian, the standard deviation is locked in both cases to $\sigma = \sqrt{\bar{x}} = \sqrt{x}$. In other words, $\sqrt{x}$ is the best estimate of the standard deviation. To illustrate this finding, if $x = 1600$, then $\sigma = 40$. The true value has a 68% chance to be in the interval [1560,1640].

For a single detector tool, reading N counts, $\sigma = \sqrt{N}$. For a dual detector tool, $\sigma_1 = \sqrt{N_1}$, $\sigma_2 = \sqrt{N_2}$. These formulæ have direct applications to density and neutron porosity logging. An interesting consequence of these algorithms is that the precision varies with the resulting density and neutron porosity. For instance, at high densities, corresponding to low gamma ray counting rates, imprecision[1] is relatively high. At low density, imprecision is relatively low. For this reason, it is easier to check the response of a density tool in a salt bed (halite whose ρ_b is 2.05 g/cm^3) than in an anhydrite bed ($\rho_b = 2.98$ g/cm^3). *Figure 10.1* shows the estimated imprecision as a function of the expected density.

---

[1]The term imprecision, the contrary of precision, may be preferred to specify the precision of a measurement as a large number corresponds to an increased imprecision (reduced precision) and a small number to a decreased imprecision (improved precision).

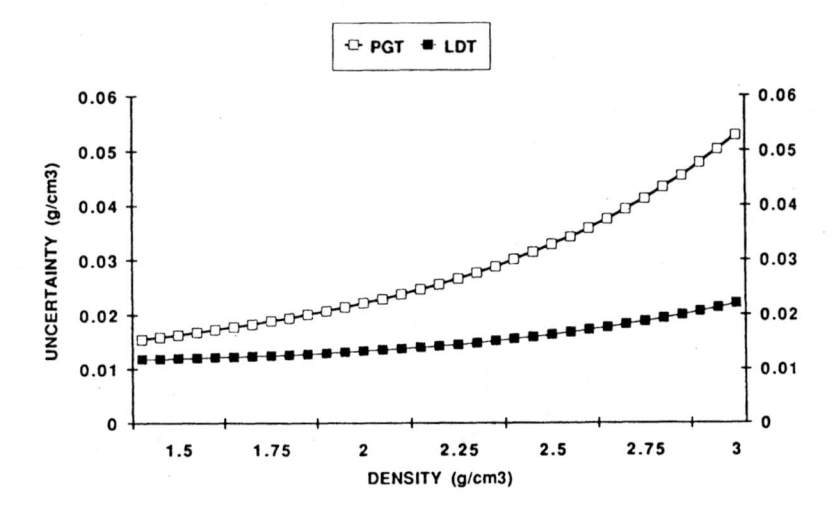

FIG. 10.1 : Imprecision of density measurements.
This chart is valid for a 1-s measurement, which corresponds to a logging speed of 9 m/min (1800 ft/h) and no filtering.

10.1.4 Nuclear magnetic resonance

A measurement cycle using nuclear magnetic resonance includes two components; wait time intervals, during which the formation spins are polarized and information acquisition periods, during which spin echoes are sensed. The wait time should be approximately three times longer than the mean T_1 of the pore fluids. A typical wait time value is 1.3 s for sandstones.

The minimum spin echo spacing is a few hundred microseconds. A cycle may include up to 1800 echoes in continuous logging. To correct for electronic offset, information is collected in phase-alternated pair. This approach doubles the measurement duration. The total cycle time is then twice the sum of wait time and of spin echo sensing time.

10.1.5 Measurement duration in logging

In almost every modern logging tool, the measurement is performed during a fixed period of time, even though sampling rates are mostly expressed in terms of depth intervals. In wireline logging, a typical sampling period would be 67 ms [4]. In MWD services, it could range from 5 to 60 s.[2]

---

[2]Communication from surface to the tool enables the change from one sampling rate, e.g., 20 s, to another one, e. g., 10 s. This allows a better adaptation to the drilling rate of penetration.

10.2 Data transmission

There are four ways to transmit data to a surface computer or system:
(a) electric transmission—analog or digital—with a cable
(b) mud-pulse transmission
(c) electromagnetic transmission
(d) direct dump of a downhole memory to a surface system.

10.2.1 Analog transmission with a cable

Signals originating downhole are distorted as they are transmitted through cables. *Fig. 10.2* shows the distortion observed on a 6100-m (20,000-ft) monocable excited by pulses of different duration. Similar effects are experienced on heptacables.

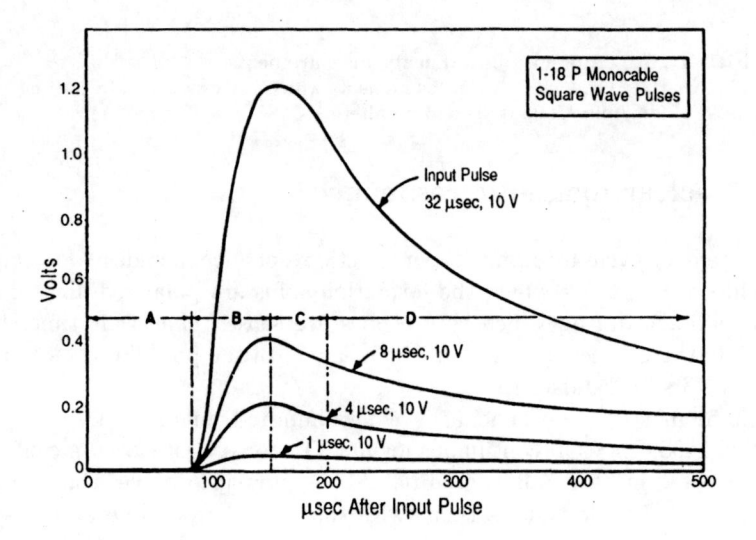

FIG. 10.2 : Output of a monocable excited by 10-V pulses.
A is the time interval required for the high-frequency content of the pulse to reach surface. B is the rise time from the beginning of the signal to its peak (peak amplitude is related to cable attenuation and to input pulse duration. C is the fall time; D is the tail time.
Courtesy of Schlumberger

10.2.2 Digital transmission with cable

Digital transmission has become a standard feature for logging companies. This technology is needed because of the ever-increasing volume of data carried down- and up-hole (*Table 10.1*). 500-kbps (kilobits per second) telemetries are now common.

TABLE 10.1
DATA RATE REQUIREMENTS

Log type	Channels	Approximate logging speed		Sampling rate	Bit resolution	Data rate
		($\frac{m}{min}$)	($\frac{ft}{h}$)	(in.)		(kbps)
IES/SN...............	2	30	6000	6.0	8	0.05
Litho-density...........	30	9	1800	6.0	16	0.48
HDT..................	5	9	1800	0.2	8	1.2
FMS...................	64	9	1800	0.1	16	62
FMI...................	192	9	1800	0.1	16	186
Combination of five tools	200	9	1800	6.0	16	40

10.2.3 Mud-pulse transmission

It is possible to use mud circulation through the bottomhole assembly (BHA) to transmit the data collected by downhole sensors. A telemetry sub is required. It controls the flow of mud downhole. Changes in the mud flow are observed as changes of standpipe pressure at surface. There are three ways to transmit information with an hydraulic transmission (*Fig. 10.3*):

(a) Positive mud pulse: Standpipe pressure increases as flow is reduced downhole.
(b) Negative mud pulse: Standpipe pressure decreases as a small portion of the mud flow inside the drill collars is diverted to the annulus.
(c) Continuous mud wave (also called "siren") [3]. *Table 10.2* gives some typical transmission rates with *Anadrill* mud-pulse telemetries.

TABLE 10.2
TYPICAL TRANSMISSION RATES WITH MWD TOOLS

Measurement	Update time (s)		
Effective data rate	3 bps		6 bps
Telemetry type	M1	M3	M10
Density and P_e	53.3	44	22
Neutron porosity	53.3	44	22
Tool face	17.8	7	11
Tool voltage	53.3	88	—
Weight-on-bit and torque	53.3	44	22
Shallow and deep resistivity	53.3	44	22
Gamma ray	53.3	44	22

Courtesy of Schlumberger

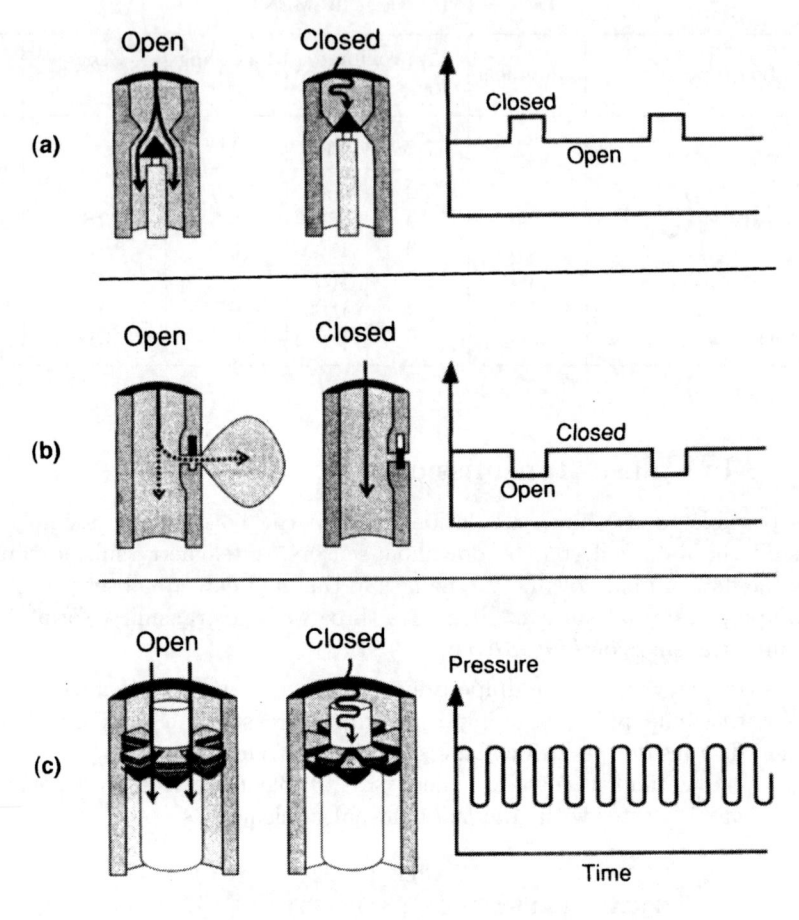

FIG. 10.3 : Mud pulse telemetries.
  a. Positive pulse.
  b. Negative pulse.
  c. Siren.
       Courtesy of IMS [2]

Effect of telemetry and of compression

Even though measurements are performed at a higher sampling rate, the mud-pulse telemetry could act as a bottle neck. This is the case with real-time MWD information. For instance, measurement duration can be 10 s, but information has an update rate of 53 s. Precision should be derived from the actual measurement duration, not from the update rate. Compression schemes, often used for large data transfer rates, are not supposed to affect the precision of the data. *Figure 10.4* displays the potential effects of a mismatch between sampling and update rates.

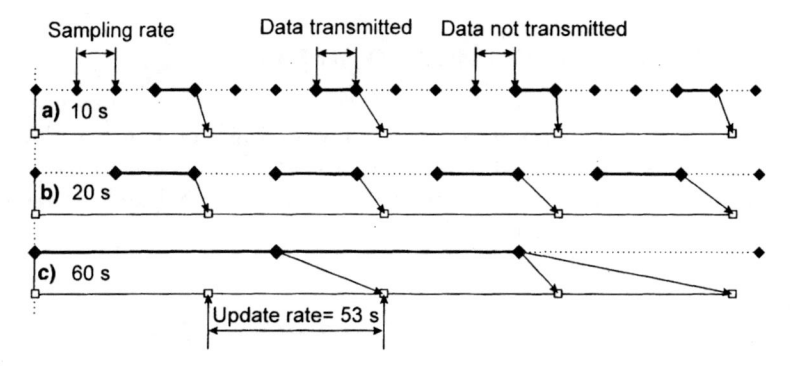

FIG. 10.4 : Interaction between sampling rate and update rate.
In all examples, the update rate is 53 s. The last value sampled is sent uphole. Data transmitted uphole is represented as a thick solid line. Data not transmitted, and therefore not available uphole, is dotted.

a. Sampling rate is 10 s. About 75% of the data is not transmitted uphole.
b. Sampling rate is 20 s. About one sample out of two is transmitted to surface.
c. Sampling rate is 60 s. In some cases, the same information is sent twice.

10.2.4 Electromagnetic propagation

Because the mud-pulse telemetry presents obvious bandwidth limitations, the capability of transmitting data directly from a downhole sensor to surface has been evaluated. It depends on the electric characteristics of the Earth, the presence of water (offshore and onshore) and the depth of the sensors.

This technique is used on shorter distances to eliminate direct electrical connections through mechanically complex components of the BHA. The data is sent from a transmitter linked to the sensor and its electronic package, to a receiver electrically connected to the mud-pulse telemetry. This approach is used when the sensor is close to the bit and is separated from the telemetry sub by a drilling motor.

10.2.5 Memory dump

Because the telemetry rates that can be obtained with mud-pulse telemetry are low, simultaneous recording to a downhole memory is performed with many logging-while-drilling tools. A far larger quantity of information, including raw and detailed quality control data, can then be stored. This data is retrieved, after the tool is back to surface, through a direct connection from a port on the tool to the surface system. Because the connection is short as compared to an entire logging cable and the information is conveyed in a digital format, no loss of information results from this process.

This can also be accomplished while the tool is still downhole, using a wireline-conveyed tool with an inductive coupling.

10.3 Conversion of time to depth

Geological information indexed to the time it was acquired is of little value. It needs to be linked to the depth at which it was acquired. This is achieved by performing a measurement of depth versus time, then combining the data indexed in time with the depth information indexed in time.

10.3.1 Relating time to depth in wireline logging

The displacement of the wireline cable is measured by a wheel located near the drum (*Fig. 10.5*). After a given length of cable is unspooled, a pulse is emitted to the central processor. After a certain number of pulses cumulate in what is called a depth sampling interval, the measurement sequence is interrupted. At a constant logging speed, there is a direct relation between sampling interval and measurement duration. Different combinations of logging speeds and sampling rates are depicted in *Fig. 10.6*.

FIG. 10.5 : The *Schlumberger* Integrated Dual Wheel spooler.
Courtesy of Schlumberger

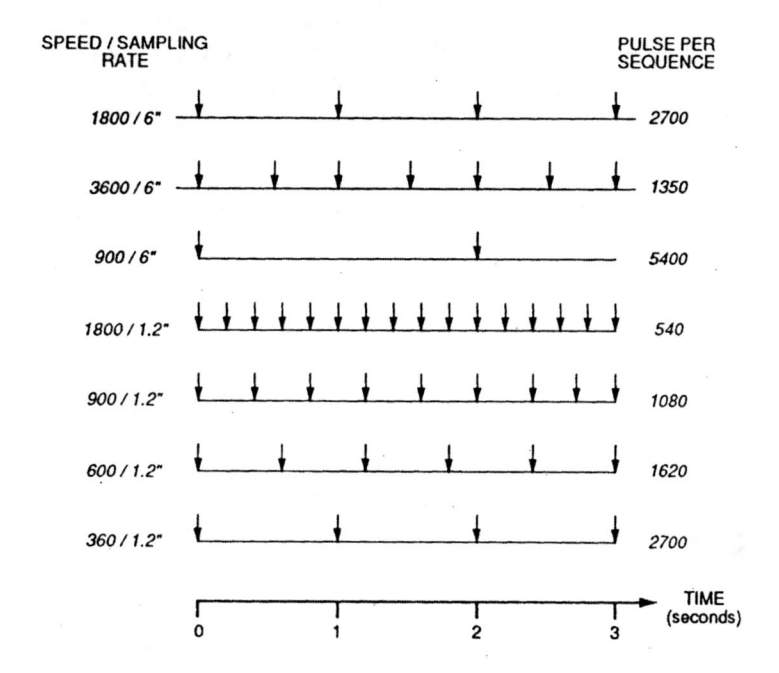

SPEED / SAMPLING RATE		PULSE PER SEQUENCE
1800 / 6"	↓ ↓ ↓ ↓	2700
3600 / 6"	↓ ↓ ↓ ↓ ↓ ↓ ↓ ↓	1350
900 / 6"	↓ ↓	5400
1800 / 1.2"	↓↓↓↓↓↓↓↓↓↓↓↓↓↓↓	540
900 / 1.2"	↓ ↓ ↓ ↓ ↓ ↓ ↓ ↓	1080
600 / 1.2"	↓ ↓ ↓ ↓ ↓ ↓	1620
360 / 1.2"	↓ ↓ ↓ ↓	2700

TIME (seconds): 0   1   2   3

FIG. 10.6 : Interrupt frequency.
At 9 m/min (1800 ft/h) and with a sampling interval of 15 cm (6 in.), an interrupt schematized by a vertical arrow is emitted every second. At 18 m/min (3600 ft/h) and the same sampling rate, there is an interrupt every 0.5 s. At 4.5 m/min (900 ft/h), a measurement sequence takes 2 s. If the sampling interval is reduced to 3 cm (1.2 in.), for 9 m/min (1800 ft/h), a sequence lasts 0.2 s. It would be necessary to reduce the logging speed to 1.8 m/min (360 ft/h) to obtain exactly the same measurement sequence as for 9 m/min-15 cm.

10.3.2 Relating time to depth in MWD

The depth measurement in MWD is described in detail in Chapter 19. The measured depth is recorded versus time in the "time-depth" file. The difference from wireline acquisition is that, because logging speed is not constant, equal time samples are not converted into equal depth samples.

Influence of the rate of penetration

If downhole data is acquired while drilling (MWD), movement of the sensors is controlled by the drilling operation. The rate of penetration (ROP) varies with rock type, bit wear and bit design. It ranges from a few meters per hour (ft/h) to several hundreds meters per hour (ft/h).

At present, the time sampling rate of the measuring equipment is fixed at the surface before each run in the well. Consultation with the geologist and drilling personnel is required so that the ROP can be estimated in advance and the sampling rate optimized to obtain the desired recording rates[3] (*Fig. 10.7*). Depending on ROP, a fixed depth interval (e.g., 6 in.) may be surveyed in drastically different times. Measurement duration is the ratio of the sampling rate and of the instantaneous ROP (*Fig. 10.8*). Consequently, precision is not a single figure as it is for a given logging speed in wireline logging, but it varies with ROP (*Fig 10.9*).

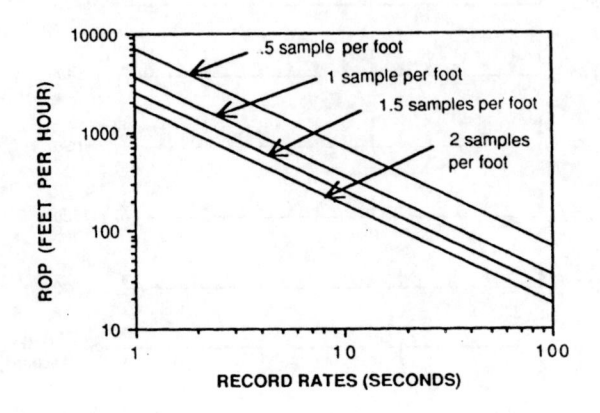

FIG. 10.7 : Relation between ROP and MWD precision.
Courtesy of the Oilfield Review [1]

10.3.3 Merging data and depth information

Once downhole data and time-depth information are available, they are merged through a merging program. This operation can be affected in many different ways by the gating that is desired. Typically several values would be available between two instants t_1 and t_2, corresponding to a measurement sequence, or one depth sample. The way these values are combined depends on the gate selected.

For instance, for a very precise measurement, such as resistivity, the last value would be selected as representative of the whole interval. Conversely, for a nuclear measurement, the average of the values available between t_1 and t_2 would be computed to represent the interval. *Table 10.3* lists the gate types used by Schlumberger.

Figure 10.10, derived from the values listed in *Table 10.4*, gives an idea of the potential difficulties linked to time-to-depth conversion. It points out that the time-indexed data contains much more information that is lost in the translation to depth. As log data is commonly referred to depth only, this information is generally lost forever.

---

[3]Sampling rate changes may be programmed to vary to avoid wasting memory space inside casing, or on highly sampled data recorded over shallow intervals that have been already been logged.

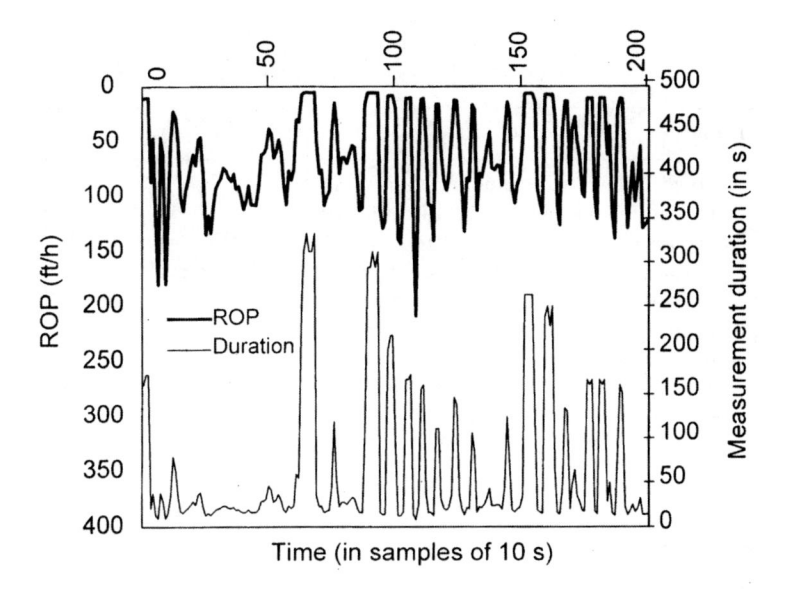

10. Effect of measurement duration on precision   127

FIG. 10.8 : Example of ROP and measurement duration.
ROP varies in a wide range, from a few feet per hour to 160 ft/h. Accordingly, measurement duration varies between a few seconds up to 300 s.

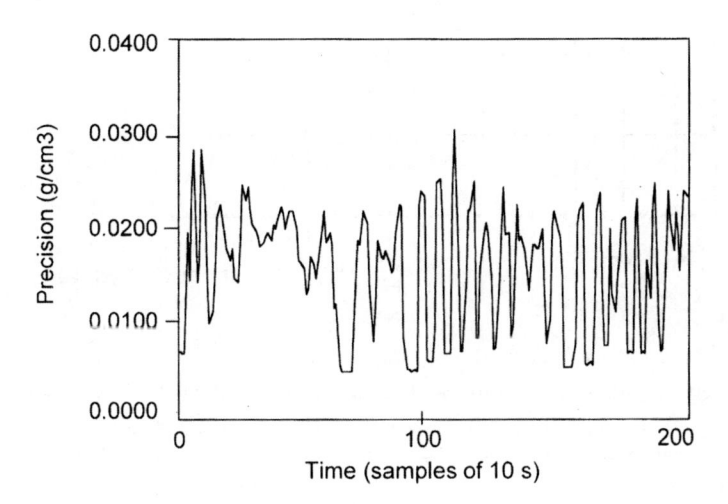

FIG. 10.9 : Variation of precision as ROP changes.
As measurement duration varies, so does precision. Intuitively, one can estimates that more precise information will be obtained when the sensors are staying a longer time in front of a geological formation.

TABLE 10.3
GATE TYPE DESCRIPTION

Name	Description
Average	average value over an interval
Count rate	converts counts to count rate
Interpolate	linearly interpolated value between two known points
Last value	most recent value encountered in an interval
Logical or	logical OR of value bits over an interval
Maximum	maximum value over an interval
Minimum	minimum value over an interval
Sum	sum of values over an interval

Courtesy of Schlumberger

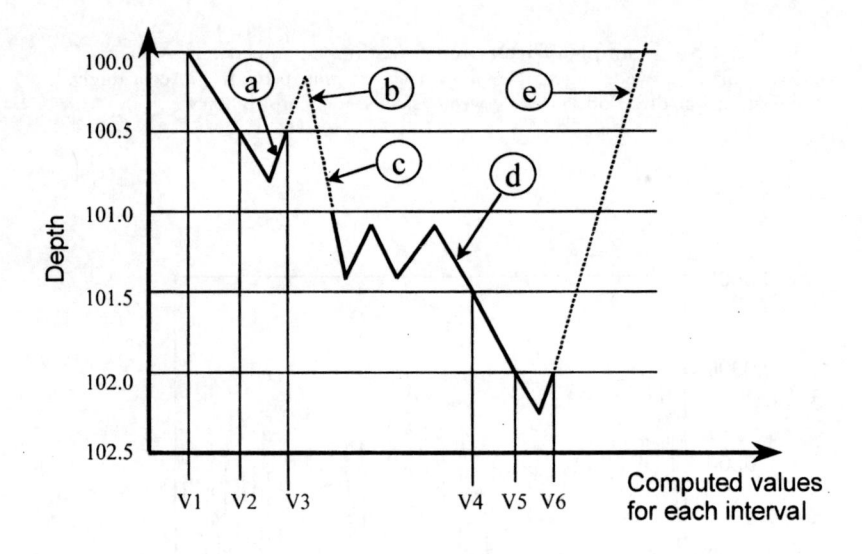

FIG. 10.10 : An example of time to depth conversion.
  a. Off bottom data is included to compute the average (and final) value.
  b. Average V3 for depth index 101.0 is computed when crossing the previous 6-in. boundary. This data is not used since V3 has already been computed.
  c. This data is not used since V3 has already been computed.
  d. All this data is used in computing V4 for the 101.5 depth index. As the sensors go up and down, the average may be misleading.
  e. V6 has been computed for depth index 102.5. This data recorded with the sensors coming up is not taken into account.

Courtesy of Schlumberger

TABLE 10.4
DATA USED FOR THE DEPTH-TIME CONVERSION

Data is recorded every 5 s in the memory, but it is decimated to 10 s when stored into the time-indexed file. Depending on the gate type, a different depth-indexed value is calculated. This value is computed for every 6-in. increment of depth 100.0, 100.5, 101.0, etc. Between 60 and 70 s, the sensors proceed by 1.4 ft, crossing three depth indices. The same final value, A3 (gate = Average) or L3 (gate = Last value) is stored at these three depth indices.

Time	Tool data	Time frame reader	Measured depth	6-in. depth	Average gate	Last value gate
5	V1					
10	V2	V2	100.0	100.0	$A1 = V2$	$L1 = V2$
15	V3					
20	V4	V4	100.1			
25	V5					
30	V6	V6	100.2			
35	V7					
40	V8	V8	100.5	100.5	$A2 = (V4+V6+V8)/3$	$L2 = V8$
45	V9					
50	V10	V10	100.6			
55	V11					
60	V12	V12	100.9			
				101.0	$A3 = (V10+V12)/2$	$L3 = V12$
				101.5	A3 (note: repeated value)	L3
65	V13			102.0	A3 (note: repeated value)	L3
70	V14	V14	102.3			
75	V15					
80	V16	V16	102.4			
				102.5	$A4 = (V14+V16)/2$	$L4 = V16$
85	V17					
90	V18	V18	102.6			
95	V19					
				103.0	$A5 = V18$	$L5 = V18$
100	V20	V20	103.1			
105	V21					
110	V22	V22	103.4			
115	V23					
120	V24	V24	103.5	103.5	$A6 = (V20+V22+V24)/3$	$L6 = V24$

Courtesy of Schlumberger

10.4 Effect of logging speed/rate of penetration and sampling rate on measurement precision

The way logging speed and sampling rates affect measurement precision depends on the nature of the measurement.

10.4.1 Induction, laterolog, imaging

The measurements themselves are fast and are not directly affected by measurement duration.

10.4.2 Nuclear tools

Density

N_{ls} and N_{ss} are the counts in the long- and short-spacing detectors for a given measurement; α and β are constants. Precision is expressed as

$$\sigma_{\rho_b} = \sqrt{\alpha/N_{ls} + \beta/N_{ss}}.$$

Neutron porosity

The neutron porosity precision σ_{ϕ_N} can be derived in a similar manner:

$$\sigma_{\phi_N} = \gamma \frac{N_{ls}}{N_{ss}} \sqrt{1/N_{ls} + 1/N_{ss}}.$$

In both cases, precision is highly dependent on time as the counts are equal to counting rates multiplied by measurement duration.

Wireline logging

At 9 m/min (1800 ft/h) with a 15-cm (6-in.) sampling rate, one depth sample is surveyed by a wireline logging tool in 1 s. In that case, the counts accumulated are equal to the counting rate or counts per second. Precisions on ρ_b and ϕ_N can be computed from the formulæ above. For the density, $\sigma_{\rho_b} = 0.017$ g/cm^3 at 2.70 g/cm^3. For the neutron porosity, $\sigma_{\phi_N} = 1.7$ pu at 29 pu.

Assume the logging speed is doubled and the same sampling rate is used. The counts accumulated in one simple measurement represent half the counts considered for the reference—noted with the subscript 0. It can be quickly derived that

$$\sigma_{\phi N} = \sigma_{\phi N_0} \sqrt{2},$$

$$\sigma_{\rho_b} = \sigma_{\rho_{b0}} \sqrt{2}.$$

In general, if the actual logging speed is a times the reference speed and the sampling rate used during logging is b times the reference sampling rate expressed in inches, then

$$N_{ls} = N_{ls_0} \times \frac{b}{a}$$

and

$$N_{ss} = N_{ss_0} \times \frac{b}{a}.$$

Therefore,

$$\sigma_{\phi N} = \sigma_{\phi N_0} \sqrt{\frac{b}{a}}$$

and

$$\sigma_{\rho_b} = \sigma_{\rho_{b0}} \sqrt{\frac{b}{a}}.$$

Similar equations relating speed, sampling rate and raw data can be expressed for all nuclear tools. The effects of speed and sampling rate on the imprecision of the density, neutron-derived porosity and other nuclear measurements are shown in *Table 10.5*.

TABLE 10.5

VARIATION OF IMPRECISION WITH SPEED AND SAMPLING RATE

Speed (ft/h)	Sampling (in.)	Imprecision multiplication factor		$\sigma_{\phi N}$ (pu)	σ_{ρ_b} (g/cm^3)
1800	6.0	1	1.000	1.70	0.017
3600	6.0	$\sqrt{2}$	1.414	2.40	0.024
900	6.0	$\sqrt{2}/2$	0.707	1.20	0.012
1800	1.2	$\sqrt{5}$	2.236	3.80	0.038
900	1.2	$\sqrt{5/2}$	1.581	2.68	0.027
600	1.2	$\sqrt{5/3}$	1.291	2.19	0.022

This table is valid for the whole range of measurement of a nuclear logging device. The imprecision at the nominal speed and sampling rate varies substantially with the measured value. This means that the same final imprecision will be obtained for the observation of n counts at 9 m/min (1800 ft/h) and the measurement of $2n$ counts, at twice this logging speed—18 m/min (3600 ft/h).

The nominal speed, 9 m/min (1800 ft/h), is a reference, a compromise for all logging conditions. In practice, less dense formations corresponding to the accumulation of a large number of counts could be logged faster, while denser formations ought to be surveyed more slowly. Design of an optimal logging program accounts for this factor and adapts logging speed to meet the desired precision (Chapter 28).

MWD

The same rationale applies to MWD. When the precision is known at a reference ROP for a given sampling rate, it is possible to derive the imprecision for a different set of conditions using the described formulæ.

Exercise 15*

The precision specification for the MWD density is 0.015 g/cm^3 at a rate of penetration of 100 ft/h. What is the rate of penetration needed for a precision of 0.020 g/cm^3?

10.4.3 NMR tools

In continuous logging, the tool speed is adjusted to ensure a full measurement cycle is completed during each depth sample interval. For a sample rate of 6 in., the maximum logging speed is

$$v_{\text{maximum}} = 300 \times \frac{\text{sample rate}}{\text{cycle time}}.$$

References

1 Allen, D., Bergt, D., Clark, B., Falconer, I., Hache, J. M., Kienitz, C., Lesage, M., Rasmus, J., Roulet, C., Wraight, P., "Logging while drilling," pp. 4-17, *Oilfield Review*, Vol. 1, No. 4, 1989.

2 International MWD Society, "State of the art in MWD," 1993.

3 Montaron, B., Hache, J. M., Voisin, B., "Improvement in MWD telemetry: the right data at the right time," SPE 25356, Singapore, 1993.

4 Schlumberger, *CMR Combinable Magnetic Resonance Tool User's Guide*, Houston, 1997.

11
Signal processing: filtering

11.1 Precision of a measurement over a thick homogeneous interval

A homogeneous formation with constant characteristics is logged over an interval l, with
$$l = n \times s.$$
s is the length of one sample; n is the number of samples covered by interval l. The n values of the measurement are equal, and so is σ, the precision of each individual measurement. The precision of the measurement over the whole interval is
$$\sigma_l = \frac{\sigma}{\sqrt{n}}.$$

If the individual measurements are not strictly equal, but are known to represent the same true value of the formation, then the average of the measurements is selected to represent the best evaluation of the formation characteristic. The precision of this averaged value is described by the same formula as above.

From this formula, it is observed that excellent precision, several magnitudes better than observed over a single sample, can be obtained when thick and homogeneous formations are logged. It is interesting to note that the majority of current interpretation programs fail to use this improvement of precision as they are based on a level-by-level analysis instead of considering one homogeneous formation at a time.

11.2 Combination of successive measurements

From the previous section, it can be inferred that improved precision can be obtained through the combination of successive measurements.

11.2.1 Precision of a weight-averaged measurement

The measurement x_d made at depth d is replaced by the sum of the measurements performed at levels located above and below. The measurement x_d is indeed included. If the same weight is given to all the measurements, for three levels at depths $d - s$, d and $d + s$ (s being one depth sample), the filtered value x_f is

$$x_f = \frac{1}{3}(x_{d-s} + x_d + x_{d+s}).$$

Using the transmission of error formula, we obtain

$$\sigma_{x_f}^2 = (1/9) \times (\sigma_{x_{d-s}}^2 + \sigma_{x_d}^2 + \sigma_{x_{d+s}}^2).$$

The errors at successive levels can be assumed to be of the same magnitude:

$$\sigma_{x_{d-s}} = \sigma_d = \sigma_{x_{d+s}}.$$

Consequently

$$\sigma_{x_f} = \frac{1}{\sqrt{3}} \times \sigma_{x_d}.$$

Filtering is improving precision by a factor of 1.73. Similarly, for five levels the improvement would be by a constant equal to 2.23. With n different weights, n being generally an odd number such that $n = 2p + 1$, we have the following formula (with $a_1 + a_2 + \cdots + a_n = 1$):

$$x_f = a_1 x_{d-ps} + a_2 x_{d-(p-1)s} + \cdots + a_p x_{d-s} + a_{p+1} x_d + a_{p+2} x_{d+s} + \cdots + a_{2p+1} x_{d+ps}.$$

Considering that $\sigma_{x_{d-ps}} = \cdots = \sigma_{x_d} = \cdots = \sigma_{x_{d+ps}}$,

$$\sigma_{x_f} = \sqrt{a_1^2 + a_2^2 + a_3^2 + \cdots + a_{2p+1}^2} \times \sigma_{x_d}.$$

Numerical example

$$a_1 = 1/4, \quad a_2 = 1/2, \quad a_3 = 1/4.$$

Then

$$\sigma_{x_f} = 0.61 \times \sigma_{x_d}.$$

Table 11.1 indicates the resulting improvement in precision.

TABLE 11.1
IMPROVEMENT OF PRECISION THROUGH FILTERING

Levels	Weights	Precision	Improvement factor
1	1	$1.00 \times \sigma_{x_d}$	1.00
3	1/3, 1/3, 1/3	$0.58 \times \sigma_{x_d}$	1.73
3	1/4, 1/2, 1/4	$0.61 \times \sigma_{x_d}$	1.63
5	1/12, 1/4, 1/3, 1/4, 1/12	$0.50 \times \sigma_{x_d}$	2.00
5	1/16, 1/4, 3/8, 1/4, 1/16	$0.52 \times \sigma_{x_d}$	1.91
5	1/5, 1/5, 1/5, 1/5, 1/5	$0.45 \times \sigma_{x_d}$	2.24

This table shows that the number of levels as well as the weight distribution impacts the improvement in precision. Systematic errors and accuracy are not changed through filtering.

11.2.2 Reduction of vertical resolution

The filtering, so beneficial to the precision, has negative effects on the definition of the geological beds. Examples of the results of averaging are shown in *Fig. 11.1*.

11.3 Optimal filtering

How far can filtering be pushed so that only noise, in other words, incoherent data, and no relevant information is removed?

An increase of sampling rate corresponds to a multiplication of the precision by a factor equal to the square root of the increase (Section 10.1). Starting from a sampling rate s, expressed as a length, an increase k times the sampling, giving a new sampling length $s' = s/k$, results in a precision $\sigma_{s'}$:

$$\sigma_{s'} = \sqrt{k} \times \sigma_s,$$

where σ_s is the precision of the measurement sampled with the sampling length s, while $\sigma_{s'}$ is the one associated to s'.

Filtering, therefore, has the same impact on data as changing sampling rate. As filtering improves precision, the sampling length is **increased**.

Example

Taking the weights listed on page 182, $\sigma_{xf} = 0.61\sigma_{x_d}$, with an original measurement x_d sampled every 15 cm (6 in.), the filtered data is equivalent to the same raw data, unfiltered, but sampled with a sampling interval of 41 cm (16 in.).

The question **How much filtering?** is transformed into: **What is the largest sampling length that can be used?**

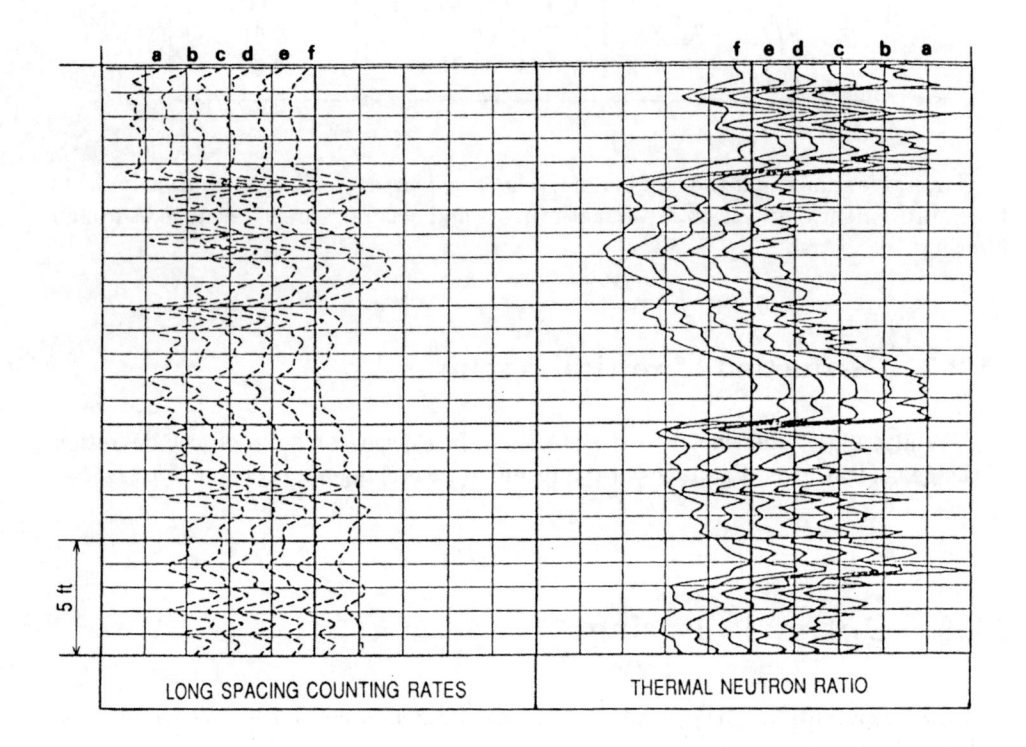

FIG. 11.1 : Effect of filtering on bed definition.
Depth scale is 0.3 m (1 ft) between adjacent horizontal lines.
  a. No averaging. Raw data only.
  b. Three-level filtering with 1/3, 1/3, 1/3 weights.
  c. Three-level filtering with 1/4, 1/2, 1/4 weights.
  d. Five-level filtering with 1/12, 1/4, 1/3, 1/4, 1/12 weights.
  e. Five-level filtering with 1/16, 1/4, 3/8, 1/4, 1/16 weights.
  f. Five-level filtering with 1/5, 1/5, 1/5, 1/5, 1/5 weights.

11.3.1 Incidence of sampling interval

The sampling length impacts the vertical resolution [2 and 6]. A more frequent sampling interval does not improve resolution by itself. For instance, a 2.5-cm (1-in.) sampling rate would not improve the value of an induction measurement that has a 240-cm (8-ft) resolution. Nevertheless, better vertical resolution commands higher sampling intervals. The basic sampling rates for *Schlumberger* logging services are listed in *Table 11.2*.

TABLE 11.2
SELECTION OF SAMPLING LENGTHS

Mode	Sampling interval	
	(cm)	(in.)
Standard	15.0	6.0
Dipmeter	0.5	0.2
Images	0.25	0.1
High resolution	3.0	1.2
Electromagnetic propagation, microspherical ...	5.0	2.0

An example of a density-neutron log recorded with 15-cm (6-in.) and 3-cm (1.2-in.) sampling rates is shown in *Fig. 11.2*. The same interval was surveyed several times with the high sampling rate.

Information theory[1] tells us the sampling frequency must be at least twice as high as the highest frequency contained in the measured signal. In well logging, this means the sampling interval must be at least twice as small as the vertical resolution of the logging tool. As tools may have different vertical resolutions, from millimeters to meters, from fractions of an inch to several feet, variable sampling rates are available from modern acquisition software. As early as the 1960s, fast and slow channels coexisted on the high-resolution dipmeter tool.

From these implications, *Table 5.3* can be transformed into a table of optimal sampling lengths for the most common logging tools.

Example

For the Litho-Density tool, the vertical resolution listed in *Table 5.3* is 38 cm (15 in.). Therefore, the effective sampling rate should be less than 19 cm (7.5 in.). An actual sampling interval of 15 cm (6 in.) is suitable if no filtering at all is applied.

---

[1]The theorem of Shannon [3]; performing a Fourier transform on the data and analyzing the frequency content would lead to the same conclusion.

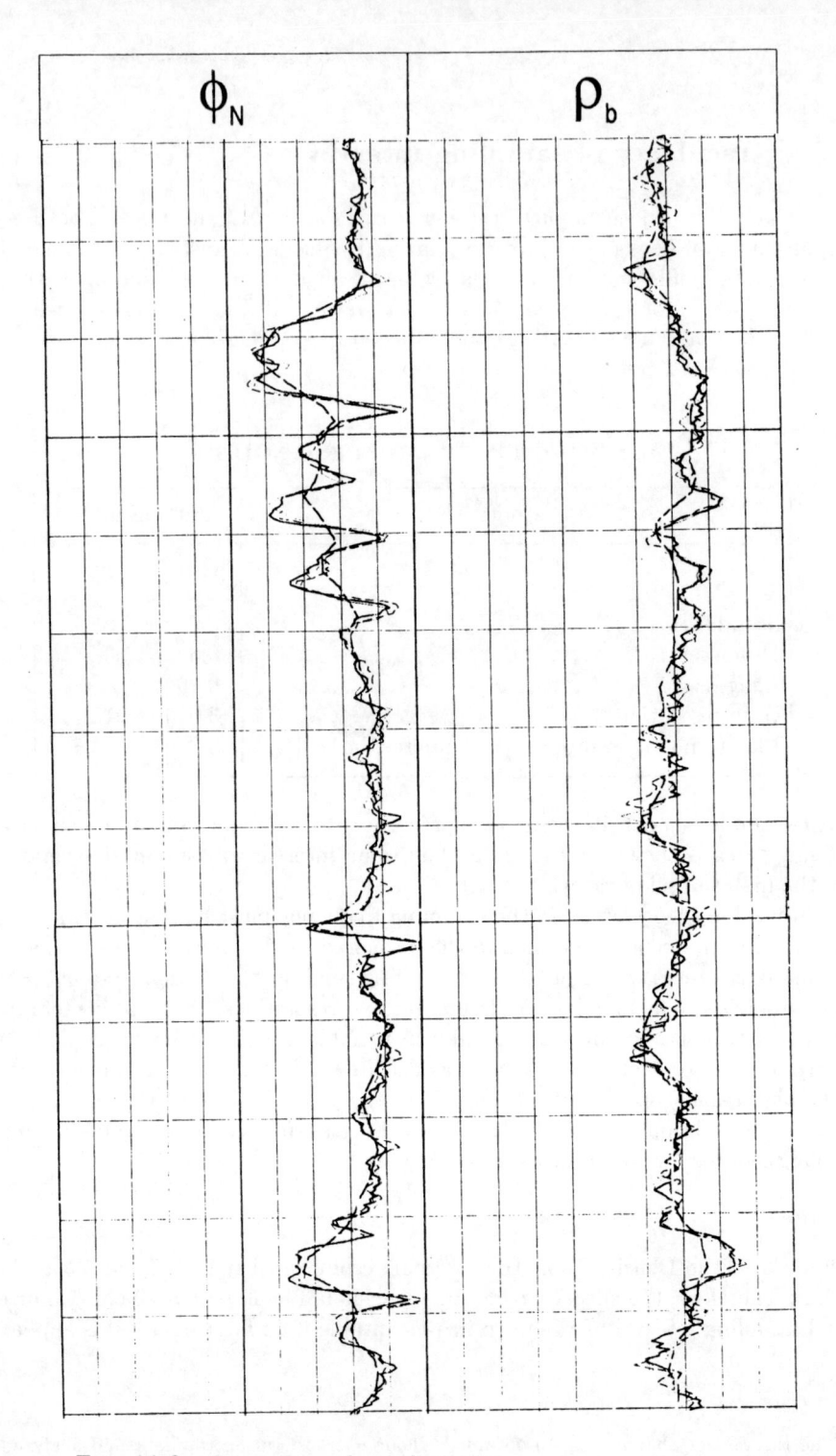

FIG. 11.2 : Comparison of sampling rates.
The 6-in. sampled data corresponds to the dashed curves. The thin variations of porosity/density are not observed on these curves. The 3-cm (1.2-in.) passes highlight porosity anomalies that repeat well.

Exercise 16

1. Find the optimal sampling rate for the deep induction, medium induction and MicroSFL* tools?

2. If the acquisition software had total flexibility, what would be the optimal sampling interval for the combination of the three measurements?

Remarks

1. The sampling rate should be a round divider of the vertical resolution, a requirement that is impractical when combining many measurements.

2. Combined logging speed/rate of penetration and sampling interval determine the measurement duration (Section 10.3). A short sampling interval, allowing many measurements over the same formation, seems desirable, but, considering the constraint on the measurement duration, which controls in many instances the measurement precision, short sampling intervals require very low logging speeds.

11.4 How is filtering controlled in the logging environment?

Raw data is usually subject to some kind of filtering. For instance, in the *Schlumberger* surface acquisition system, filtering is performed in two stages:

1. Raw data is smoothed into a calibrated output. The smoothing is controlled by a constant. One option corresponds to no smoothing. Another option corresponds to a three-level averaging with equal weights.

2. The calibrated output is filtered a second time for optical display. This is done on three depth intervals with weighting coefficients of 1/4, 1/2 and 1/4.

The combined filtering with a smoothing constant of 1 corresponds to five windows with weights 1/12, 1/4, 1/3, 1/4 and 1/12. The overall statistical precision is consequently improved by a factor of 2. This corresponds to an effective sampling length of 60 cm (2 ft) for data originally sampled every 15 cm (6 in.).

11.5 Median averaging

Median averaging is an efficient way to remove spikes. *Figure 11.3* shows how it performs. A window of n samples is considered, with n an odd number. The measurements included in this window are sorted in order of magnitude, from the smallest to the largest. The median measurement, of rank $(n + 1)/2$ is selected as the final "median-averaged" or "despiked" value. This type of filtering has been used for the borehole-compensated sonic log.

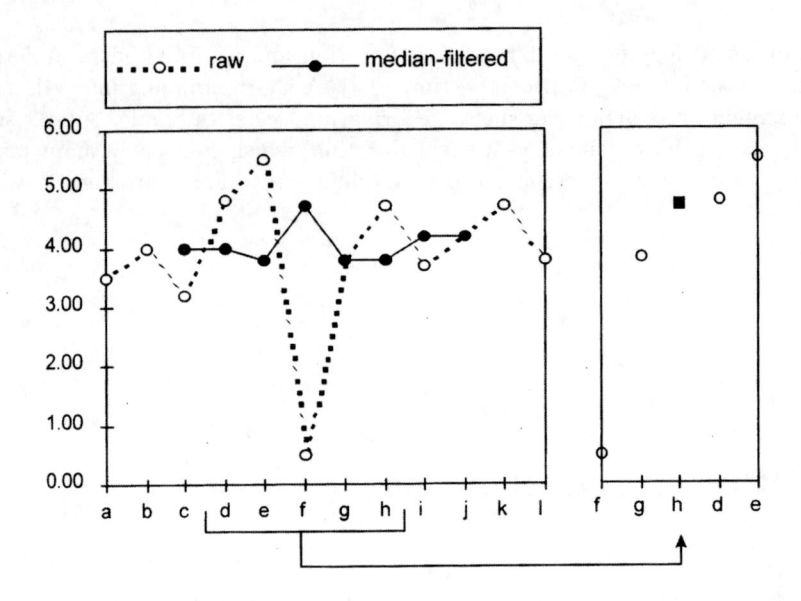

FIG. 11.3 : Median averaging.
The averaging window includes five samples. Samples d, e, f, g, h are ranked by size. The median sample, h, is selected as the final value.

11.6 Stacking

Is there a way to improve precision without sacrificing vertical resolution? Stacking is the answer. Stacking consists of running several passes over the same interval, then averaging the measurements taken at the same depth.

Example

Four passes of neutron porosities are recorded over the same interval (*Fig. 11.4*). Recorded values are ϕ_1, ϕ_2, ϕ_3 and ϕ_4, associated with precisions σ_1, σ_2, σ_3 and σ_4. The resulting porosity is

$$\phi = \frac{\phi_1 + \phi_2 + \phi_3 + \phi_4}{4}$$

with the precision

$$\sigma = \sqrt{\frac{\sigma_1 + \sigma_2 + \sigma_3 + \sigma_4}{n-1}}.$$

Careful depth-matching of the contributing passes needs to be performed before stacking.

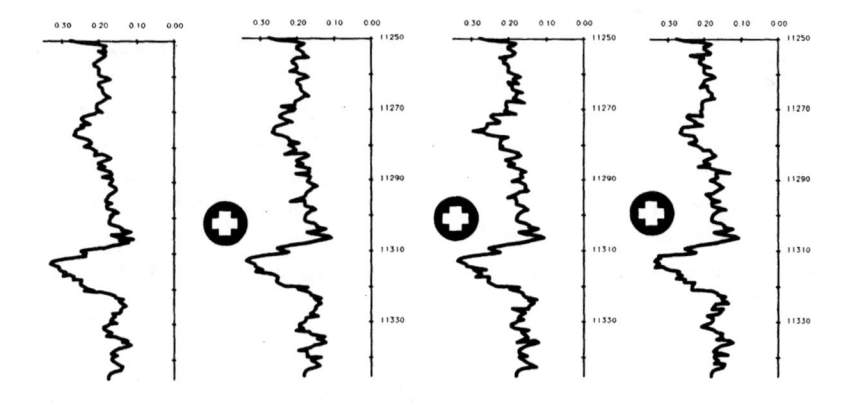

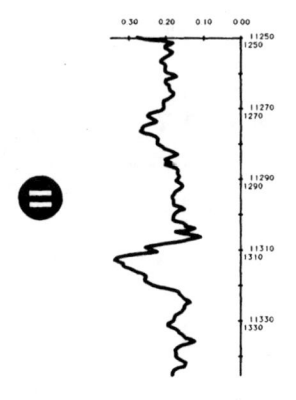

FIG. 11.4 : Stacking of four passes.

11.7 Active filtering

Sophisticated filtering methods have been developed to obtain the best compromise between the conservation of the resolution and the reduction of uncertainty.

Kalman filtering applied on most spectral gamma ray measurements is one example [1 and 4]. In a few words, the software based on such a method first differentiates zones of low activity (homogeneous beds) and of high activity (transition from one facies to the other). It then tunes the amount of filtering to these two conditions, heavy filtering on zones of the first type and soft filtering on zones of the second type. Complex filters are numerous and each has different characteristics.

11.7.1 Spectral gamma ray logging

Natural radioactivity originates from thorium, potassium and uranium. The knowledge of these three elemental contents in the formation allows some inferences on the clay typing, geochemistry, etc. Th, K and U contents cannot be measured directly. They are calculated from the counting rates accumulated in several energy windows. The *Schlumberger* NGS* Natural Gamma Ray Spectrometry tool, for instance, has five windows ranging from 0 to 3 MeV. The direct utilization of the five window counting rates to obtain the three elemental yields through a least-square method creates anticorrelations (called Mae Westing) and, occasionally, some negative readings on the elemental curves.

An adaptive filtering technique has been developed to decrease these effects and improve the appearance of the log.

The equation describing the theoretical response of the tool is shown below:

$$\begin{bmatrix} W_{1n} \\ W_{2n} \\ W_{3n} \\ W_{4n} \\ W_{5n} \end{bmatrix} = \begin{bmatrix} A_{11} & A_{12} & A_{13} \\ A_{21} & A_{22} & A_{23} \\ A_{31} & A_{32} & A_{33} \\ A_{41} & A_{42} & A_{43} \\ A_{51} & A_{51} & A_{53} \end{bmatrix} \begin{bmatrix} Th_n \\ U_n \\ K_n \end{bmatrix} + \begin{bmatrix} \epsilon_{1n} \\ \epsilon_{2n} \\ \epsilon_{3n} \\ \epsilon_{4n} \\ \epsilon_{5n} \end{bmatrix}$$

In a shorter expression

$$W_n = A\,\Theta_n + \epsilon_n.$$

Geological continuity—that characteristics within the same facies do not change rapidly—can be expressed by

$$\hat{\Theta}_n = (1 - k_n)\,\hat{\Theta}_{n-1} + k_n\,\Theta_n,$$

in which k_n is an adaptive gain based on the total gamma ray counting rate.

When there is a strong link between the values of the successive yields, k_n is small and the standard deviation on the gamma ray distribution in a fixed depth window is low. When the estimation at a given depth is not linked to the value at the previous depth, k_n is large and the standard deviation is large. In the second case, a boundary between facies is likely to have been crossed.

Figure 11.5 displays logs corresponding to a simple least-squaring technique and to an adaptive technique.

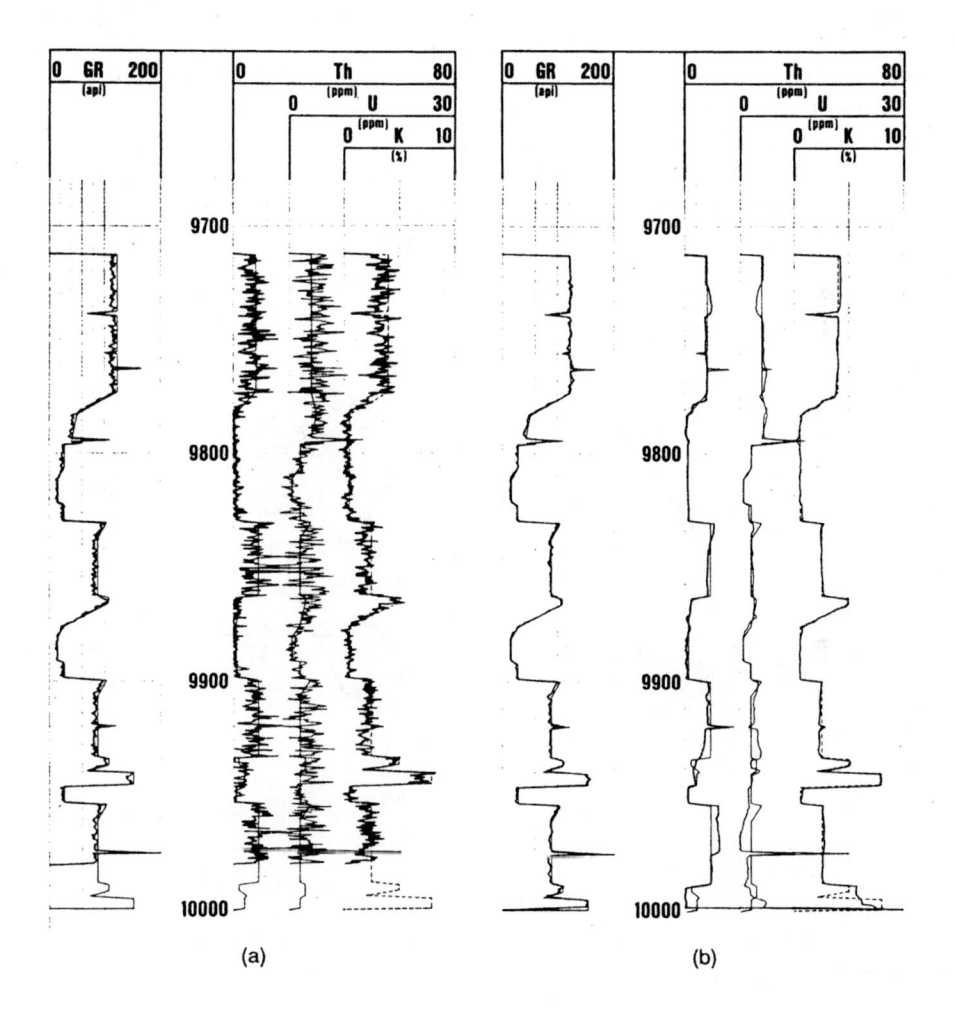

FIG. 11.5 : Comparison of raw and filtered data on a synthetic formation model.
   a. Data after least-squares computation.
   b. Data after adaptive filtering.

Courtesy of Schlumberger [5]

11.8 Azimuthal averaging

After describing the common filtering techniques, it is useful to highlight the presence of a less known effect, that is linked to the geometry of the logging sensors and to the development of deviated wells.

In a vertical well crossing horizontal beds, the geometry of the sensors is matching the geometry of the geology. In other words, the spherical or cylindrical symmetry of the transmission of energy and signals corresponds well to the symmetry of the formation. The sensors scan the formation over an interval that is identical to the thickness of this formation.

In a deviated well crossing a planar bed, such a match is no longer present. The sensors are scanning the formation over an interval that is not equal to the thickness of the bed. As measurements average the characteristics of the formation over the volume of investigation of the sensors, misleading measured values may result.

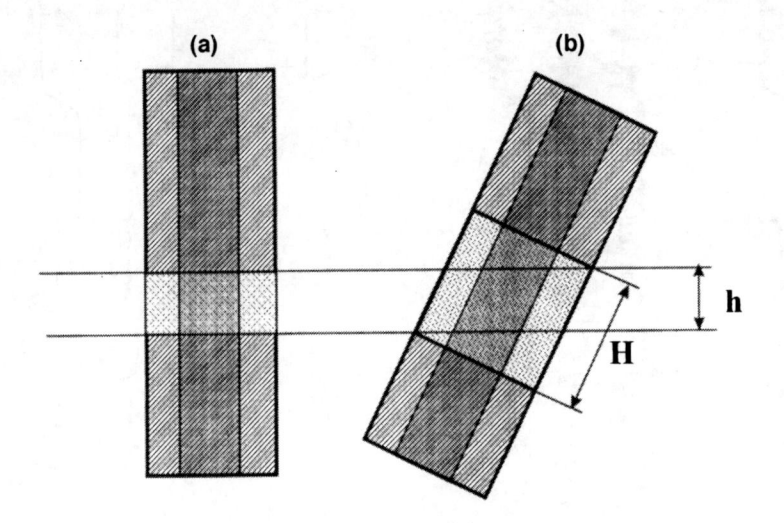

FIG. 11.6 : Comparison of a vertical well and a deviated well.
The geological bed has a thickness h.
   a. Vertical well. The sensors scan the bed over the same thickness h.
   b. Deviated well. The sensors are affected by the same bed over a distance H.
Courtesy of Schlumberger

Figure 11.6 compares the vertical and deviated wells geometries, while *Fig. 11.7* gives an example of the distortion caused by the apparent dip between the formation and the well trajectory.

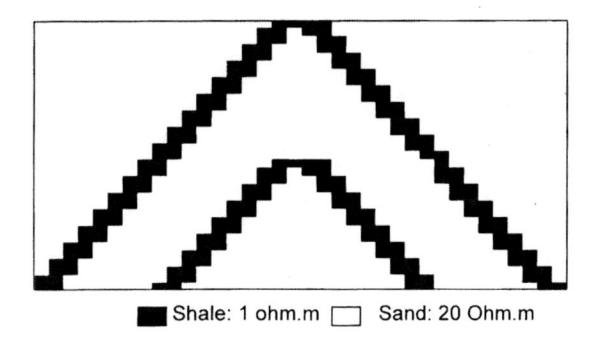

FIG. 11.7 : Distortion caused by azimuthal averaging.
The shale bed represents only 4 of the 36 quadrants over the whole interval. The oil-bearing sand bed represents the remaining 32 quadrants. This zone is scanned by a conductivity device that averages conductivities—4 quadrants at 1 Ω/m or 1000 millimhos and 32 quadrants at 20 Ω/m or 50 millimhos. The measurement is constant over the interval at 6.4 Ω/m or 156 mmhos.

References

1 Lifermann, J., *Les principes du traitement statistique du signal, I. Les méthodes classiques*, Masson, Paris, 1981.

2 Mathis, G. L., Gearhart, D., "The vertical resolution of P_e and density logs," pp. 150-161, *The Log Analyst*, Vol. 30, No. 3, 6-1989.

3 Max, J., Audaire, L., Berthier, D., Bigret, R., Carré, J. C., Chevalier, H., Escudié, B., Hellion, A., Lacoume, J. L., Martin, M., Miquel, R., Peltié, P., Trottot, M., Valette, S., Vergne, R., *Méthodes et techniques de traitement du signal et applications aux mesures physiques*, Masson, Paris, 1985.

4 Ruckebush, G., "A Kalman filtering approach to natural gamma ray spectroscopy in well logging," *IEEE Trans. in Automatic Control*, Vol. AC-28, No 3, 3-1983.

5 Schlumberger, *Essentials of N.G.S. interpretation*, Schlumberger, Paris, 1981.

6 Torres, D., Alberty, M., Jackson, C., "Real-time frequency domain filtering maximizing vertical resolution while minimizing noise," paper U, *Trans*. SPWLA 29th annual logging symposium, 1988.

12

Enhancement of vertical resolution through processing

In the early 1950s, many bankers were able to qualitatively interpret the Microlog, an electrical log with electrodes spaced 2.54 mm (1 in.) and 5.08 mm (2 in.) from each other [7].[1] The microlog could distinguish permeable from impervious formations with great detail. In the formula yielding the hydrocarbon volume, the microlog was helpful in evaluating the net-to-gross ratio (N/G).

This chapter reviews the reasons why such a detailed knowledge of the formation, relative to such a minute proportion of the reservoir, could be of significance.

12.1 Effect of vertical resolution

12.1.1 Hydrocarbon volume estimation

Failure to identify fine variations of lithology or facies may lead to major errors in the evaluation of hydrocarbon volumes [2, 6, 14 and 15]. The problem stems from the lack of linearity of many formation evaluation equations, in particular saturation and shaliness correction equations. Furthermore, a measurement with low resolution often makes a linear averaging of the characteristics of the formation.

This problem is highlighted in a sequence of thin laminations of sands and shales. For simplicity, the shale index is assumed to swing completely from 0 to 100%. The

---

[1] Admittedly, the resolution on the film was hardly better than 0.3 m (1 ft) in spite of the very short spacings.

intrinsic characteristics of the formation are listed in *Table 12.1*.

TABLE 12.1
FORMATION CHARACTERISTICS

	Units	Sand	Shale
Resistivity	(Ω/m)	25	1
Density, ρ_b	(g/cm^3)	2.34	2.5
Thermal neutron porosity, ϕ_N	(pu)	20	35
Apparent porosity, ϕ_A	(pu)	20	22.3
Water saturation, S_w	(su)	20	100

High-resolution logs

The measurements are able to extract the intrinsic value. The hydrocarbon volume is computed as 8%.

Low-resolution logging

Figure 12.1 shows the resistivity values are smeared into a common value of 1.92 Ω/m. Similarly, ρ_b and ϕ_N are averaged and yield respectively the values of 2.42 g/cm^3 and 27.5 pu. The resulting porosity and water saturation[2] are 10 pu and 64 su, bringing the oil volume to 3.6%, or 55% less than that calculated with high-resolution tools.

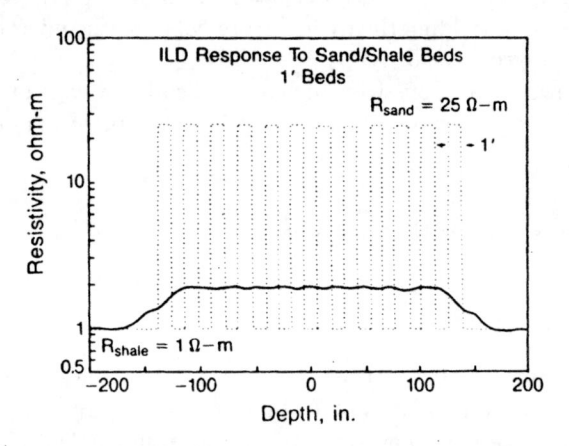

FIG. 12.1 : Response of induction in thin beds.

Courtesy of Schlumberger [1]

---

[2]The water saturation is computed through a total shale relation,

$$\frac{1}{R_t} = \frac{\phi^2 S_w^2}{R_w(1 - V_{sh})} + \frac{V_{sh} S_w}{R_{sh}},$$

in which V_{sh} and R_{sh} are assumed to be known. If a laminated model is used (when the lamination geometry is known) then the error on the saturation determination would be much smaller.

The quasi-**miraculous** increase of oil volume in thin laminations discovered by high-resolution tools has actually been observed in different areas (*Figs 12.2* and *12.3*).

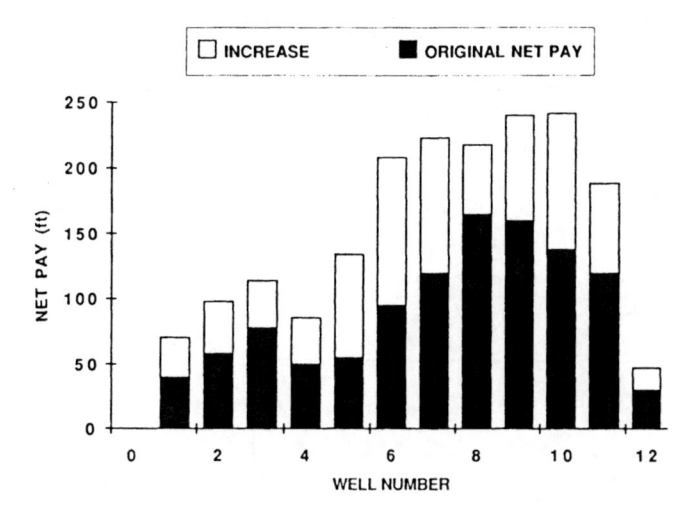

FIG. 12.2 : Example of increased oil in place estimated from dipmeter logs.
Courtesy of Journal of Petroleum Technology [13]

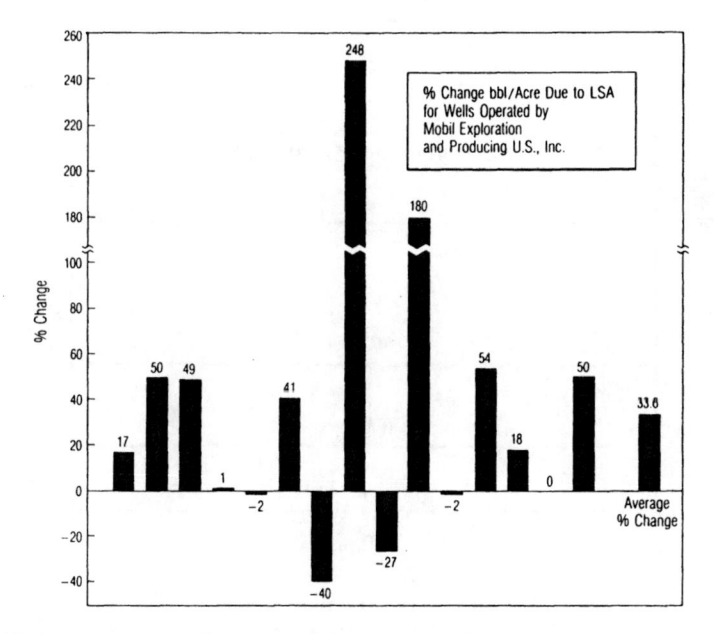

FIG. 12.3 : Differences of oil in place estimation resulting from high-resolution evaluation.
Courtesy of Schlumberger [3]

12.1.2 Recovery factor

The identification of changes of permeability is critical for optimal completion. This is illustrated with the layered model shown in *Fig. 12.4a*. The whole reservoir is 10 ft thick. Without high-resolution tools, it can be misinterpreted as homogeneous with a constant permeability of 40 mD. With high-resolution tools, several individual layers can be recognized. They are 1.5 m (5 ft), 0.9 m (3 m) and 0.6 m (2 ft) thick. If the whole zone is perforated and swept by a water front (*Fig. 12.4b*) only 40% of the zone would be covered at the time the water front reaches the producing well. Selective perforating, avoiding high permeability zones in the first phase, would allow a much higher recovery. Such a program requires the input information provided by high-resolution logging.

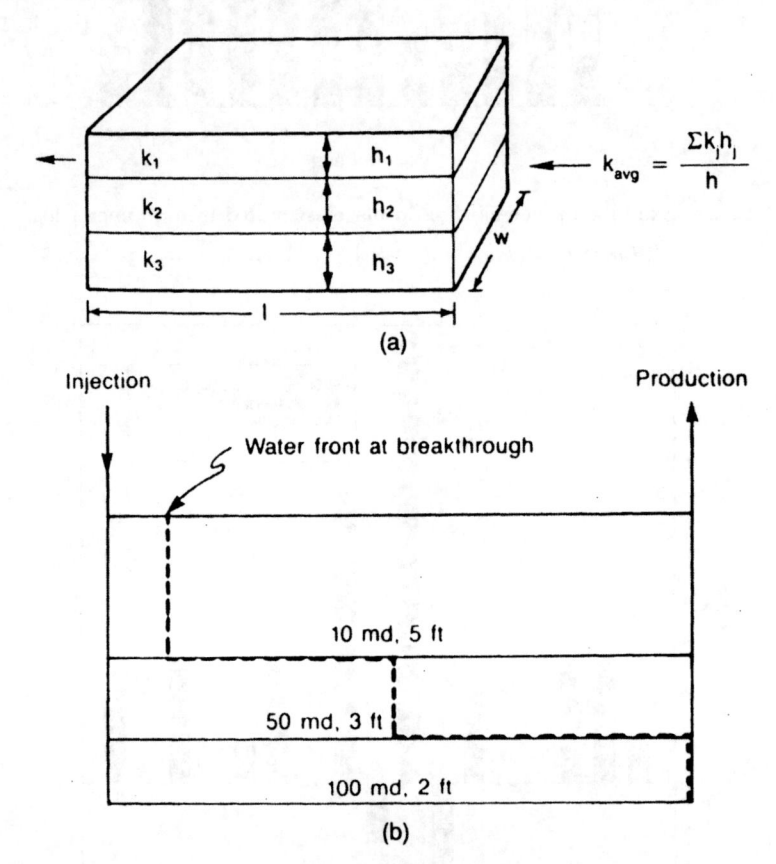

$$k_{avg} = \frac{\Sigma k_j h_j}{h}$$

FIG. 12.4 : Layered reservoir model.
  a. Reservoir model characteristics.
  b. Position of the water front at breakthrough.
Courtesy of PennWell Publishing Company [XVII]

12.2 High-resolution sensors

To obtain high-resolution information it is necessary to start from a set of sensors with the highest resolution possible. These sensors collect data used in the signal processing methods explained at a later stage. Most of the high-resolution sensors have been designed specifically to yield sharp measurements. In some cases, the modification of an existing tool has allowed the recording of a sharper log. The high-resolution sonic is shown as an example of such a development [VI and 16].

High-resolution sonic

It is possible to rewire a standard 0.9- to 1.5-m (3- to 5-ft) borehole compensated sonic sonde in order to obtain a much shorter span between receivers. A resolution of 10 cm (4 in.) can be obtained (*Fig. 12.5*). The borehole compensation, a critical issue for this device, is performed in a way similar to the 0.6-m (2-ft) spacing sonic.

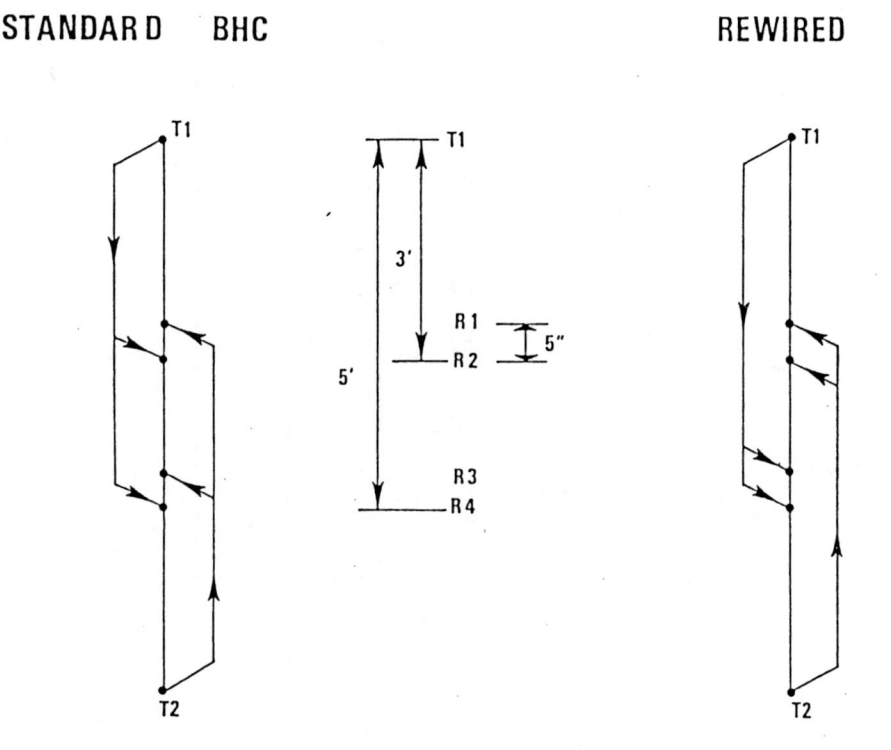

FIG. 12.5 : High-resolution sonic.
Sketch of the sonde configuration.
Courtesy of Schlumberger [VI]

152   12. Enhancement of vertical resolution through processing

The improvement of resolution is clearly shown in *Fig. 12.6* where the high-resolution sonic is compared with the standard tool.

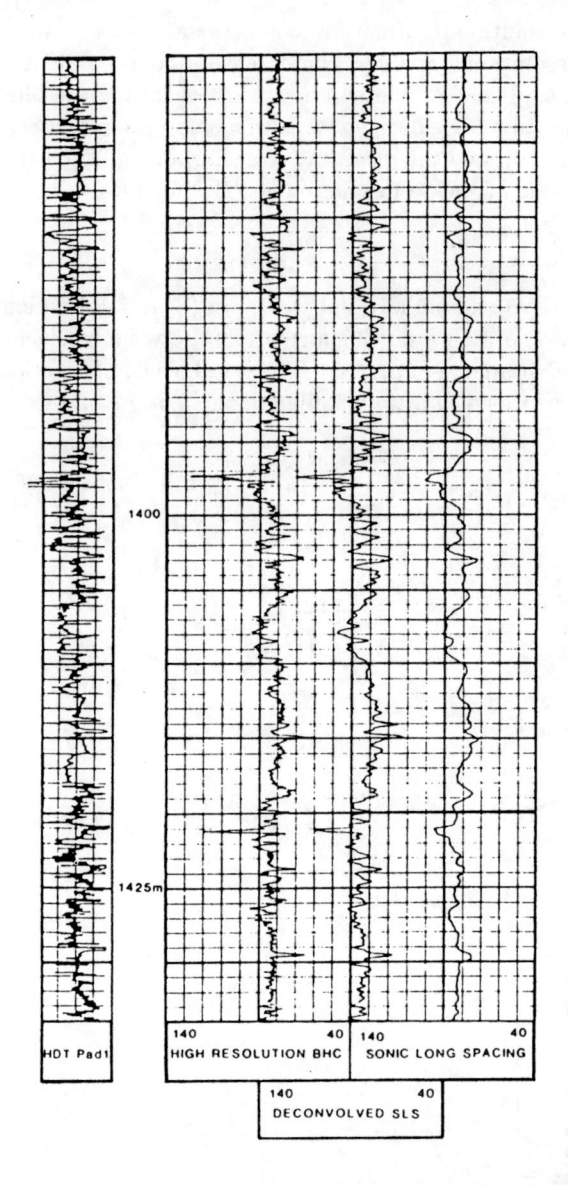

FIG. 12.6 : Comparison of the vertical resolution of different sonic curves.
On the left track, the microresistivity curve recorded by the high-resolution dipmeter is also shown.

Courtesy of Schlumberger [VI]

12.3 Signal processing

12.3.1 Deconvolution

The signal collected by any sensor, S is the **convolution** product of the formation response, f, and of a filter corresponding to the geometrical and dynamic characteristics of the sensor, s:

$$S = f * s$$

f is really what we are interested in. It can be recovered by designing a filter s^{-1}, called an inverse filter, that compensates s. The process of canceling the effect of a filter with a second filter designed to be its inverse is called **deconvolution**:

$$f = S * s^{-1}$$

Application to sonic

When both high resolution and long spacing are required, the hardware solution developed in Section 12.2 is no longer possible. A signal processing technique is required [11 and 12]. It uses a mean-square deconvolution controlled by a converter. This converter, when convolved with one sonic tool vertical response, produces a response with a different span. The initial 0.6-m (2-ft) response is deconvolved into a 0.15-m (0.5-ft) response. A standard sonic, a high-resolution sonic and a deconvolved sonic are shown together in *Fig. 12.6*.[3]

12.3.2 Combination of measurements

For most of the recently introduced technology, the logging curve printed on films and prints is generally the result of the combination of several raw measurements. For instance, the density measurement is the combination of short- and long-spacing counting rates and the thermal neutron porosity is derived from the combination of short- and long-spacing counting rates.

---

[3]The sonic transit time is the convolution of the slowness of the formation, S, by a 2-ft window: DT = S * $\boxed{\text{2 ft}}$. A 6-in. span sonic, DT', is the convolution of the same slowness, S, with a 6-in. window: DT' = S * $\boxed{\text{0.5 ft}}$. Assume there is an operator O that can convert a $\boxed{\text{2-ft}}$ window into a $\boxed{\text{0.5-ft}}$ window such that $\boxed{\text{0.5 ft}}$ = $\boxed{\text{2 ft}}$ * O; then, DT' = S * $\boxed{\text{2 ft}}$ *O = DT*O. The 6-in. deconvolved sonic is obtained from the original sonic provided the operator O can be determined.

The general approach to enhanced resolution is to use the raw measurement with the highest resolution to correct the main measurement [4].

Example

Measurement A is a function of u and v, with v being the raw measurement with the highest resolution. The classical tool response would be

$$A = f(u, v).$$

The enhanced resolution log yields A' linked to A as follows:

$$A' = A + g(v).$$

It can be anticipated that this approach requires that v is actually sensitive to small variations in the formation—that its field of investigation is not limited to the borehole wall. Otherwise, this technique would just superimpose borehole noise on the classical measurement.

Application to nuclear tools: alpha processing

Alpha processing is the application of the above technique to nuclear tools [8 and 9]. It is described here for the density measurement. The short and long spacings of the density tools—depending on the manufacturer—are approximately 15 cm and 38 cm (6 and 15 in.).

The short spacing is linked to a shallow depth of investigation, a primary requirement to correct for mudcake effects, and an excellent resolution, around half the spacing distance. This is because only gamma rays with a much lower energy than the source are detected. These lower energy gamma rays cannot be found less than a couple of inches from the source. In the favorable case where a large percentage of the formation contributes to the overall signal seen by the detector, the short-spacing detection can be duly used to enhance the resolution. Assume that a density ρ_{SS} can be defined from the short-spacing detector counting rates. ρ_{SS} is a function of ρ_b and ρ_{mc}, respectively the formation and mudcake densities:

$$\rho_{SS} = f(\rho_b, \rho_{mc})$$

$$\rho_{SS} = g(\rho_b) + h(\rho_{mc}, \text{environment}).$$

Alpha processing works if function h varies at a much slower rate than function g. In order to find the offset resulting from the environment, ρ_{SS} is averaged over a distance similar to the vertical resolution of the original density measurement. This yields $\rho_{SS\text{ average}}$. By subtracting this average from the original measurement and keeping the assumption on a slowly changing environment, a high-resolution, formation-dependent component is obtained:

$$\delta\rho = (\rho_{SS\text{ original}} - \rho_{SS\text{ average}}).$$

This is added to the original measurement:

$$\rho_{b \text{ enhanced resolution}} = \rho_b + \delta\rho.$$

It could be said that this seems to be a neat stunt. The short-spacing detector is primarily designed to compensate for mudcake and rugosity effects and it is now used to scan the formation with high resolution. It has to be kept in mind that the short-spacing detection can do one of the two in given conditions and not the two simultaneously. An example of an alpha-processed density is shown in *Fig. 12.7*.

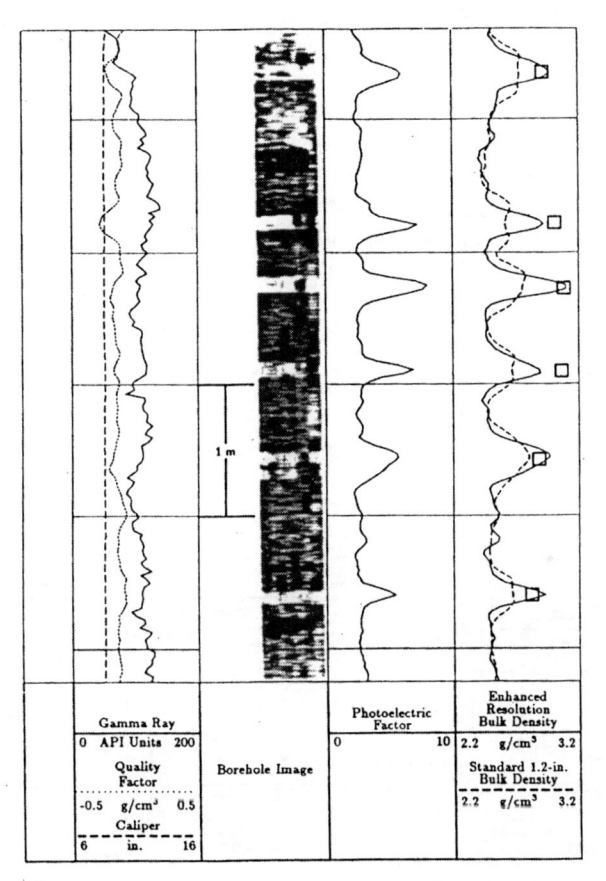

FIG. 12.7 : Comparison of standard 1.2-in. and alpha-processed bulk densities.
In the middle track, the borehole image derived from the Formation MicroScanner tool is shown. The open squares displayed in the same track as the alpha-processed bulk density represent core-derived densities. In track 1, the quality factor indicates the magnitude of the environmental corrections. In general, if the quality factor deviates by more than 0.07 g/cm^3 from the average baseline, the enhanced resolution result is suspect.

Courtesy of SPWLA [8]

Enhanced resolution of the induction

Several companies have proposed processing methods to reduce the induction vertical resolution from 2.1 m (7 ft) to 0.9 m (3 ft) [1]. Like alpha processing, the method combines two measurements at different depths of investigation.

Deconvolution methods have been attempted in the past. They have been hampered by the so-called blind frequencies of the induction tool, at which it is unable to see resistivity changes every 75 and 45 cm (2.5 and 1.5 ft).

The technique developed by *Schlumberger* fills the **missing frequencies** of the deep measurement by substituting them with the medium induction information. This second measurement has no blind frequencies because of the shorter spacing between coils. The method preserves the deep induction depth of penetration while injecting the medium induction superior resolution.

Because the improved resolution is obtained from shallow information, the sensitivity of the enhanced log to rugose boreholes filled with relatively salty muds is increased [5 and 10]. The best enhancement of resolution is obtained when the resistivity of the invaded zone exceeds that of both the uninvaded zone and the adjacent shale layer. *Fig. 12.8* shows an example of the processing method.

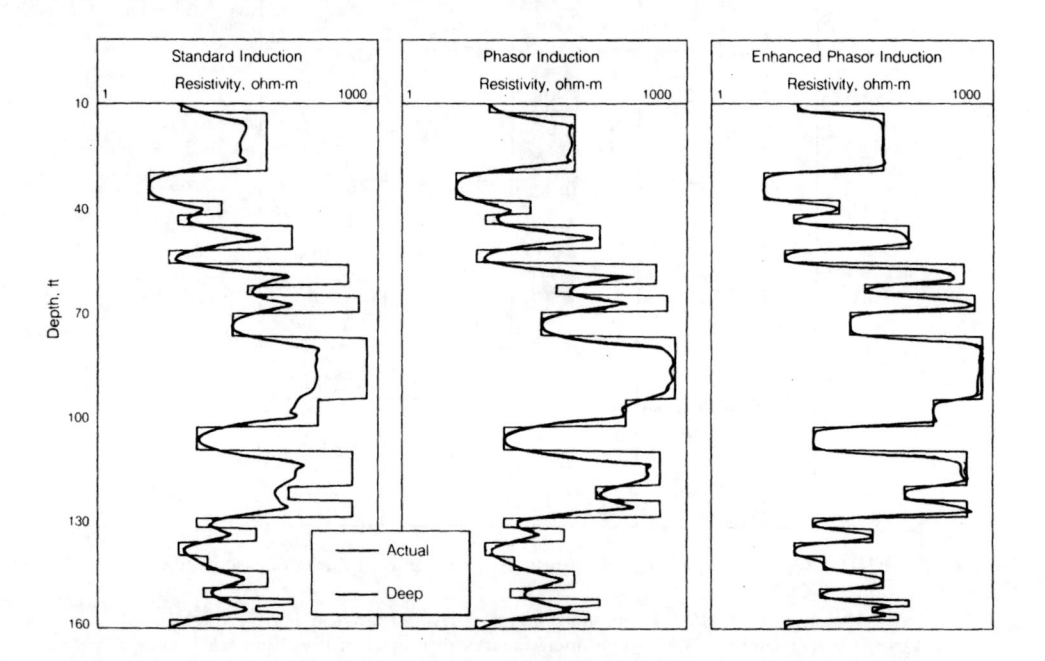

FIG. 12.8 : Comparison of standard and enhanced resolution induction logs.

Courtesy of Schlumberger [3]

References

1 Alberty, M., Epps, D., Strickland, R., "Field test results of the high resolution induction," paper P, *Trans.* SPWLA 29th annual logging symposium, 1988.

2 Allen, D. F., "Laminated sand analysis," paper XX, *Trans.* SPWLA 25th annual logging symposium, 1984.

3 Allen, D., Anderson, B., Barber, T. D., des Ligneris, S., Everett, B., Flaum, C., Hemingway, J., Morriss, C., "Advances in high resolution logging," pp. 4-14, *The Technical Review*, Vol. 36, No. 2, 4-1988.

4 Allen, D., Cafiero, M., Cheshire, S., Flaum, C., Hart, R., McGann, G., Spaeth, R., Wilson, D., A., "Strategies for thin-bed evaluation," pp. 28-42, *Oilfield Review*, Vol. 1, No. 3, 10-89.

5 Barber, T., "Induction vertical resolution enhancement, physics and limitations," paper O, *Trans.* SPWLA 29th annual logging symposium, 1988.

6 Chugbo, A., Roux, G., "Seven years of production of a thin oil column at Obagi," *Trans.* Seminar on recovery from thin oil zones, hosted by the Norwegian petroleum directorate, Stavanger, 1988.

7 Doll, H., "The microlog, a new electrical logging method for detailed determination of permeable beds," pp. 155-164, *Trans. AIME*, Vol. 189, 1950.

8 Flaum, C., Galford, J. E., Hastings, A., "Enhanced vertical resolution processing of dual detector gamma-gamma density logs," paper M, *Trans.* SPWLA 28th annual logging symposium, 1987. Also pp. 139-149, *The Log Analyst*, Vol. 30, No. 3, 1989.

9 Galford, J. E., Flaum, C., Gilchrist, W. A. Jr., Duckett, S., "Enhanced resolution processing on compensated neutron logs," SPE 15541, *Trans.* SPE 61st annual technical conference and exhibition, New Orleans, 1986. Also pp. 131-137, *SPE Formation Evaluation Journal*, Vol. 4, No. 2, 6-89.

10 Minette, D. C., "Thin bed resolution enhancement: potential and pitfalls," paper GG, *Trans.* SPWLA 31st annual logging symposium, 1990.

11 Peyret O., Private communication, 1986.

12 Runge, R. J., Powell, N. J., "The effect of sampling rate on sonic log span adjustment," paper D, *Trans.* SPWLA 8th annual logging symposium, 1967.

13 Sallee, J. E., Wood, B. R., "Use of microresistivity from the dipmeter to improve formation evaluation in thin sands, Northeast Kalimantan, Indonesia," pp. 1535-1544, *Journal of Petroleum Technology*, Vol. 36, No. 9, 9-1984.

14 Schulze, R., Ives, G., Etter, T., "Thin bed analysis in Central Oklahoma," paper LL, *Trans.* SPWLA 26th annual logging symposium, 1985.

15 Suau, J., Albertelli, L., Cigni, M., Gragnani, U., "Interpretation of very thin gas sands," paper A, *Trans.* SPWLA 25th annual logging symposium, 1984.

16 Theys, P., "Advances in logging and perforating technology for improved evaluation and production of thin reservoirs," *Trans.* Seminar on recovery from thin oil zones, hosted by the Norwegian petroleum directorate, Stavanger, 1988.

13
Tool response

Tool response is the link between the raw measurements and usable formation characteristics. Tool response could be the functional relationship between gamma ray detector count rates and formation electronic density. Anything affecting the raw measurement so that it is no longer possible to have a direct transform between this measurement and the related formation characteristic is called environmental effect. These effects are reviewed in the next chapter.

Tool response and environmental effects are sometimes difficult to uncouple. Nuclear tools equipped with several detectors are able to scrutinize formations at different distances from the borehole wall. One detector reads the formation and the borehole environment (mudcake, borehole wall, invaded zone). The second one primarily observes the environment. For practical reasons, a single algorithm is generally used to relate the detector counting rates to formation characteristics while simultaneously eliminating the environmental effects.

Beyond this practical aspect, another difficulty may occur when formation and environmental parameters are coupled—when the environmental correction varies with changes in the formation. Thermal neutron detection can be taken as an example. The salinity effect varies with the lithology and a different correction applies whether the formation is mainly sandstone or dolomite.

There are four general methods for determination of the tool response. Chronologically, they are

(a) physical simulation
(b) experimental measurements
(c) scaled experiments
(d) mathematical modeling.

13.1 Physical simulation

13.1.1 The resistivity network

The experimental assessment of the performance of a resistivity tool such as the laterolog is a practical impossibility because of the large dimensions investigated. In addition, evaluation is complicated by the number of parameters involved and related to mud, annulus, flushed and virgin zones, adjacent beds.

An efficient way to understand the performance of a given tool in such a complex situation is to simulate its response by postulating the actual values of the different zones involved. This was initially achieved with a resistor network (*Fig. 13.1*). For long tools with many electrodes, this procedure, although successful and accurate, took several days to produce the response of the tool to a single set of formations [7].

FIG. 13.1 : The resistor network.
Thousands of resistors were used to derive the tool response of the first electrical logging tools. The resistors are grouped in clusters representing R_t, R_{xo}, shoulder bed and borehole resistivities.

13.2 Experimental measurements

Intuitively, the most reliable and controllable method to derive tool response is to use real formations whose characteristics are determined independently of the logging tool. As seen before, this approach could not be followed for deep investigating resistivity tools. Conversely, it is the most popular method for nuclear tools whose volumes of investigation are in the range of a tenth of a cubic meter (or a cubic foot). *Figures 13.2, 13.3* and *13.4* give examples of experimental setups.

FIG. 13.2 : Calibration blocks for nuclear tools.
Setup used in North America in the early 1970s.

162   13. Tool response

(a)

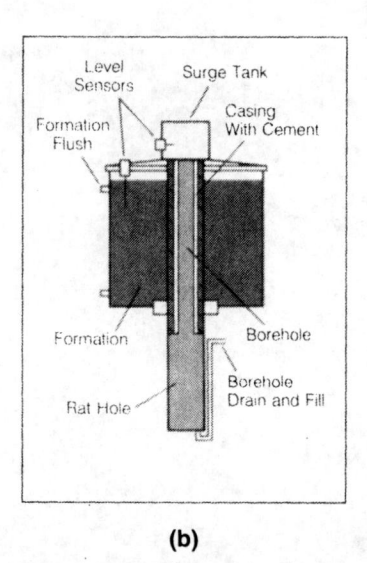

(b)

FIG. 13.3 : New facilities at Houston for measuring the response of nuclear tools in a wide variety of formation and borehole environments.

a. Artificial formation. A logging tool is introduced into the borehole.

b. Sketch of the block depicted above.

Courtesy of The Technical Review [6]

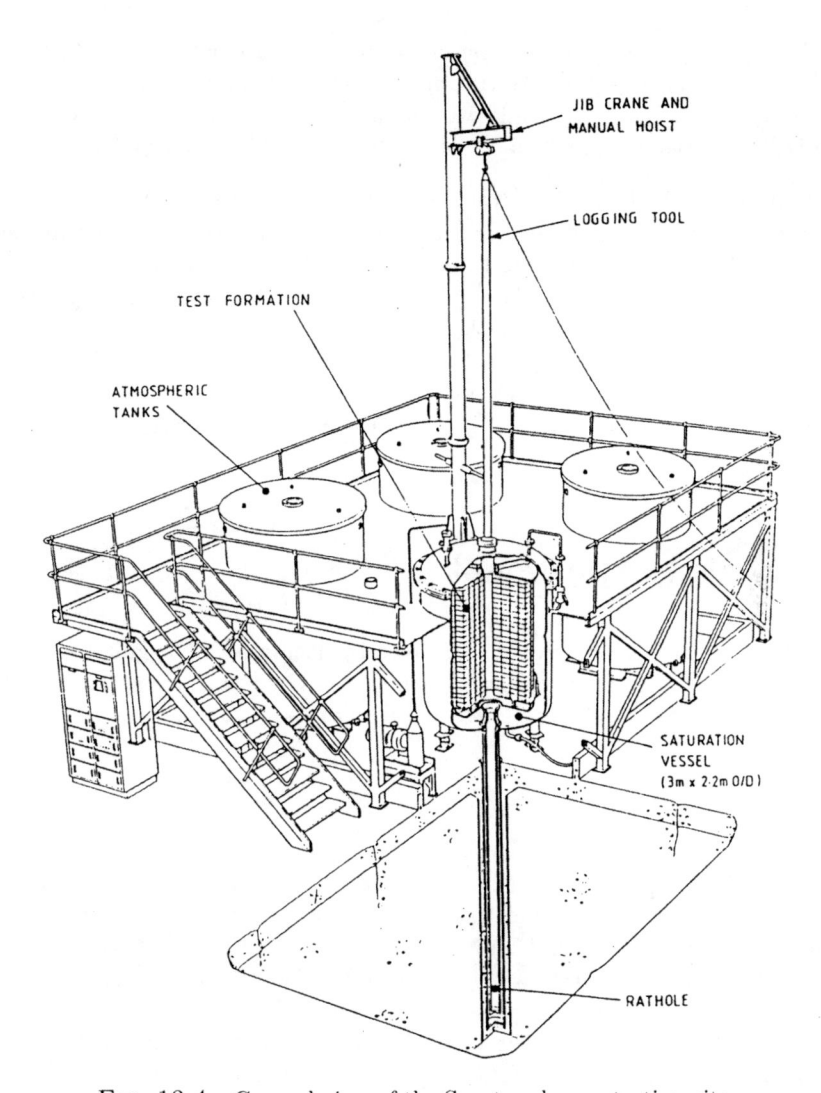

FIG. 13.4 : General view of the Spartan demonstration pits.
Courtesy of the Offshore Supplies Office, UK Department of Energy [3]

13.2.1 Test formation design considerations

The following problems are inherent in the building of experimental formations [3 and 11]:

- Traces of elements may affect the measurement being characterized. For instance, for the thermal neutron, the presence of few parts per million of boron (or, worse, gadolinium) introduces an important bias on the tool response.

- Homogeneity of the rock sample is desired. Homogeneity can be checked with huge scanning facilities. An alternative is to build the formation from small blocks of material that could be checked individually. However, in this case, discontinuities at the interfaces of the blocks could cause problems.

- Natural quarried rocks require complex handling. Several days are necessary to remove all the original moisture. All the remaining air is then flushed out before deaerated water is used to impregnate the block under rigorous pressure control. Rock samples are then taken at regular intervals to check the axial and radial homogeneity of the block.

Typically, the characteristics of the artificial formations can be known to within 0.003 g/cm^3 for the density and to 0.3 pu for the neutron porosity.

13.3 Scaled experiments

One way to solve the size problem involved by deep-reading logging tools is to use reduced scale mock-ups. This approach has been used by the University of Houston [9] for laterolog tools and by *Schlumberger* for sonic tools [2 and 8].

13.3.1 Resistivity

The scale model taken as an example (*Fig. 13.5a*) analyzes the response of the shallow laterolog tool. The model tool is scaled down by 20 to 1. The dark areas represent the fiberglass mandrel and the white areas with the labels are the brass electrodes. The overall length of the tool is around 50 cm (20 in.). *Figure 13.5b* shows the reduced scale setup. Porous concrete is used in a fiberglass tank filled with saline water. The scaled tool is moved through a borehole drilled in the formation.

Figure 13.5c displays the results of the scaled experiment. Good agreement is obtained between the actual resistivity value (10 Ω/m) and the measured values.

13.3.2 Sonic

Laboratory experiments using scale models were performed to verify the Stoneley wave attenuation by a horizontal fracture crossing the borehole. The experimental setup is shown in *Fig. 13.6* and consists of two aluminum cylinders stacked on the top of each other in a fluid tank.

A transmission coefficient across the fracture could be measured and compared with theoretical predictions (*Fig. 13.7*).

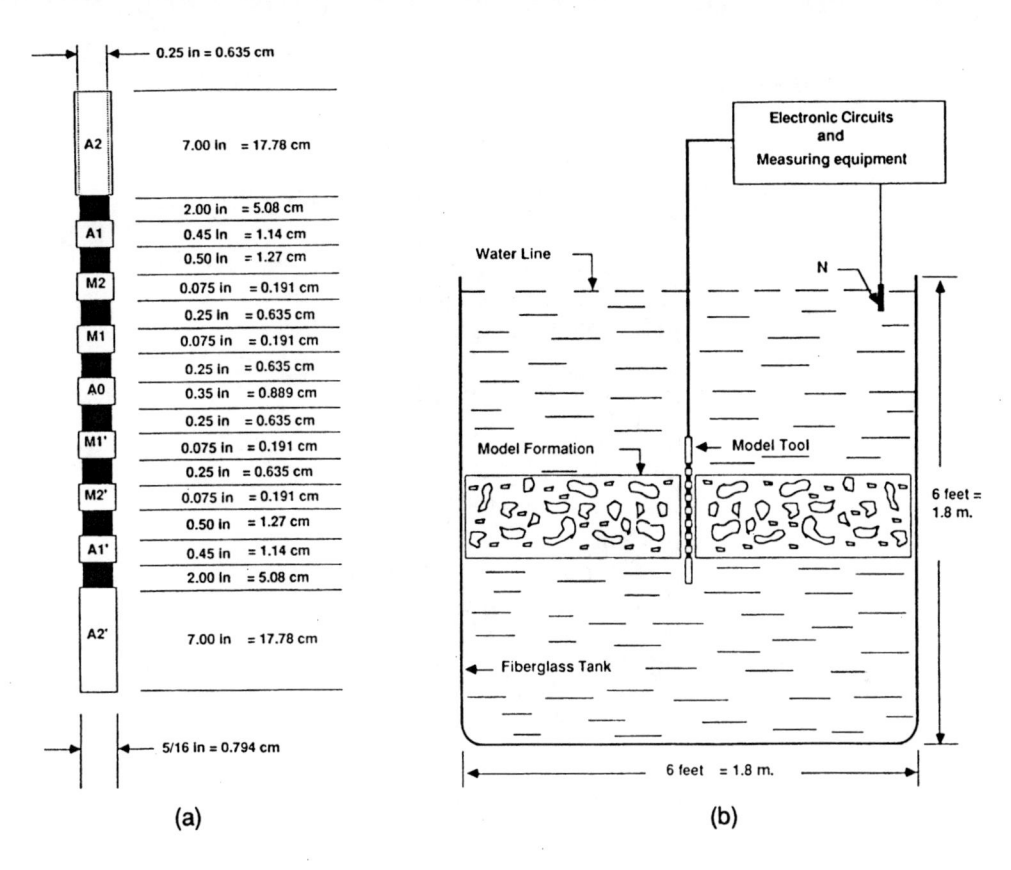

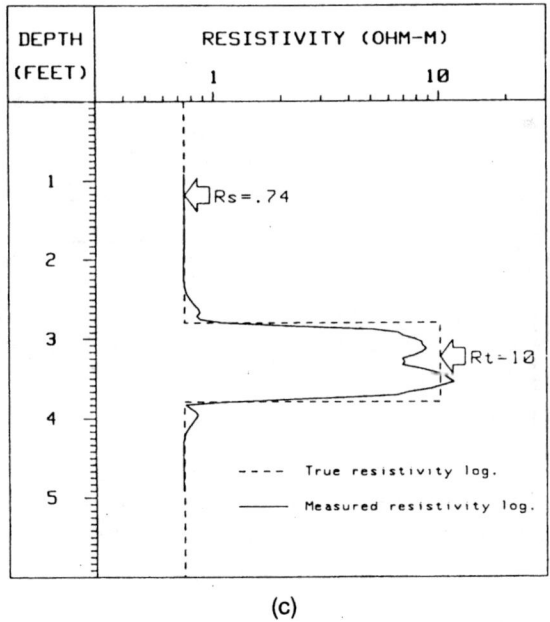

FIG. 13.5 : University of Houston setup for resistivity scaled experiments. *Courtesy of SPWLA [9]*

   a. Model of the tool.

   b. Complete setup overview.

   c. Results of the experiments.

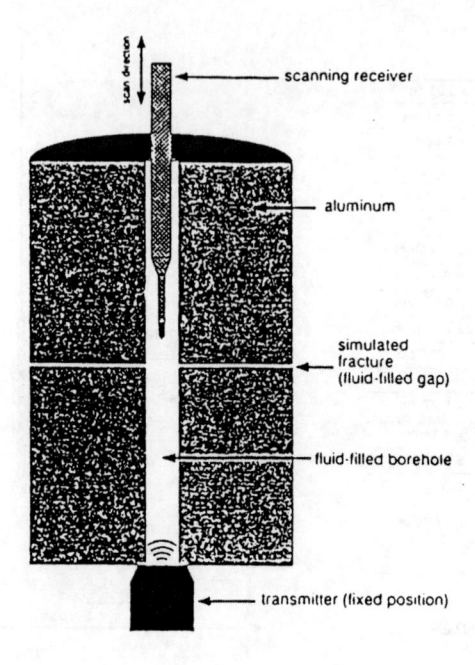

FIG. 13.6 : Scaled setup for sonic tool response definition.

Courtesy of SPWLA [2]

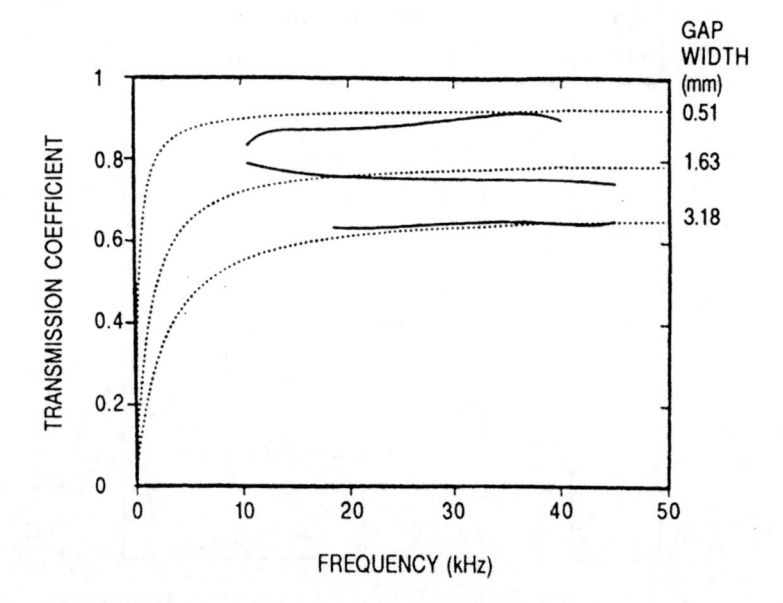

FIG. 13.7 : Theoretical and measured Stoneley wave attenuation.

Courtesy of SPWLA [2]

13.4 Mathematical modeling

13.4.1 General

Mathematical modeling has become a necessary complement to laboratory experiments performed during the design and development of logging tools.

Determination of accurate tool response needs to be established under numerous formation and environmental conditions. Different matrices, porosities, and fluids are considered over a wide range of hole sizes, mudcakes, tool positions, temperature and pressure. Collection of such data necessitates vast facilities that unfortunately have limitations despite their size. For instance, high temperatures commonly encountered in normal logging conditions are difficult to reproduce.

Mathematical modeling extends the experimental database beyond laboratory capabilities and allows more parameters to be considered. Another benefit is the separation of effects that cannot be distinguished under normal laboratory conditions, such as borehole and formation temperatures. It is important to note that mathematical modeling cannot completely replace the experimental use of laboratory formations, which are needed to check the soundness of the model and, sometimes, to calibrate it. The study of the effect of temperature on neutron porosity illustrates this point: the simple assumption that the reduction of neutron porosity results from a decrease of density with higher temperature. Data obtained using this simple model fits poorly with that obtained in the laboratory. However, by including in the model the variation of thermal capture cross sections with temperature, good agreement with experimental data is obtained (*Fig. 13.8*).

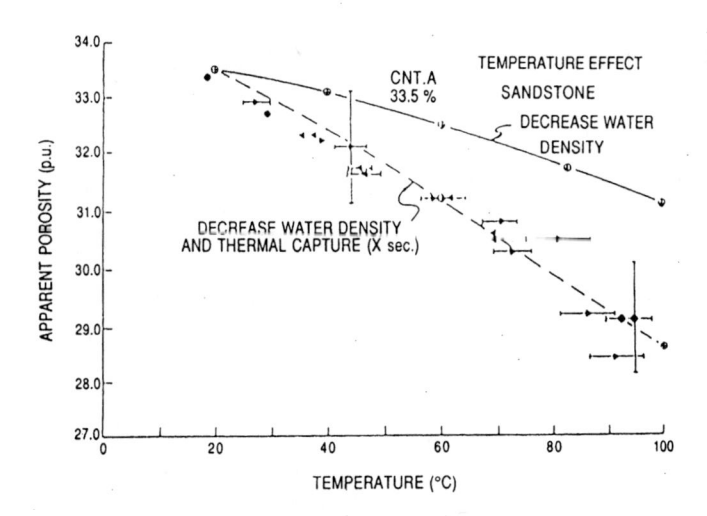

FIG. 13.8 : Variation of neutron porosity with temperature.

The cost of mathematical modeling has decreased considerably. This has resulted in the development of models that are more complex and hence more realistic.

13.4.2 Sonic modeling

Acoustic propagation in the borehole is extremely complex owing to the presence of headwaves,[1] trapped fluid modes,[2] and surface waves[3] [5]. The sonic waveform includes all these. In openhole conditions, the compressional component is generally well recognized but later arrivals are difficult to identify. When casing is present, even compressional arrivals can be smeared.

Mathematical modeling expands the understanding of acoustic propagation and optimizes design of the receiver array and the choice of frequency ranges. There are two types of modeling, analytic and finite.

The **analytic model** assumes an acoustic point source on the axis of a perfectly cylindrical, fluid-filled borehole that penetrates a homogeneous, isotropic and elastic formation. It yields waveforms quite similar to those obtained from laboratory experiments in nonrugose boreholes (*Fig. 13.9*).

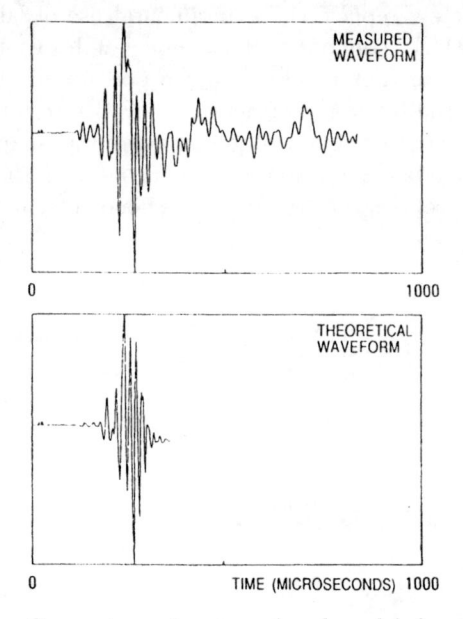

FIG. 13.9 : Comparison of measured and modeled acoustic waves.

---

[1] Headwaves are wavefronts that develop from pressure wavelets in the mud and shear wavelets in solids. Typically, the compressional wavefront accompanies a headwave.

[2] When acoustic energy is reflected from the borehole walls, a family of propagations called trapped modes is created.

[3] Surface waves are associated to packets of energy that propagate along the borehole wall. A surface wave is related to the Stoneley wavefront.

In the **finite elements** method, space is divided into small circular rings of rectangular or triangular cross sections. The motion of every element at each time step is simulated by using the minimum energy principle. The method handles the presence of bed boundaries and rugose boreholes. Numerical results have shown excellent agreement between the analytical and finite element models for geometries where both can be used (*Fig. 13.10*).

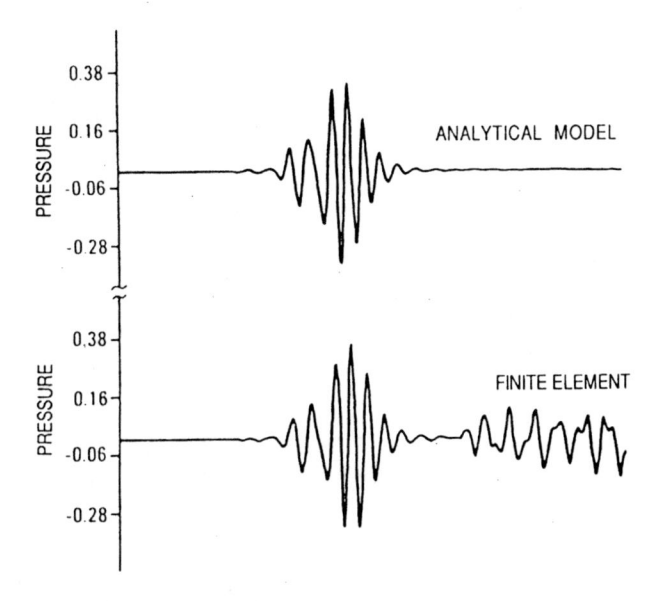

FIG. 13.10 : Comparison of analytical and finite elements models.

13.4.3 Resistivity modeling

Because the resistor network solution is time consuming as well as restrictive, direct simulation of the network has been implemented. Resistivity modeling has evolved in parallel with the development of computer technology [1]. In the early eighties, only simple formation models were considered. Typically, the model would involve the inversion of matrices with about 40,000 elements. Comparison with network measurements was excellent for reasonable resistivity contrasts (*Fig. 13.11*). The direct simulation extended the field of investigation to very large ratios of resistivities (R_t/R_m=10,000). Producing the response of the tool to a single set of formations took 1 h.

Computer codes have increased in number and complexity. They are capable of handling more realistic invasion profiles and successions of beds with different resistivities. Dipping and anisotropic formations can also be handled. *Figure 13.12* summarizes the early programs for induction modeling. Section 15.6 documents the recent progress in resistivity modeling.

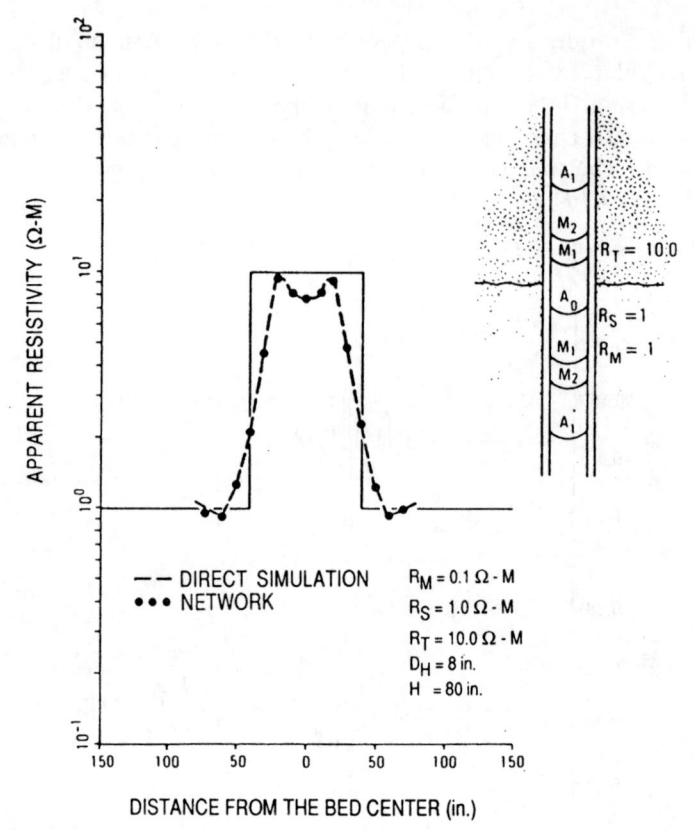

FIG. 13.11 : Comparison of computer simulation and values given by the resistor network.

13.4.4 Nuclear modeling

Nuclear modeling describes the behavior of neutrons or gamma rays interacting with atoms of the formation, borehole and tool. The calculated flux of particles can be used to predict counting rates without the need for experimental normalizing factors.

Neutron detection illustrates this approach. The system consists of a source, shielding and detectors. Formation porosity is derived from the counting rates of the detectors. These counting rates depend on the gas in the detectors and the cross sections of the formation elements, which are themselves dependent on the energy of the incident particle. This is referred to as radiation transport and is described by the Boltzmann equation.

General analytic solutions of this equation do not exist. However, two mathematical techniques have been developed to solve this equation numerically:

In the **multigroup** transport method, the energy range of interest is divided into a finite number of individual groups, as are the directions. Interaction cross sections are replaced by averaged values for each group.

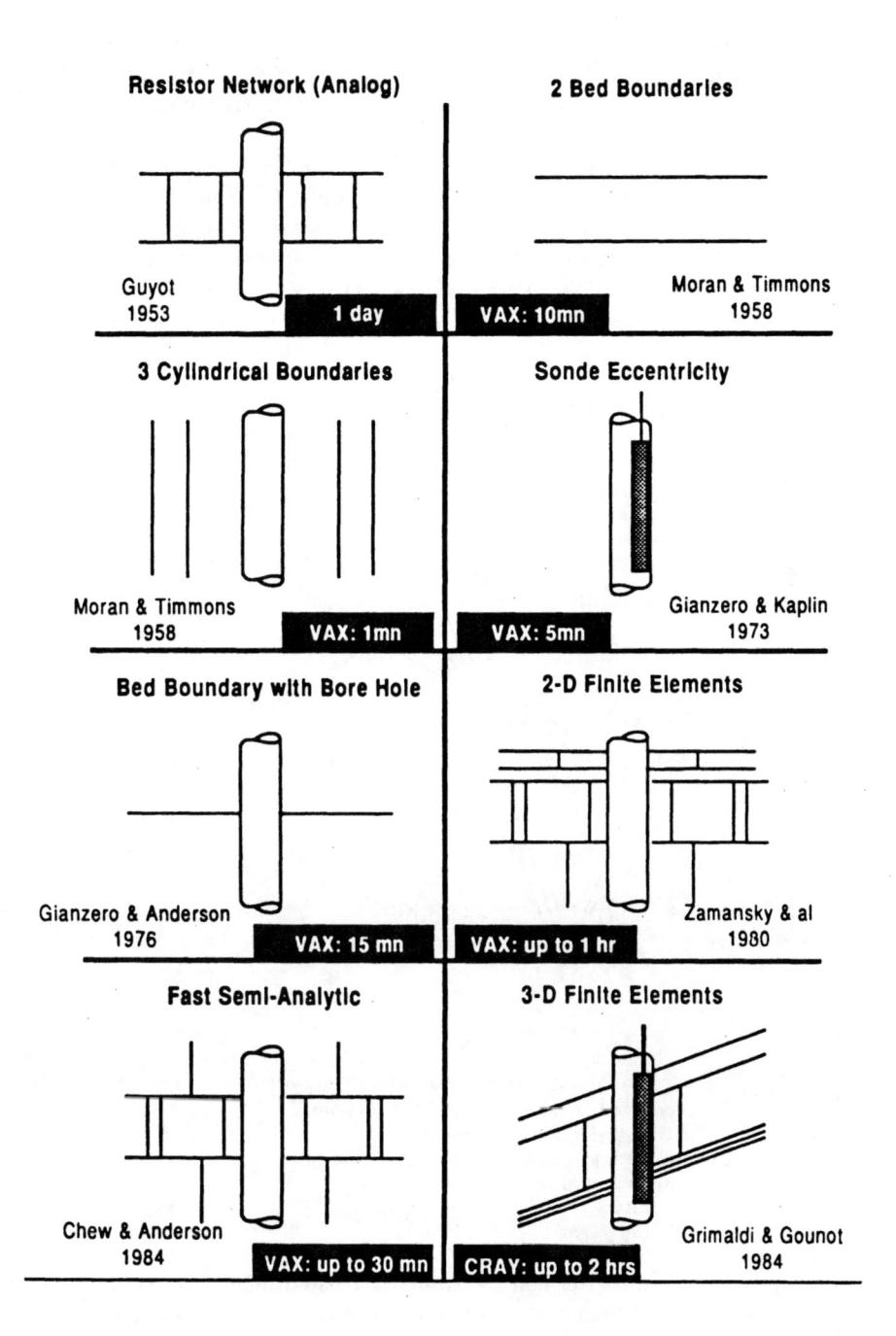

FIG. 13.12 : Computer codes for induction simulation.
The CPU times required today are two orders of magnitude shorter.
Courtesy of C. Clavier

Three-dimensional (3D) models based on multigroup transport theory are now within the practical capacity of present-day computers [4]. Alternatively, a two-dimensional (2D) approximation can be made that assumes azimuthal symmetry in cylindrical geometries.

The **Monte Carlo** treatment does not directly solve the transport equation, but simulates individual particle histories. Each particle emitted by the source in a random direction follows a succession of straight-line paths between collisions. Such distances are treated as random variables constrained by the geometry and the cross sections of the medium. The type of particles produced, their energy and direction, are specified with their respective probabilities. *Fig. 13.13* represents the history of a particle.

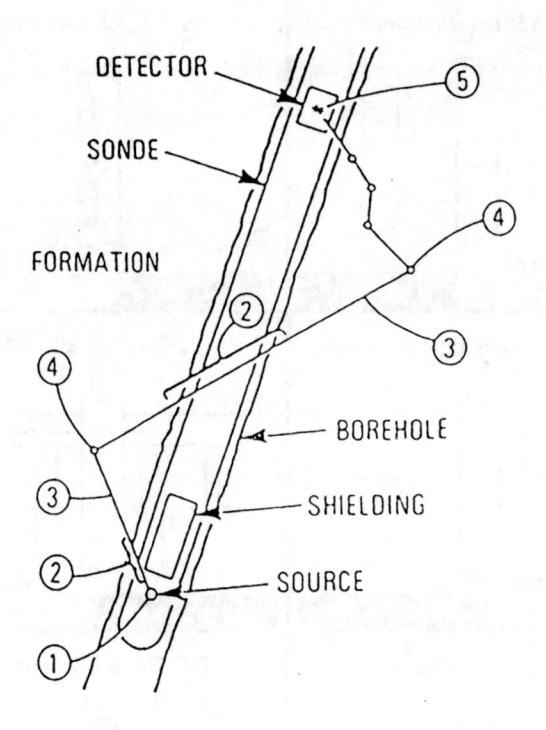

FIG. 13.13 : History of a neutron particle emitted by the logging neutron source.
1. Emission from the source.
2. Path through the logging tool and mud.
3. Path through the formation.
4. Interaction with formation atoms.
5. Detection.

Three-dimensional geometries can be explicitly treated but numerous particle histories are needed to yield a realistic simulation, which can result in excessive computer time. In a practical calculation, particular attention is given to **important** particles, in the sense that they are likely to interact with a detector sometime during their random walk.

For instance, it is reasonable that particles traveling away from the detector just after their emission from the source are minimized, and particles emitted in the direction of the detector are emphasized. Appropriate weighting of these **important** particles prevents a distortion of the final result while reducing computer time. *Fig. 13.14* represents a few histories simulated by the Monte Carlo method. All the depicted histories end up at the detector, which is not the case in general.

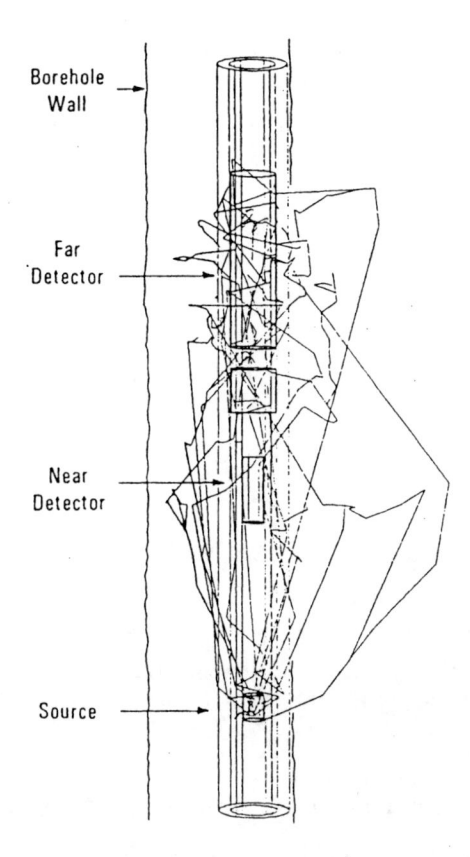

FIG. 13.14 : Graphic representation of few simulated particles histories.

Courtesy of Schlumberger [10]

An important feature of Monte Carlo simulation is the ability to edit the detailed histories of the particles. For example, it is possible to take a census of particles that went farther than a given distance or which spent most of their flight inside the tool.

Considerations of depth of investigation, vertical resolution and quality of the tool design are easily approached by Monte Carlo simulation even if they cannot be sorted out by laboratory measurements. Experimental and calculated values of the response of neutrons are in good agreement without the need for normalization.

References

1 Anderson, B., Barber, T. D., Singer, J., Broussard T., "ELMOD-putting electromagnetic modeling to work to improve resistivity log interpretation," paper M, *Trans.* SPWLA 30th annual logging symposium, 1989.

2 Brie, A., Hsu, K., Eckersley, C., "Using the Stoneley normalized differential energies for fractured reservoir evaluation," paper XX, *Trans.* SPWLA 29th annual logging symposium, 1988.

3 Butler, J., Locke, J., Bradshaw, K., "A new facility for the investigation of nuclear logging tools and their calibration," paper LL, *Trans.* SPWLA 10th European formation evaluation symposium, 1986.

4 Butler, J., Kemshall, P., Locke, H., Armishaw, M., "PC2—a porosity code for personal computers," paper Q, *Trans.* SPWLA 33rd annual logging symposium, 1992.

5 Chang, S., van der Hijden, J., Orton, M., "Acoustic waveforms explained," pp. 16-21, *The Technical Review*, Vol. 35, No. 1, 1-1987.

6 Flaum, C., Galford, J., Olesen, J. R., Scott, H., "New environmental corrections for the compensated neutron porosity log," pp. 23-31, *The Technical Review*, Vol. 36, No. 2, 4-88.

7 Gouilloud, M., Theys, P., "L'informatique appliquée au logging," Schlumberger, 1981.

8 Hornby, B. E., Johnson, D. L., Winkler, K. W., Plumb, R. A., "Fracture evaluation from the borehole Stoneley-wave arrivals," *Trans.* SEG 57th annual international meeting, 1987. Also pp. 1274-1288 in *Geophysics*, Vol. 54, No. 10, 10-1989.

9 Shattuck, D. P., Bittar, M. S., Shen, L. C., "Scale modeling of the laterolog using synthetic focusing methods," pp. 357-369, *The Log Analyst*, Vol. 28, No. 4, 7-1987.

10 Tsang, J. S. K., Evans, M. L., "Monte Carlo computational mode for the gamma ray spectroscopy tool," SPE 12052, *Trans.* SPE 58th annual technical conference and exhibition, 1983.

11 Van Baaren, J. P., Heller, H. K. J., Visser, R., "Construction of reservoir rock samples for acoustic research," paper FF, *Trans.* SPWLA 10th European formation evaluation symposium, 1986.

14
Environmental corrections

14.1 Definition

As discussed in [4] and Section 4.2, every measurement has imperfections that give rise to an error viewed as having random and systematic components. Systematic errors can often be reduced. If a systematic error arises from a recognized effect, the effect can be quantified. If the effect is significant in size relative to the required accuracy of the measurement, a correction can be applied to compensate for the effect.

Accordingly, a measurement that has not been corrected cannot be expected to meet its accuracy specifications.

Petrophysical example

At 30 pu, the borehole size correction of the neutron porosity provided by a MWD system is 11 pu for a 254-mm (10-in.) holesize. It is easy to understand that failure to apply a correction would completely invalidate this measurement.

Environmental effects belong to this specific class of effects that varies with the well environment. The response of a logging tool is derived in narrow conditions, called reference conditions, that almost never reflect the real conditions encountered in a well environment. Environmental corrections enable the log user to obtain valuable information in these real conditions.

Environmental corrections require two ingredients:

- the availability of an algorithm or processing method defined beforehand and linking the parameters that influence the measurement to a quantity called correction

- a measurement or an evaluation of these parameters, also called "influence quantities."

Example

Neutron porosity is measured. The raw measurement is ϕ_0, equal to 25 pu. That would be the final porosity value if the borehole size were h_0, equal to 203 mm (8 in.), corresponding to the reference condition. The actual borehole size is h, equal to 10 in. Corrected porosity is

$$\phi_{corr} = \phi_0 + f(h_{actual}, h_0).$$

Defining the algorithm and measuring a valid input, in this example respectively f and h, are two prerequisites to sound corrections and hence to accurate measurements.

14.2 Correcting for the environmental effects

Environmental corrections are performed at different steps of the data cycle:

(a) Some corrections cannot be separated from the direct tool response. This is the case for the density tool, which is always corrected through the counting rates of different energy windows. This also applies to the focusing of the dual laterolog current beam.

(b) Some corrections are made in real time when the tool is downhole. This is the case for the caliper correction for the density and neutron porosity logs. The validity of the input commanding the correction needs to be carefully checked, and, in any case, raw data needs to be recorded simultaneously so that the correction can be removed. These corrections differ from the first group in that they can be more easily disallowed.

(c) Some corrections are performed on the surface at the wellsite, after logging. Service companies generally provide some processing packages that can be performed right at the conclusion of logging. Such corrections require that the logging engineer has received proper training and customer has specified his/her requirements. Typical corrections are the temperature and pressure effects on the neutron log. More elaborate packages allow the application of corrections on spectral gamma ray, array induction logs and imaging measurements.

(d) Corrections are made at the computing center, with standard processing. Some corrections may need more computing power than is available at the wellsite, but could be classified as automatic in that they demand only a simple input parameter selection. *Schlumberger* preinterpretation modules of the Preplus* family fall into this category.

(e) Interpretive and iterative processing at the computing center requires several passes and the intervention of an expert log analyst.
For resistivity measurements, shoulder bed, invasion, bed thickness and dip corrections cannot be applied simultaneously. It is necessary to assume a formation model, compute the logs associated with this model, compare the modeled with the actual log information and try again until an acceptable match is found. This exercise consumes extensive computing power and requires a high experience level.
(f) Intensive modeling necessary to account for the combination of the environmental effects. Such corrections may require an in-depth knowledge of the code to enable software modifications. Such corrections are further developed in the next chapter. They are no longer routine work and pertain to the class of special projects.

In some cases, evaluation of the formation is required before a proper correction can be implemented. Some iterations are needed to ensure thorough environmental corrections prior to formation evaluation. The salinity correction for the thermal neutron porosity log falls in this category [3]. Different charts and algorithms apply depending on lithology.

14.3 Determination of environmental corrections

14.3.1 Anticipating the need for corrections

Environmental corrections have often been developed because the uncorrected measurements were not giving the expected results. *Table 14.1* gathers a number of effects that have been recognized in the history of logging.

14.3.2 Methods of determination of correction algorithms

Most methods developed for determination of tool response are also used for quantification of environmental corrections. This quantification is formidable because it must take into account far more parameters than those affecting tool response in a simple formation.

Table 14.2 is a nonexhaustive list of corrections desired for the most common logging tools.

TABLE 14.1
ERRONEOUS READINGS LINKED TO DOWNHOLE EFFECTS

Log	Measurement	Effect on the log	Origin
Spontaneous potential..........	Electrochemical potential	Parasitic DC noise	Foreign interference
Early induction ...	Circumferential resistivity	Low resistivities too low Reduced penetration	Shoulder effect Skin effect
Laterolog..........	Radial resistivity	Reduced penetration	Shoulder effect
Sonic..............	Axial slowness	Fastest travel path	Lack of additiveness
Density...........	Density	Low density in H rich rock	(Z/A) not additive
Lithology..........	Photoelectric cross section	Bias towards higher P_e P_e too high	P_e is not additive Foreign interference (Ba)
Thermal neutron...	Neutron porosity	Porosity is too high	Absorbers (Gd, Bo, Cl, Fe)
Early time decay...	Neutron capture cross section	Bias towards the longest τ	No additiveness
Gamma ray spectroscopy.......	Elemental concentration	Shifts in elemental ratios	Cross-coupling

Courtesy of C. Clavier

14.3.3 Challenges linked to environmental corrections

14.3.3.1 More complex environments

The standard well used to be drilled vertically, with simple muds. The technological advances have made deviated or complex trajectory wells, drilled with sophisticated muds and bottomhole assemblies a common occurrence. While the implementation of chartbook corrections took a few lines of software codes, most "modern" corrections take several thousand lines.

14.3.3.2 Combination of environmental corrections

As a consequence, many corrections may be required. As an example, invasion, dip, shoulder bed and borehole corrections may be required for resistivity logs. The classical vision of environmental corrections is that all these effects are independent and can therefore be treated sequentially by simple well-behaving algorithms (*Fig. 14.1*). In reality, most environmental effects interfere one with the other (the correction of two effects is not exactly the sum of the corrections) and need to be corrected simultaneously.

TABLE 14.2
ENVIRONMENTAL CORRECTIONS FOR COMMON LOGGING TOOLS

Measurement	Correction	Origin
Neutron porosity	Borehole size Temperature Pressure Standoff Lithology Salinity Cement/casing	Experimental and extension by mathematical modeling
Laterolog	Oval hole centered Cross corrections Shoulder/no invasion Shoulder/invasion Dip Groningen	Mathematical modeling 3D mathematical modeling Mathematical modeling Mathematical modeling Mathematical modeling Approximation
Induction	Borehole conductivity Shoulder/no invasion Shoulder/invasion Dip Cross corrections	Mathematical modeling Mathematical modeling Mathematical modeling Mathematical modeling Mathematical modeling
Density	Mud cake P_e on density Hole size Pad tilt Mudcake on P_e Boundary shape	Experimental and extension by mathematical modeling
Spectral gamma ray	Borehole, no barite Casing, water in annulus no barite Barite Casing, barite Potassium mud	Experimental
Microlaterolog	Mudcake Hole size Tilt/channeling Anisotropy	Mathematical modeling

Courtesy of C. Clavier

14.3.4 Modeling

This is why (forward and inverse) modeling has taken such an important part in the preparation of petrophysical data for final interpretation. The corresponding software include minimization techniques and several iterations are often required.

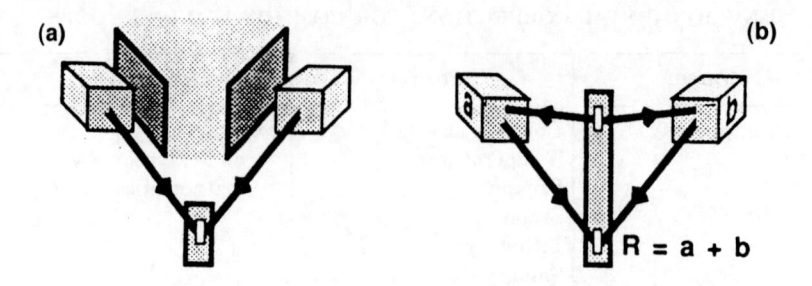

FIG. 14.1 : Idealistic combination of corrections.
  a. Effects A and B do not interfere one on the other.
  b. Response at sensor is the sum of the contributions effects A and B.
      Courtesy of C. Clavier

14.3.5 How can environmental corrections be checked?

One of the best ways to evaluate the accuracy of the corrections is to log the same formation twice in different environments. The corrected logs should be identical except for statistical variations.

Such an approach has been followed to check the validity of the thermal neutron porosity correction algorithms when they are extended to very large holes (*Fig. 14.2*). A small diameter hole was first drilled, then logged. Then the hole was enlarged and logged again. Corrected logs read similar values, which shows the environmental modeling, the environment data base and the correction inputs were all valid.

Field studies are also a good way to check the validity of the environmental corrections. While hole and mud characteristics may vary as the field is developed, it is often possible to find geological markers that are stable throughout the entire area. In these markers, the environmentally corrected data should show less dispersion than the raw data.

14.4 Error propagation from the input to environmental corrections

Once the charts and algorithms [XII, 1 and 6], enabling the performance of the corrections have been assembled, the work is far from complete:

(a) The correction input has to be valid.
(b) The correction input must be on depth with the data it is supposed to modify.

Uncertainties of the environmental correction inputs are transmitted through the correction algorithm to the logging parameters. This is reviewed with two practical examples, the induction borehole and the thermal neutron corrections.

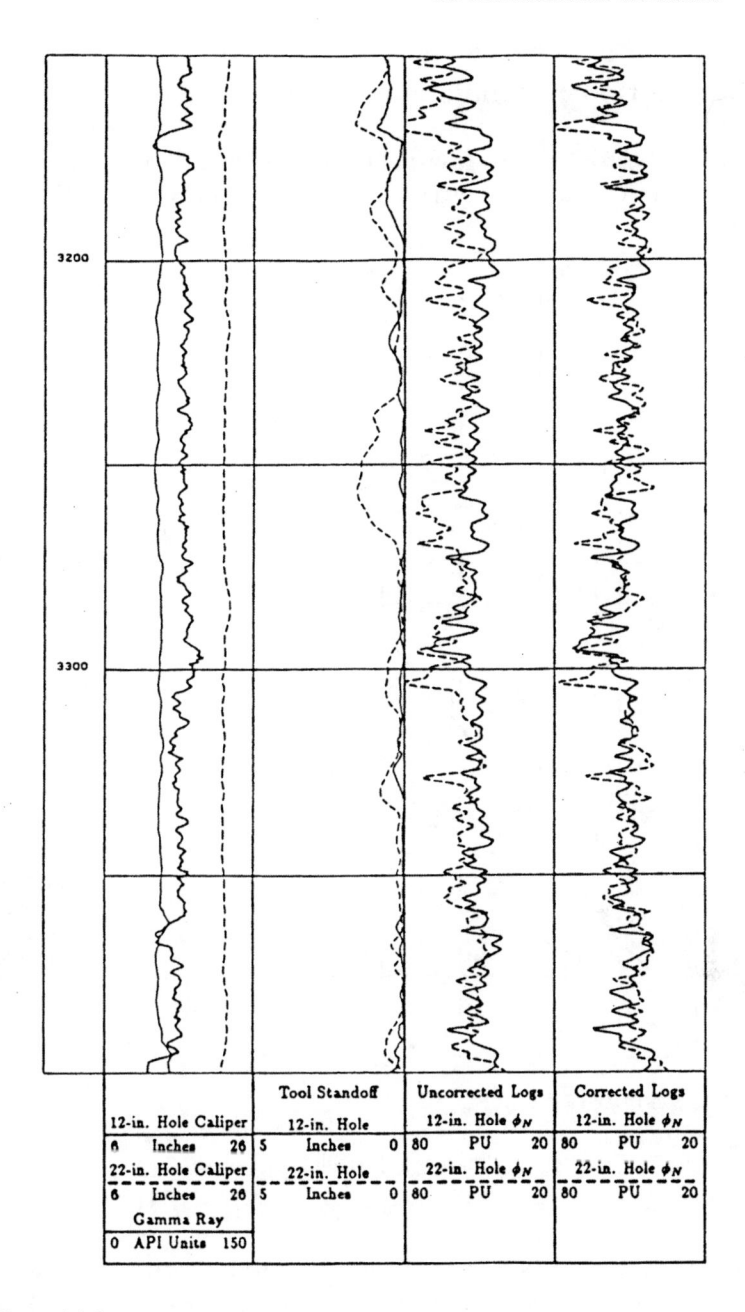

FIG. 14.2 : Logging the same formation with different environments. The neutron porosity log is run two times in the same formation, first in a 30.5-cm (12-in.) hole, then in a 56-cm (22-in.) hole. The standoff varies from one run to another. While the uncorrected logs display large discrepancies, environmentally corrected logs match closely.

Courtesy of Schlumberger [5]

14.4.1 Induction borehole correction

An important correction of the raw induction reading is the removal of the borehole conductivity contribution. The correction can be performed manually through the chart shown (*Fig. 14.3*).

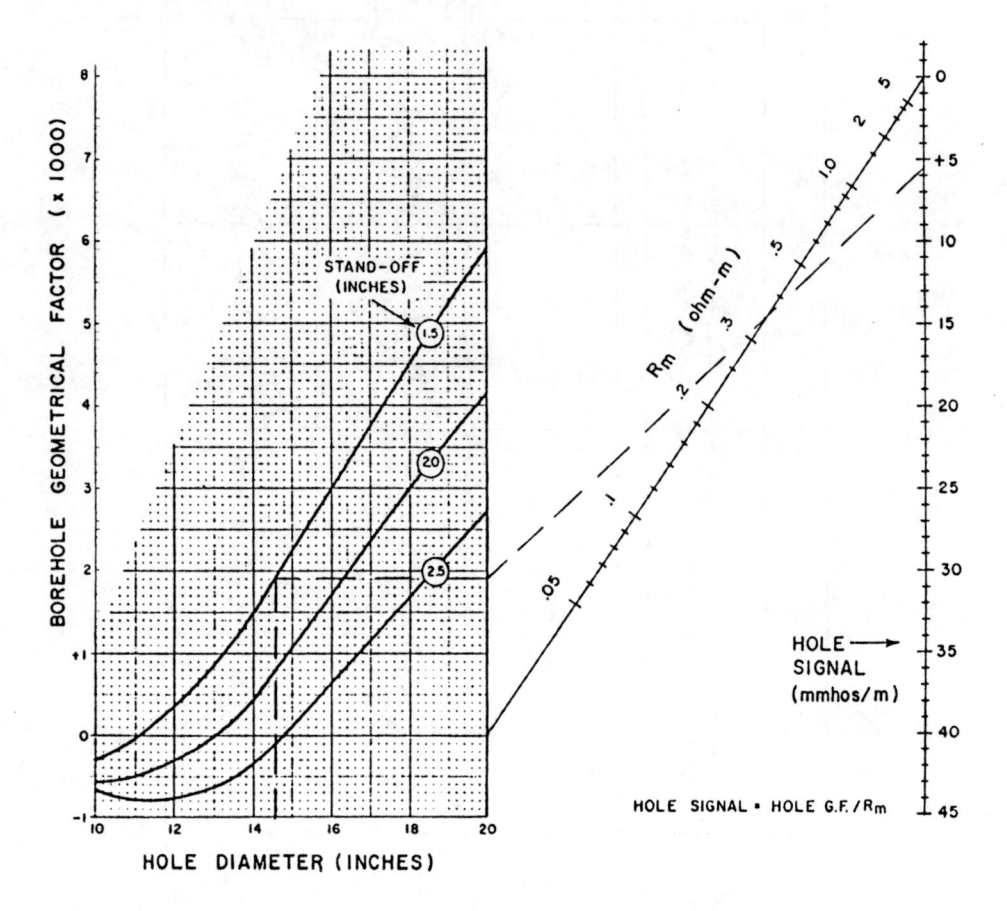

FIG. 14.3 : Borehole conductivity correction chart.
Courtesy of Schlumberger [XII]

A formation of actual resistivity R_t is logged. The log reads 4.65 Ω/m, equivalent to 215 mmho/m. In order to compensate for the borehole effect on the logging device, three assumptions are made: a 40.6-cm (16-in.) borehole diameter, a standoff of 3.8 cm (1.5 in.) and a mud resistivity of 0.2 Ω/m equivalent to a mud conductivity, C_m, of 5000 mmho/m.

From the chart (*Fig. 14.3*), we find that the borehole geometrical factor, G_b, is 30×10^{-4}.

The nominal borehole contribution is $G_b \times C_m$ or 15 mmho/m to be subtracted from the log conductivity to obtain the formation conductivity, 200 mmho/m, which is equivalent to 5 Ω/m.

The impact of invalidity of the input assumption is evaluated in the following cases:

(a) **The borehole is not circular.** A dual-axis caliper yields the long and short axes, a and b, as 40.6 cm (16 in.) and 35.6 cm (14 in.) As the geometrical factor is proportional to the area of the hole section, the value $d' = \sqrt{ab}$ is input into the chart. The resulting G_b is 22×10^{-4}. The borehole conductivity is 11 mmho/m. The actual formation conductivity is 204 mmho/m, a difference of 2% from the value derived from a superficial analysis of the environmental inputs.

(b) **The uncertainty of the caliper measurement is 1.27 cm (0.5 in.).** Maximum uncertainty of formation resistivity is computed in two cases:

$$d = 15.5 \qquad G_b = 26 \times 10^{-4}$$

$$d = 16.5 \qquad G_b = 34 \times 10^{-4}.$$

(c) **The hole geometry is such that standoff** is not the nominal 3.8 cm (1.5 in.) value but **6.4 cm (2.5 in.)** (*Fig. 14.4*).

(d) Because of tension, **the caliper is off-depth** and the hole geometry changes with depth. **The actual caliper value is 35.6 cm (14 in.)**, but 40.6 cm (16 in.) is input.

(e) **Mud composition is varying** because of segregation. **The actual mud resistivity is 0.166 Ω/m**, which is equivalent to 6000 mmho/m.

(f) **Actual standoff is 5 cm (2 in.) and the actual caliper is 38 cm (15 in.)** The resulting geometrical factor is 11×10^{-4}.

The shifts resulting from a limited knowledge of the environment are indicated in *Table 14.3*.

TABLE 14.3
EFFECT OF INCORRECT INPUT

Anomaly	Stand-off	Borehole diameter	Borehole geometrical factor	Conductivity (mmho/m)			R_t	Difference
	(in.)	(in.)	($\times 10^{-4}$)	Mud	Borehole	Formation	(Ω/m)	(%)
None.......	1.5	16.0	30	5,000	15.0	200.0	5.0	0
Elliptical...	1.5	16.0-14.0	23	5,000	11.5	203.5	4.9	2
Cal +0.5 in.	1.5	15.5	26	5,000	13.0	202.0	4.9	2
Cal −0.5 in.	1.5	16.5	34	5,000	17.0	198.0	5.1	2
Standoff...	2.5	16.0	6	5,000	3.0	212.0	4.7	6
Sticking....	1.5	14.0	15	5,000	7.5	207.5	4.8	5
R_m wrong.	1.5	16.0	30	6,000	18.0	197.0	5.1	2
SO+Cal...	2.0	15.0	11	5,000	5.5	209.5	4.8	5

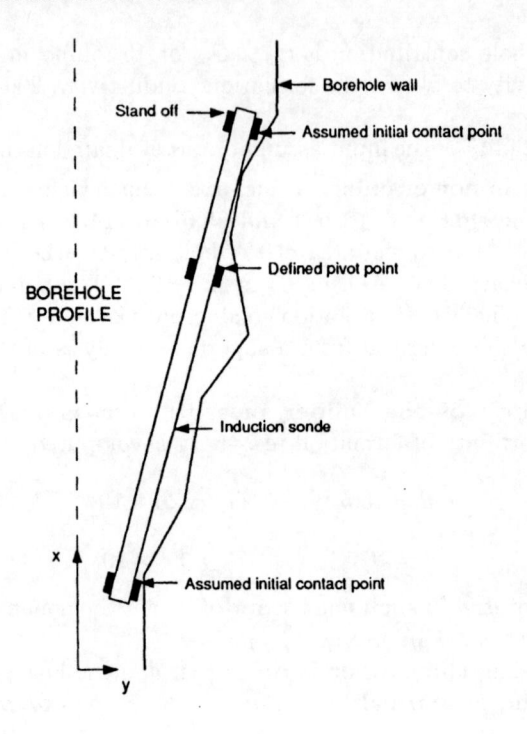

FIG. 14.4 : Modeling of standoff behavior.
The actual standoff affecting the measurement may not be exactly equal to the size of the rubber accessories added to the induction sonde. Though rubber fins of 3.8 cm (1.5 in.) are used, a standoff as large as 10 cm (4 in.) is observed in this example.

A general formula for the transmission of errors owing to borehole conductivity is

$$C_t = \frac{C_{\log} - C_m G_b}{1 - G_b},$$

where G_b is a function $f(d_h, SO)$ of the hole diameter d_h and of the standoff SO. In other words,

$$\sigma^2_{C_t} = \left(\frac{1}{1-G_b}\right)^2 \sigma^2_{C_{\log}} + \left(\frac{G_b}{1-G_b}\right)^2 \sigma^2_{C_m} + \left(\frac{C_{\log} - C_m}{1-G_b}\right)^2 \sigma^2_{G_b},$$

with

$$\sigma^2_{G_b} = \left(\frac{\partial f}{\partial d_h}\right)^2 \sigma^2_{d_h} + \left(\frac{\partial f}{\partial SO}\right)^2 \sigma^2_{SO}.$$

G_b is around 1%, and so the formula above can be simplified to

$$\sigma^2_{C_t} = \sigma^2_{C_{\log}} + G_b^2 \sigma^2_{C_m} + (C_{\log} - C_m)^2 \sigma^2_{G_b}.$$

14.4.2 Thermal neutron correction

The thermal neutron porosity correction depends on many environmental variables:
$$\phi_N = f(\Sigma_w, d_h, SO, P, T)$$
where Σ_w is the formation salinity, d_h is the hole diameter, SO is standoff, P is pressure and T is temperature.

$$\sigma_{\phi_N}^2 = \left(\frac{\partial f}{\partial \Sigma_w}\right)^2 \times \sigma_{\Sigma_w}^2 + \left(\frac{\partial f}{\partial d_h}\right)^2 \times \sigma_{d_h}^2 + \left(\frac{\partial f}{\partial SO}\right)^2 \times \sigma_{SO}^2 + \left(\frac{\partial f}{\partial P}\right)^2 \times \sigma_P^2 + \left(\frac{\partial f}{\partial T}\right)^2 \times \sigma_T^2.$$

Each term $\frac{\partial f}{\partial x}$ represents the sensitivity of the neutron porosity to a specific environmental effect x. Table 14.4 gathers the values of $\frac{\partial f}{\partial x}$ for a value of ϕ_N of 20 pu for the *Schlumberger* neutron tool [XII].

TABLE 14.4
SENSITIVITY OF THERMAL NEUTRON POROSITY
TO ENVIRONMENTAL PARAMETERS AT 20 pu

Sensitivity is expressed in pu. Conversion to the metric system is not made.

Environmental effect	Sensitivity of ϕ_N
Borehole size	1.5/in.
Mudcake thickness	3.0/in.
Borehole salinity	2.0/(10^6 ppm)
Formation salinity	12.0/(10^6 ppm)
Mud weight (no barite)	0.2/(lbm/gal)
Mud weight (barite)	0.1/(lbm/gal)
Standoff	3.0/in.
Pressure	0.2/(kpsi)
Borehole temperature	1.75/(100°F)

The most influential environmental parameters are **standoff** and **temperature**. For instance, a change of 1.27 cm (0.5 in.) in standoff implies a change of 1.5 pu in ϕ_N while the same magnitude of change in the caliper reading causes only a variation of 0.75 pu in ϕ_N.

Exercise 17*

Given *Fig. 14.5*, express the error on a 20-pu ϕ_N reading resulting from the use of a geothermal gradient instead of a real temperature measurement. The temperature is assumed to be approximately 300°F (150°C) and incorrect by 20°F (11.1°C).

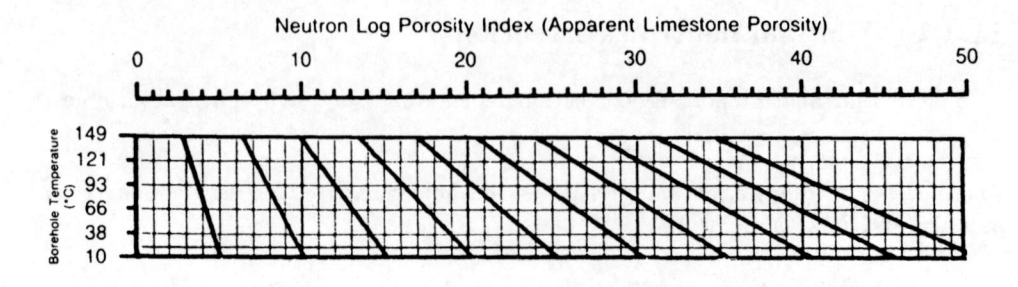

FIG. 14.5 : Thermal neutron porosity correction for temperature.
Courtesy of Schlumberger [XII]

14.5 Importance of the description of the environment

For most measurements, borehole geometry (shape, rugosity, caving) may cause large uncertainties. The position of the tool in the hole and the mud parameters may also have significant impact. It is therefore useful to obtain accurate and complete information on these three information categories. As shown later in Chapter 15, simple hole behavior (round geometry, constant temperature gradients, constant mud weight) should not be relied upon.

14.5.1 The role of sensors describing the environment

While it is glamorous to design a multienergy, multidetector device based on imaging and advanced physics, it seems there is little glory in building a sonde that will permit an accurate description of the borehole environment. Still...

14.5.2 A note of concern: caliper accuracy

Calipers are essential to the environmental correction of many curves—gamma ray, density, neutron, induction logs. Caliper information may introduce errors because of the following factors:

(a) **Substandard caliper calibrations**. For instance, calipers need to be calibrated in the proper calibrating rings. Obviously, the two rings used should be selected so that most of the recorded values fall between the diameters of these rings.

A 2-mm (0.1-in.) uncertainty on each calibration point (axis of the tool not perfectly parallel to the axis of the cylindrical rings, etc.) can be anticipated. This means that a caliper calibrated with 20-cm (8-in.) and 30-cm (12-in.) rings would read with an uncertainty of 1.25 cm (0.5 in.) at 40 cm (16 in.).

(b) **Caliper stability**. There is generally no after-survey caliper verification. Any drift may be left undetected.

(c) **Caliper tool response**. Not all calipers respond linearly. This can be checked by observing the response of the caliper in more than two rings.

(d) **Effects of depth mismatch** on the caliper can be serious. Caliper channels from other runs can be used only if precise depth matching is performed.

14.5.3 A note of optimism

Real boreholes are not smooth and the tool axis is seldom absolutely parallel to the borehole axis. There is indeed no guarantee that the borehole is round. However, Nature is kind enough to reduce the number of potential variables. The behavior of a long tool string in a hole only slightly larger than the tool will keep the angle between the tool and the borehole axis very small. For an induction tool, borehole signals in the reference 20-cm (8-in.) hole are rather small in most muds and are significant only when formation resistivities are high. Highly resistive formations are usually hard, and do not wash out much. If the caves or washouts are shorter than the tool string, the tool will remain aligned with the **in-gauge** portions of the hole and will be nearly centered over caved sections.

14.5.4 MWD and the well environment

MWD measurements are collected a few minutes after the bit has cut the formation. Invasion of the virgin formation by mud components is smaller than at wireline logging times and hole enlargement minimized. The mud in the hole is obviously the one that has been used during drilling—by contrast, formations surveyed with wireline logging have been invaded by muds with different compositions. Because invasion is limited, tools do not need to have large depths of investigation, and for this very reason, they are scanning fewer neighboring beds that need to be corrected for with deep-investigation wireline sensors. In summary, MWD measurements require fewer and smaller environmental corrections. It is fair to say that, by contrast, the mechanical environment—shocks and vibrations—is far more challenging.

14.5.5 Are environmental corrections really useful?

The fact that environmental corrections require a good database and valid environmental parameters may be disheartening. In addition, logging companies are continuing to search for better correction algorithms and it is likely that new transforms will become available for almost every logging service. Is all this trouble worth it?

There is one specific instance where this perseverance has paid off. In 1983, a new dolomite response[1] was defined for the *Schlumberger* thermal neutron tool [2]. A specific field with a mainly dolomitic lithology proved in 1990 that the previous work was not just of academic interest [7].

Figure 14.6 shows the density-neutron crossplots derived from data corrected with the *OLD* and the *NEW* transforms. The second crossplot is in better agreement with the core and field geological data. Moreover, the environmental corrections have a significant impact on the final net-to-gross ratio which is increased five-fold. Combined with the resulting effect on oil saturation, the hydrocarbon pore volume is multiplied by seven and comes in line with the current production.

14.5.6 How can environmental corrections be improved?

- **Integration of all available data**

  Each logging measurement is often corrected independently of the others. For instance, the photoelectric index from the spectral density tool is not used to correct the barite effect on the natural gamma ray spectral tool although the photoelectric curve is sensitive to the barite content. Cross corrections and integrated software using all inputs should be designed.

- **Proper environment modeling**

  The representation of the environment is often too simple. More realistic environments are described in Chapter 15.

- **Use of multisensor information to describe the environment**

  When a logging device includes a large number of transmitters and receivers that scan the **same** formation azimuthally, some sensitivity to the environment results—the shallow investigating sensors have the most sensitivity. This redundancy can be used to correct for the environment without measuring its characteristics quantitatively. This approach is the basis of the density spine-and-rib method (Chapter 16), of the multispacing induction borehole correction or of the azimuthal neutron borehole invariance.

---

[1] The behavior of the thermal neutron tool in front of dolomitic formations could be seen as a direct application of the establishment of the tool response. For thermal neutron tools, tool response and environmental effects cannot be separated.

- **Design and use of tools that measure the environment**

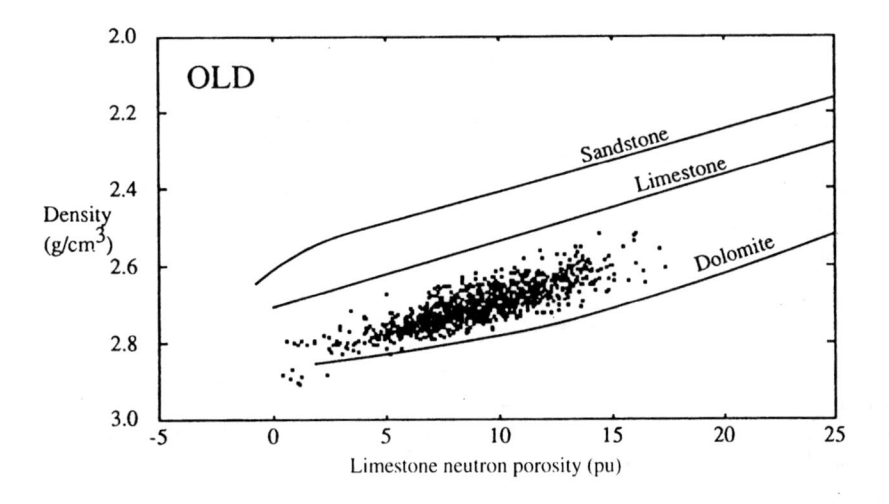

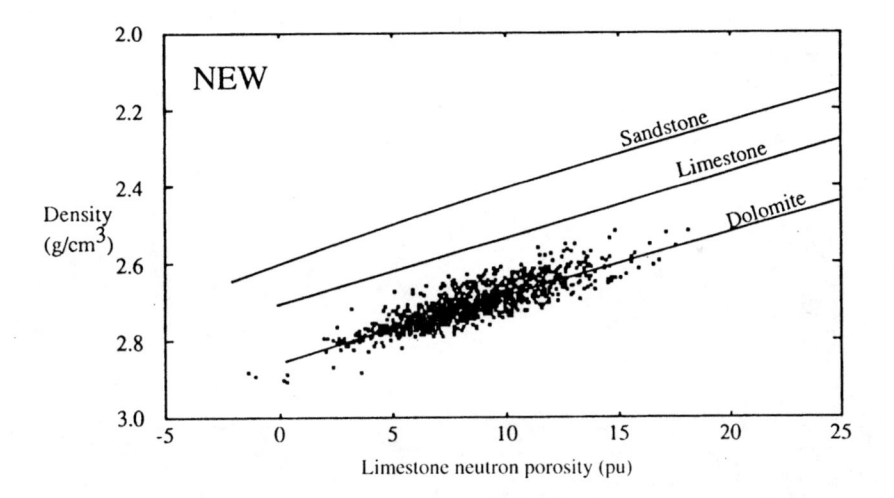

FIG. 14.6 : Comparison of data corrected with two versions of environmental corrections.
With the *OLD* transform, most of the data falls above the dolomite line, which suggests significant admixtures of quartz or calcite, inconsistent with the known lithology. With the *NEW* transform, most points fall on the dolomite line.

Courtesy of SPWLA [7]

References

1. *Atlas Wireline Services, log interpretation charts*, Western Atlas International, Houston, 1989.

2. Ellis, D. V., Case, C. R., "CNT-A dolomite response," paper S, *Trans.* SPWLA 24th annual logging symposium, Calgary, 1983.

3. Ellis, D. V., Flaum, C., Galford, J. E., Scott, H. D., "The effect of formation absorption on the epithermal neutron porosity measurement," SPE 16814, *Trans.* SPE 62nd annual technical conference and exhibition, Dallas, 1987.

4. International Organization for Standardization, *Guide to the expression of uncertainty in measurement*, ISBN 92-67-10188-9, Geneva, Switzerland, 1995.

5. Gilchrist, W. A. Jr., Galford, J. E., Flaum, C., Soran, P. D., Gardner, J. S., "Improved environmental corrections for compensated neutron logs," SPE 15540, *Trans.* SPE 61st annual technical conference and exhibition, New Orleans, 1986. Also in *SPE Formation Evaluation Journal*, pp. 371-376, Vol. 3, No. 2, June 1988.

6. *Halliburton Logging Services chartbook*, Halliburton Logging Services, Houston, 1991.

7. Pagès, G., Francisque, J. H., Flaum, C., Yver, J. P., "The impact of utilizing the new CNL transforms and environmental corrections," *Trans.* SPWLA 13th European formation evaluation symposium, Budapest, 1990.

15

The real environment

Corrections developed to account for the influence of the environment are essentially valid only when they are small. The log analyst is weary of doubling or halving the originally measured value because of the environment.

Logging sensors face particularly drastic conditions—hole sizes twice as big as bit size, solids in suspension in muds assumed to be homogeneous, temperature gradients with different slopes, horizontal wells while the logging tools are often designed to log vertical wells. Understanding these extreme conditions is essential to successful log analysis. With the fast pace of technology and the innovative trends of the oil industry, it may be anticipated that what is extreme today may be standard in a few years.

In this chapter, the environmental effects that cannot be corrected by a casual or routine approach are reviewed. They demand a solid understanding of what occurs in field conditions in addition to robust algorithms and software. While "classical" environmental corrections may require milliseconds of computation time, real environmental corrections may be the basis of extensive case studies.

Also, considering their multitude, one may always find a combination of environmental effects that can explain **any** log reading. This is dangerous as poor quality data can be qualified as valid through the opportunistic use of several corrections. The user's good judgment is critical in this matter.

15.1 Temperature

Subsurface temperatures are most often calculated by interpolation between the surface and bottomhole values. This crude approach has two weaknesses:

(a) The bottomhole temperature recorded by downhole thermometers located at the top of the logging string does not reflect formation temperature and varies with time as a function of the mud thermal conductivity.

192   15. The real environment

(b) Rock thermal conductivity varies. Temperature profiles are distorted by the presence of faults, salt and igneous intrusions, etc. *Figure 15.1* shows the formation temperature/depth profile in an hypothetical rock sequence.

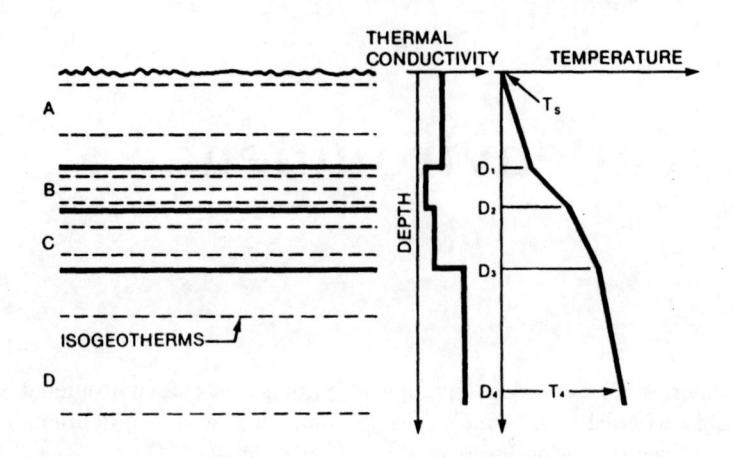

FIG. 15.1 : Isogeotherms.

Courtesy of SPE [40]

15.2 Borehole shape

The borehole is assumed to be perfectly circular, with a constant diameter along the section drilled with the same size bit and with a smooth interface between the mud and the formation wall. *Figure 15.2* shows those ideal conditions are far from reality [10].

15.2.1 Distortion of hole shape

Distortion of the borehole shape is not random. Rock mechanics gives some insight into the conditions under which holes are deformed and the shapes they assume [8].

Before drilling, a rock volume is subjected to tectonic (geostatic) stresses and to the overburden. A well-defined stress system is attached to this condition. Drilling modifies the stress system over a distance three to four times the drilling radius. The new stress system can be quantitatively described by the theory of elasticity [6].

Depending on the ranking of the original geostatic stresses and overburden, on one side, and of the hydrostatic pressure on the other side, the borehole wall will break and bits of formation will fall. These pieces of rocks would have the theoretical shapes shown in *Fig. 15.3a*.

Figure 15.3b shows a rock sample recovered from a well where hole shape distortion was confirmed by a caliper survey.

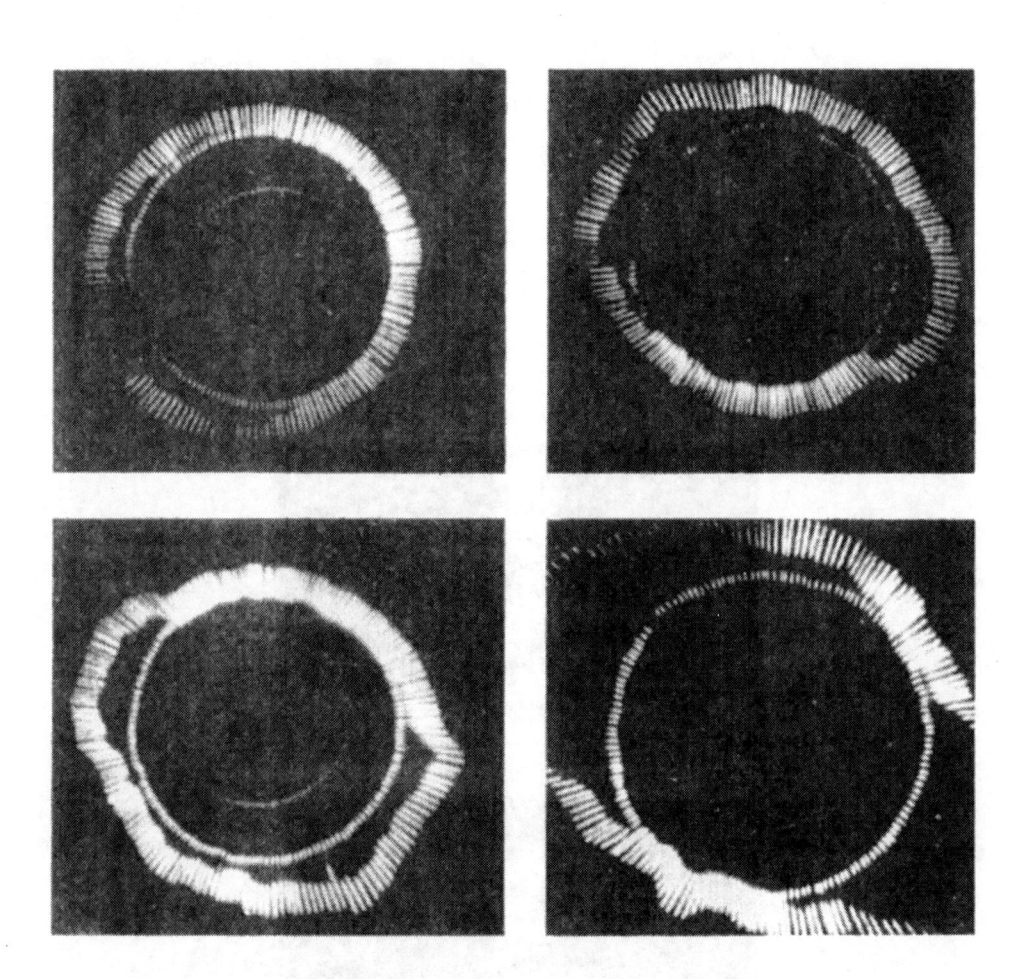

FIG. 15.2 : Real hole profiles.
Irregular hole shapes observed with a modified borehole televiewer. The dotted circle is concentric with the tool and represents a 21.6-cm (8.5-in.) diameter. Borehole walls are shown by the inner edge of the long traces. The formation is limestone; the borehole fluid is water.

Courtesy of SPE [40]

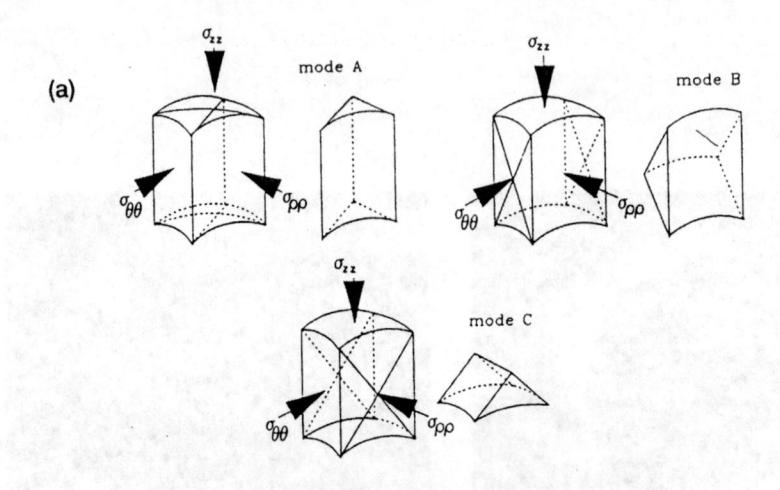

FIG. 15.3 : Possible shapes of formation chips.
  a. Theoretical shapes derived by stress analysis. There are three possible modes (A, B, C) of borehole failures.
  b. Pictures of an actual formation chip recovered from a well in Indonesia. Borehole failure occurred in mode B.
  Courtesy of Total CFP [5 and 9]

Hole shape distortion needs to be carefully monitored and understood as it can cause stuck pipe and inferior cement quality. It may also prevent logging tools from acquiring data relevant to the formation [12]. The hole would have the tendency to deteriorate and take one of the shapes shown in *Fig. 15.4*. From all the possibilities shown, it can be inferred that a logging tool with sensors facing the formation aligned with the long axis of the hole would be affected by rugosity and hole surface anomalies.

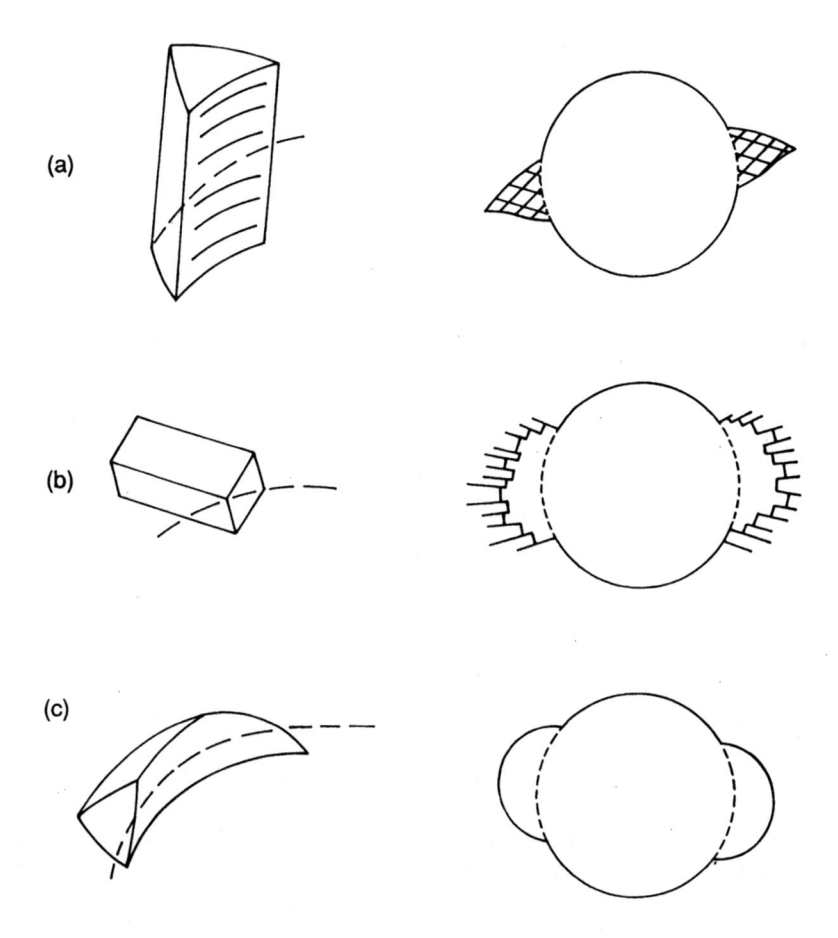

FIG. 15.4 : Relation between formation chip shape and borehole deformation.
Formation chips are shown on the left with the borehole cross sections on the right:

    a. Prismatic chips/lemon shape.

    b. French-fry shaped chips/rugose shape.

    c. Ring-shaped chips/Mickey mouse ears (in French, oreilles de Bécassine) shape.

Courtesy of Total CFP

15.2.2 Distortion from the drilling mode

In the past, a wedge at the bottom of the drilling assembly, called a whipstock, was used to drill a deviated well. Nowadays, steerable drilling systems have replaced the whipstock. They include a downhole motor, a bent sub or offset stabilizers and a navigation sub. The downhole motor (positive displacement motor or turbine) converts the force of the drilling fluids pumped through the drillstring into rotational power to drive the drillbit.

The bent sub, a short tubular section near the drill bit, sideloads the bit, causing the hole to deviate. Its angle, as referred to the drill bit, controls the azimuth and inclination of the borehole. Current steerable units enable an angle build rate of 20° per 30 m (100 ft). The real-time orientation information, generally provided by a mud pulse telemetry, allows continuous monitoring of hole trajectory (Chapter 20).

Steerable systems have a tendency to drill helical holes. This is understood to be a result of the tilt angle imposed at the bit while drilling. This helical shape may cause increased reaming time. Its impact on logging can also be significant (*Fig. 15.5*). The thermal neutron, P_e and $\Delta\rho$ curves are strongly affected. On this example, the density information is properly compensated.

15.2.3 Effects of hole shape and tool positioning on logs

The variable borehole size and shape influence the logging tool in three ways:

(a) creation of standoff
(b) pad tilting
(c) eccentering.

Effect on sonic

While the effect of caves and eccentering is reduced by normal and depth-derived borehole compensation—resulting in an offset on transit times smaller than 3 μs/m (1 μs/ft)—tilting may cause sizeable errors—up to 20 μs/m (7 μs/ft) [11]. The Stoneley wave is not affected by the eccentering of the sonic tool.

Effect on neutron-porosity

Standoff is a crucial environmental parameter for the neutron-porosity log. Solutions proposed to alleviate its impact range from running a special caliper measuring standoff (for instance a powered caliper device) to using signal processing (based on the comparison of thermal neutron short- and long-spacing counting rates).

Effect on induction

Standoff effect has a significant impact in conductive muds (Section 14.4.1).

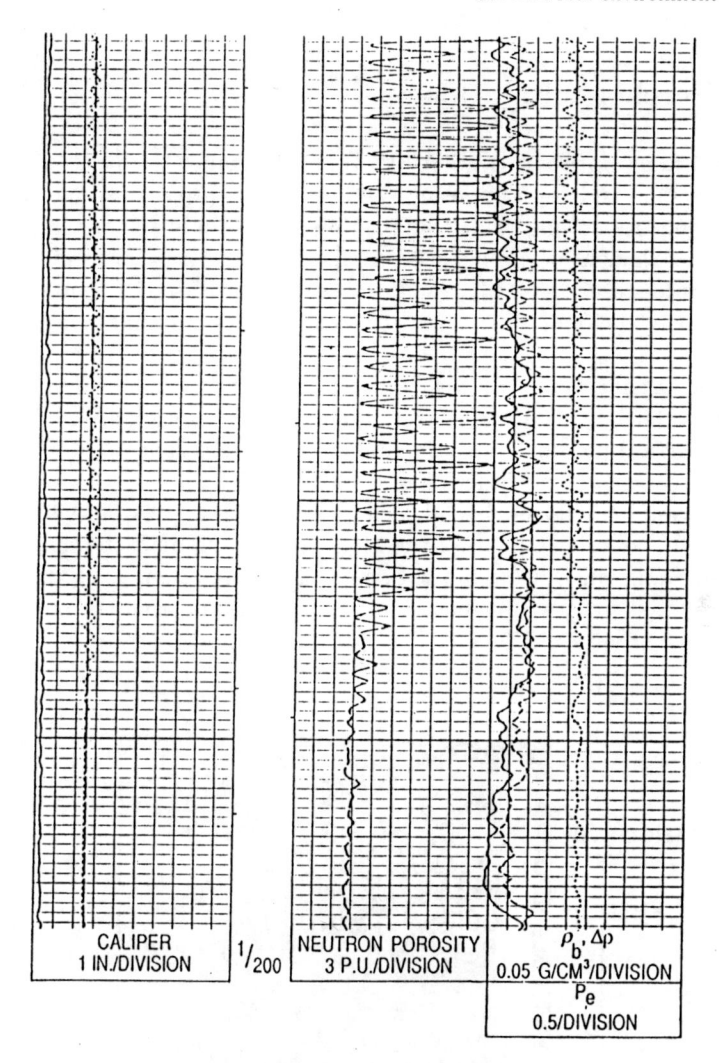

FIG. 15.5 : Distortion of logs resulting from helical borehole.
A variation of caliper size of 1.2 cm (0.5 in.) amplitude occurs every 0.8 m
(2.5 ft) of hole. The most affected measurement is the P_e curve (long-dash).
ϕ_N and $\Delta\rho$ are also affected. Note that the borehole compensated ρ_b is the
least affected.

Courtesy of Total CFP

Effect on laterolog

Eccentering affects mainly the shallow laterolog [37]. With the tool eccentered, lower resistivity readings are collected. The magnitude of the effect depends heavily on tool design, with new generations of tools being more immune. As the correction can be quantified by mathematical modeling and canceled, it is essential to constrain the tool to one of the two, the centered or the eccentered configuration.

15.2.4 Borehole alteration

Even if drilling does not degrade the borehole shape, the process of excavating the drilled volume and superimposing the hydrostatic pressure would alter the rock mechanic properties around the wellbore. Hole shape could remain essentially round but substantial changes of properties can take place in formations able to stay in an elastic or plastic state. Typically, shales behave in this fashion. *Figure 15.6* shows a model including two zones with different characteristics, an altered zone and a virgin zone.

The sonic logging tool run with different transmitter-to-receiver spacings, and therefore with different depths of investigation, allows determination of the mechanical and geometrical parameters of the altered and virgin zones. The radial thickness of the altered zone is on the order of 25 cm (10 in.).

The compressional transit time of the altered zone is approximately 30 to 45 μs/m (10 to 15 μs/ft) larger than that of the virgin formation. This shows that the short-spacing sonic tool, which has a limited depth of investigation, gives the sonic transit time of the altered zone, with values significantly different from the ones obtained by seismic surveys.

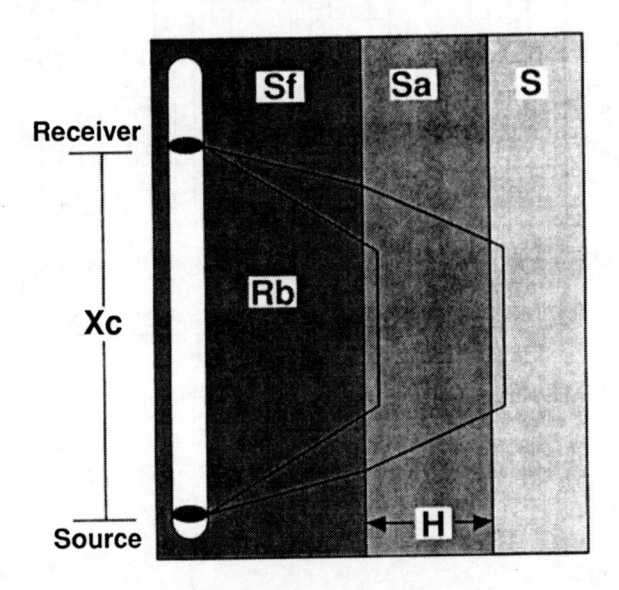

FIG. 15.6 : Borehole alteration.
Depending on the source-to-receiver spacing, X_C, the sonic energy, will travel at different rates in the undamaged formation (slowness S) than in the altered zone (slowness S_a and thickness H). The borehole has a slowness S_f. The sonic tool lies at a distance R_b from the borehole wall.
Courtesy of SPWLA [7]

15.3 Borehole trajectory

Most environment models assume a vertical hole crossing horizontal beds (or at least beds perpendicular to the wellbore axis). This assumption is particularly convenient as it gives a cylindrical symmetry to the formation. Such a geometry simplifies considerably the computation of tool response.

More and more wells are drilled with high deviation and horizontal wells are routine. The axis of the borehole no longer being perpendicular to the bed boundaries affects the logging data. These effects have not been investigated until recently.

For instance, an apparent dip would affect induction logging data two ways: First, the thickness of the bed as seen on the log would be significantly different from reality. Second, the magnitude of the measured resistivity value would be substantially different from the formation value (*Fig. 15.7*).

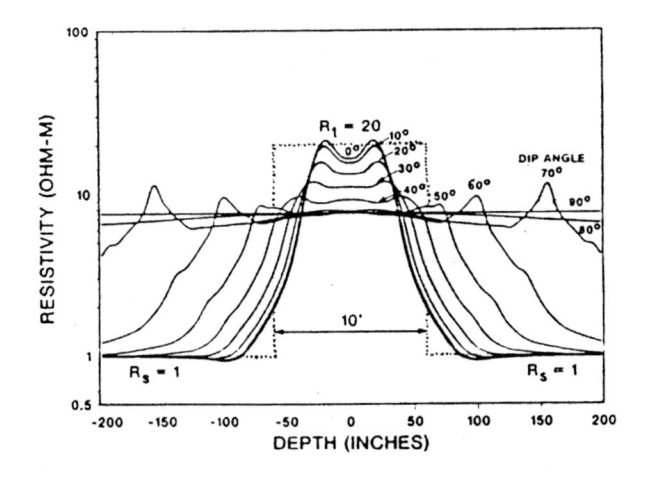

FIG. 15.7 : Effect of dip on deep induction response in a 10-ft resistive bed.

Courtesy of SPWLA [29]

In order to assess the impact of apparent dip, a simple formation model is assumed. The zone of interest is made of a single homogeneous bed with very thick shoulder beds. The well is supposed to be drilled with oil-base mud and therefore invasion is regarded as negligible. In these conditions, there is a direct relation between dip and the correction factor used to derive the water saturation:[1]

$$S_{w\text{ true}} = C \times S_{w\text{ Ild}}.$$

The coefficient C is computed for three levels of formation and shoulder resistivities, R_t and R_s and listed in *Table 15.1*.

---

[1] An Archie's equation with $m = n = 2$ is used.

Table 15.1
Effect of dip on induction-derived saturation

Dip	$R_t = 10\ \Omega/\mathrm{m},\ R_s = 1\ \Omega/\mathrm{m}$ Bed thickness (ft)											
	1	2	3	4	5	6	8	11	14	18	24	40	70
0	0.36	0.42	0.50	0.58	0.66	0.74	0.80	0.83	0.86	0.88	0.92	0.95	0.97
15	0.36	0.43	0.51	0.57	0.65	0.71	0.77	0.81	0.84	0.87	0.90	0.94	0.97
30	0.38	0.45	0.52	0.59	0.64	0.68	0.72	0.77	0.81	0.85	0.88	0.93	0.96
45	0.40	0.48	0.54	0.58	0.62	0.65	0.69	0.75	0.79	0.83	0.87	0.93	0.96
55	0.42	0.50	0.54	0.58	0.61	0.63	0.68	0.73	0.78	0.82	0.86	0.92	0.96
60	0.43	0.50	0.54	0.57	0.60	0.63	0.68	0.73	0.77	0.82	0.86	0.92	0.96

Dip	$R_t = 100\ \Omega/\mathrm{m},\ R_s = 2\ \Omega/\mathrm{m}$ Bed thickness (ft)											
	1	2	3	4	5	6	8	11	14	18	24	40	70
0	0.16	0.19	0.22	0.27	0.32	0.37	0.43	0.47	0.50	0.55	0.60	0.72	0.83
15	0.16	0.19	0.23	0.27	0.31	0.36	0.40	0.45	0.49	0.53	0.59	0.71	0.83
30	0.18	0.21	0.25	0.30	0.32	0.35	0.39	0.43	0.47	0.52	0.58	0.70	0.82
45	0.20	0.25	0.29	0.31	0.33	0.35	0.38	0.42	0.46	0.51	0.57	0.70	0.82
55	0.24	0.29	0.31	0.32	0.33	0.35	0.38	0.42	0.46	0.51	0.58	0.70	0.83
60	0.26	0.30	0.31	0.33	0.34	0.36	0.39	0.43	0.47	0.52	0.58	0.71	0.83

Dip	$R_t = 500\ \Omega/\mathrm{m},\ R_s = 4\ \Omega/\mathrm{m}$ Bed thickness (ft)											
	1	2	3	4	5	6	8	11	14	18	24	40	70
0	0.10	0.12	0.14	0.16	0.19	0.21	0.24	0.27	0.30	0.33	0.37	0.48	0.64
15	0.10	0.12	0.14	0.16	0.19	0.21	0.24	0.27	0.29	0.32	0.37	0.48	0.64
30	0.11	0.13	0.16	0.18	0.20	0.21	0.23	0.26	0.29	0.32	0.37	0.48	0.64
45	0.13	0.16	0.18	0.20	0.21	0.22	0.24	0.26	0.29	0.32	0.37	0.49	0.65
55	0.16	0.19	0.20	0.21	0.22	0.23	0.24	0.27	0.30	0.33	0.38	0.50	0.66
60	0.18	0.21	0.21	0.21	0.22	0.23	0.25	0.28	0.30	0.34	0.39	0.51	0.67

Courtesy of SPWLA [16]

In relation to laterolog devices, *Fig. 15.8* gives a model of the distortion introduced by the presence of an apparent dip on the shallow laterolog current lines.

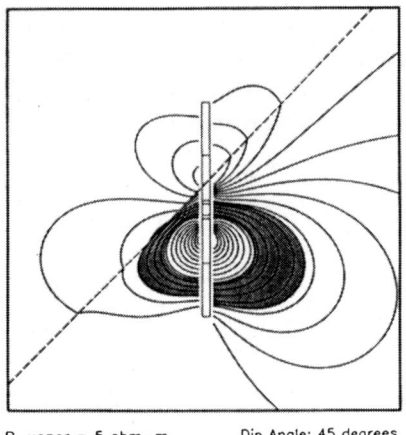

R-upper = 5 ohm—m   Dip Angle: 45 degrees
R-lower = 1 ohm—m

FIG. 15.8 : Current patterns of the shallow laterolog near a dipping boundary, computed with an analytic model.
$R_{upper} = 5$ Ω/m; $R_{lower} = 1$ Ω/m; dip angle = $45°$.

Courtesy of SPWLA [15]

15.3.1 Anisotropy

An anisotropic formation has properties that vary according to the direction in which they are measured.

Of particular interest in the production of hydrocarbons is permeability. Measuring permeability parallel to layers of sedimentary rocks may give a value that is different from the one measured perpendicular to the layer (*Fig. 15.9*).

For petrophysicists, anisotropy to current flow has been under scrutiny since wells have been drilled with an apparent dip angle to the formation [1 to 4]. Before then, the current flows enabling the measurement of formation resistivity were mostly circulating in a plan perpendicular to the axis of the borehole—and of the logging tool—and had constant characteristics in this plan.

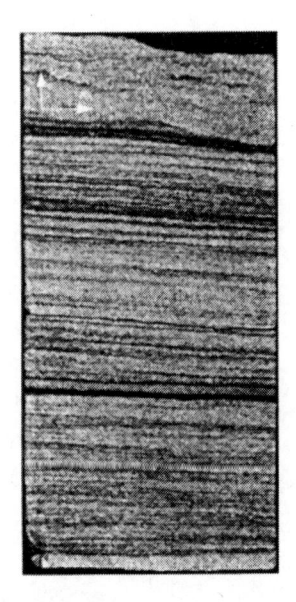

FIG. 15.9 : Permeability anisotropy in a sandstone. Thin layers of shale and quartz block most of the vertical flow, making horizontal permeability much higher than the vertical permeability.

Courtesy of Oilfield Review [1]

Running a resistivity device in a deviated well enables the measurement of two distinct formation characteristics, the horizontal resistivity, R_h, and the vertical resistivity, R_v. *Figure 15.10* shows how resistivity anisotropy allows the understanding of some mismatches between resistivity readings.

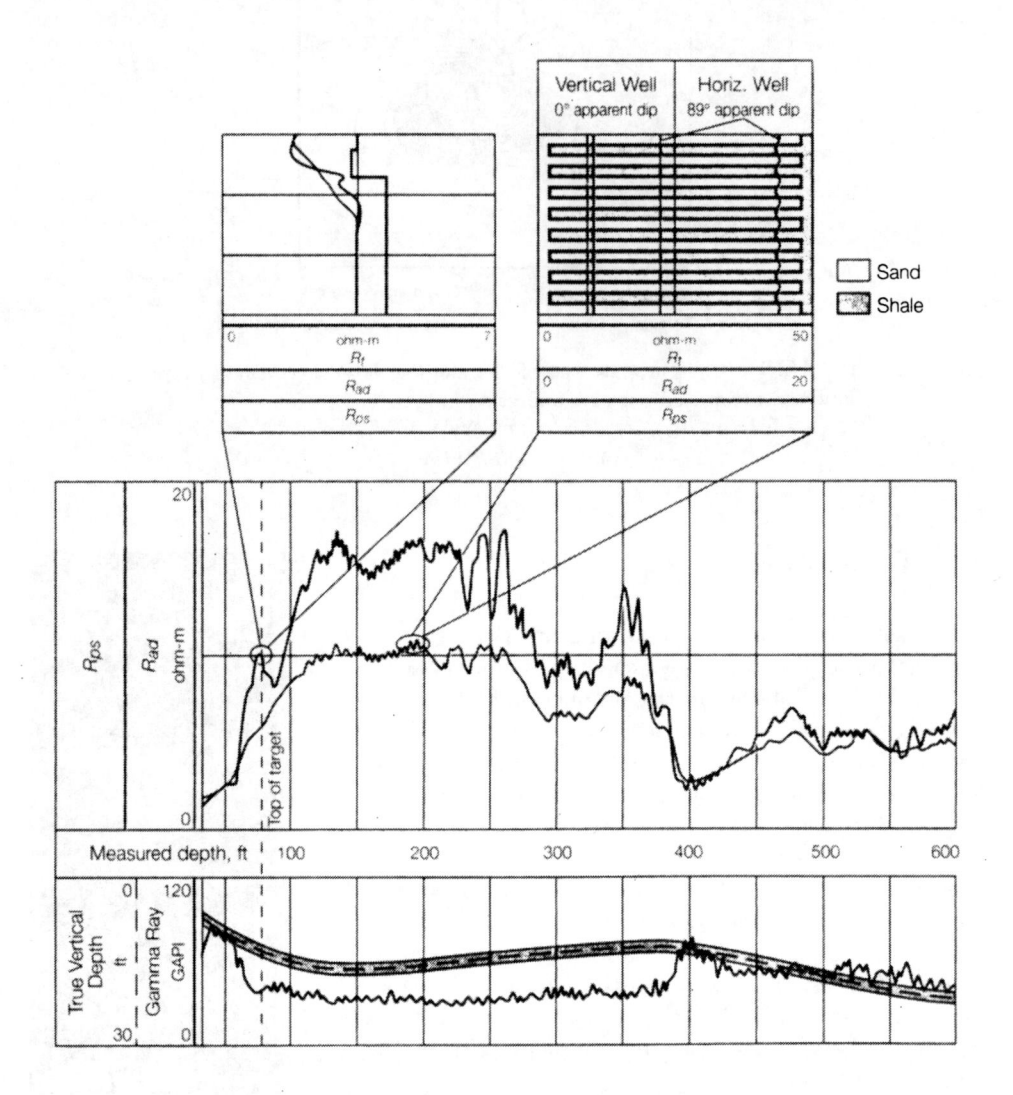

FIG. 15.10 : Resistivity anisotropy explains why the induction log, R_{ps} and R_{ad} do not agree.

The top right model represents a formation model with a series of low and high resistivity streaks. The models with the formation traversed with a 0° and a 90° apparent dip reproduce the measured log responses—4 Ω/m in the vertical well and 9 to 10 Ω/m in the horizontal well.

Courtesy of Oilfield Review [13]

15.4 The mud

15.4.1 Mud characteristics that impact logs

Mud properties are monitored several times per day. Standard techniques established by API [18] many years ago are used.

The properties [20] that need to be reported in relation with logging are:

(a) physical: density, temperature, viscosity (plastic or funnel), filtrate loss
(b) chemical: KCl equivalent, chlorides, pH
(c) mass balance: % oil, % water, % barite.

Historically, drilling muds have been composed mostly of salt water. But mud composition today could be quite complex. Presence of oil and solids, unusual chemical composition, lack of homogeneity have effects on logs. Above all, the invasion process has also a dramatic incidence on logs.

15.4.2 Oil-base mud

Shales react with water-base muds by swelling. This results in a considerable deterioration of the borehole, causing difficulties in running casing and cementation. Use of oil-base mud minimizes these effects.

The complete name for oil-base mud is *inverted oil emulsion mud*. In such mud, the continuous liquid phase is oil. Water is present as minute droplets in emulsion with the oil and is used to control the viscosity of the mixture. Water emulsifiers and various thickening, wetting and weighting agents are also added. An example of oil-base mud composition is 50% oil, around 20% water, 2% emulsifiers, 3% salts (NaCl or $CaCl_2$), 13% barite and 12% filtrate producer.

The main reason for using oil-base mud is to increase speed and efficiency [25]. The second objective is hole control. Oil-base mud tends not to cause swelling shales to cave as water-base muds do.

Oil-base mud precludes use of a number of logs, such as spontaneous potential and laterolog. It may also affect some logging measurements, for instance the sonic transit time and amplitude and formation tester response [17, 21 and 23]. As a trade-off, hole condition is generally far superior. A number of tools have been designed or adapted to collect information in oil-base mud. For instance, *Schlumberger* has developed the OBDT* Oil-Base Mud Dipmeter Tool, an induction-type device. *Gearhart Industries* has developed a tool for the same purpose but based on sonic emission. Similarly, the *Schlumberger* EPT* Electromagnetic Propagation Tool is adapted to the oil-base mud environment to derive an S_{xo} value.

Industry dogma has held that little or no invasion results from oil-base mud. In fact, oil filtrate invasion is becoming a more widely recognized fact. The conclusions drawn for water-base mud invasion need to be adapted to the oil-base mud environment, but only a few studies have been conducted so far [24].

15.4.2.1 Pseudooil-base mud

The original oil-base muds had several shortcomings; they could be environmentally unfriendly and they could have undesired behavior in temperature—degradation of rheological characteristics at low temperature. These limitations have induced the development of pseudooil-base muds that compensate for these drawbacks—at a cost.

15.4.2.2 Silicate mud

Drilling operators can also replace oil-base mud with water-base fluids that contain soluble silicates to stabilize shale and clay sections [26]. This allows them to comply with increasingly stringent environmental regulations while maintaining borehole stability.

Silicate agents react with the calcium and magnesium ions in the formation to form a gel that inhibits the penetration of water in the formation and limits the swelling of shales.

Sodium silicates have a high pH—sometimes above 13—and can affect oilfield equipment [19]. The high pH causes the attack of certain elastomers used downhole and in surface pumps; failure from mechanical or chemical damage results. In addition, when allowed to dry, the silicate forms a cement-like crust on the surface of the equipment; abrasive action is required to clean the surface.

15.4.3 Effect of mud solids

Mud is composed not only of fluids but up to several percent solid particles; commonly, bentonite and barite. Particle size is limited at surface by special screens that remove solids larger than 100 or 200 microns (or μm) in diameter, but debris produced by drilling and originating from adjacent beds may be forced back into the formation by the differential pressure between the mud column and the formation.

Mud solids invasion depends on the size of the plugging particle and of the bridging pores (*Fig. 15.11*). Mud solid invasion is confirmed in cores by the presence of barite, a mineral seldom found in rocks [22].

The effects of particle impregnation on logs are as follows:
(a) Density is reduced if the mud solids are lighter than the formation.
(b) Density is increased if weighting material is used.
(c) Neutron porosity is reduced.
(d) Microresistivity devices may be affected by montmorillonite impregnation. It modifies the electrical behavior of the near borehole region.
(e) P_e may be drastically affected by the use of barite.
(f) Deep resistivity and sonic readings are generally unaffected.

Lost circulation material

Lost circulation material is used when the formation has been fractured because of excessive mud weight. Light plugging material (e.g., nut shells) is used to seal the fractures and restore circulation. This material is unfortunately unfriendly to the turbine and pulser used by MWD tools and may cause a loss of information.

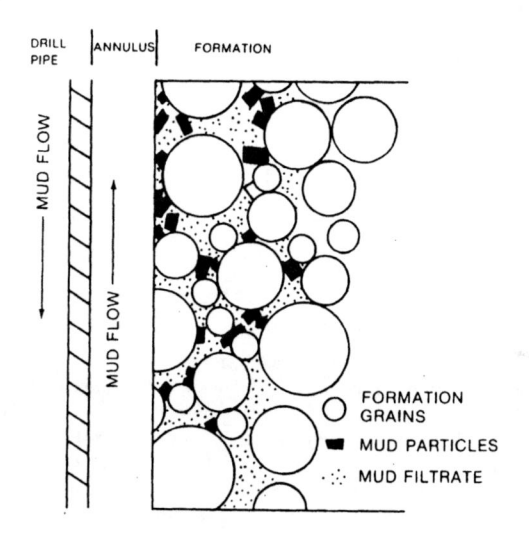

FIG. 15.11 : Mud solids invasion.

Courtesy of SPE [40]

15.4.4 Mud homogeneity

Regardless of its composition, mud is often assumed to be homogeneous. This is hardly achievable as it may take several weeks of drilling to reach total depth. During this period, mud may undergo important changes in chemistry and density.

Figure 15.12 depicts data underlining the segregation of mud at the bottom of the well.

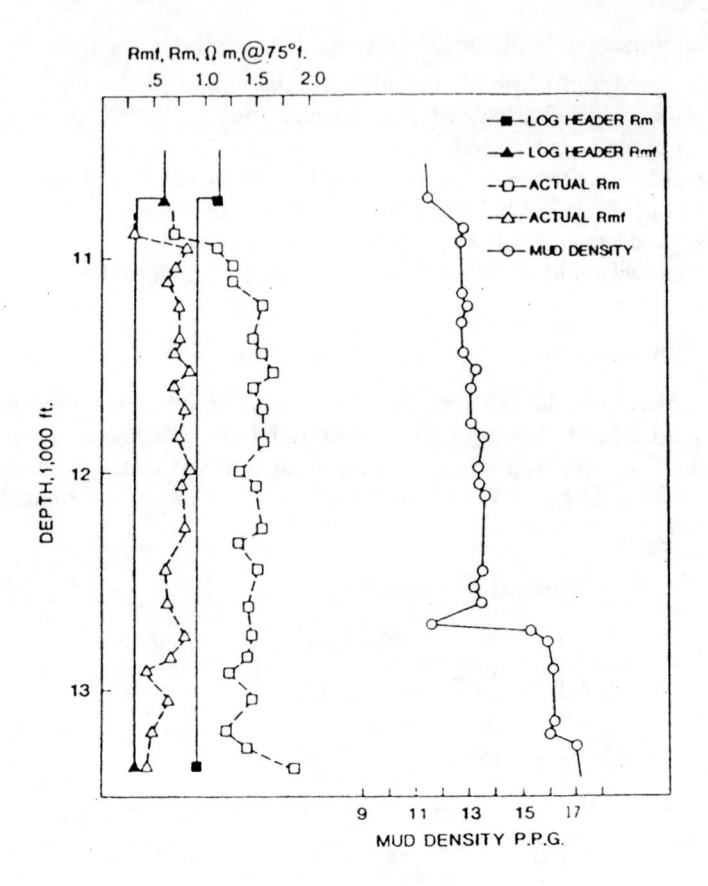

FIG. 15.12 : Example of mud heterogeneity.
Comparison of log heading information (in black) with measured values (in white). The triangles represent R_{mf} and the squares R_m. The resistivity scale is on the top of the diagram. The depth is plotted every 1000 ft on the y-axis. Variations of the mud resistivities can be tied back to changes of mud density as indicated by the curve on the right (scale in lbm/gal labeled as ppg at the bottom). Higher density is linked to mud segregation.
Courtesy of SPWLA [27]

15.4.5 Invasion

Because the pressure of borehole fluids exceeds that of the formation fluids, there is always some degree of fluid invasion in the formation. Invasion is often a nuisance as the invaded zone is a screen, more or less opaque, that blurs the picture of the virgin formation. Invasion, however, is not all that negative as, during this phenomenon, fluid motion takes place that can help diagnose the mobility of formation fluids.

The following paragraphs discuss the resistivity profiles that can arise from invasion of a hydrocarbon zone by water-base filtrate.

Filtrate from a water-base mud moves into an oil zone with a connate water saturation. Capillary and diffusion forces can be assumed negligible (though, in general, they are not). If the formation is **clean**, then a typical invasion profile develops (*Fig. 15.13*). The exact shape of the saturation profile depends on the fluid relative permeabilities and viscosities. The filtrate concentration is the volume fraction of filtrate in the aqueous phase; thus the product of filtrate concentration and water saturation is the volume fraction of filtrate in the pore space.

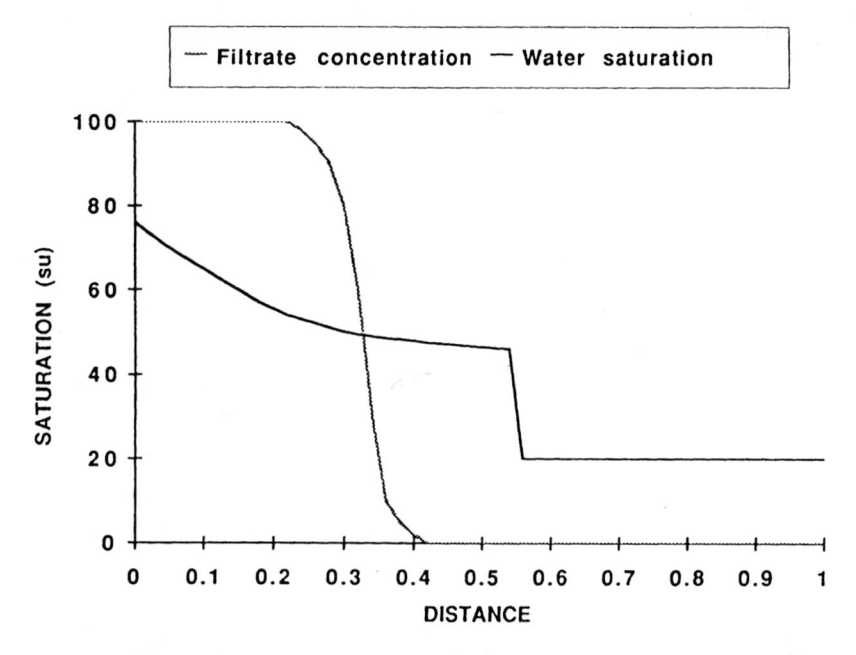

FIG. 15.13 : Saturation profile.

Courtesy of Schlumberger [24]

In *Fig. 15.13*, the filtrate front and the water saturation front do not coincide. Further, since the filtrate concentration is 100% behind the concentration front, connate water is being flushed out of the near wellbore zone and is banking in the region between the saturation front and the concentration front. This complete flushing of connate water is an idealization, as in practice a certain fraction of connate water will remain in some dead-end pores and micropores, although diffusion will eventually balance the salt content in these locations and the adjacent ones.

If the salinity of the formation and filtrate waters is known, along with the saturation exponent n and formation factor F, the resistivity profile corresponding to *Fig. 15.13* can be computed. Such a profile is shown in *Fig. 15.14*, where the filtrate is less saline than the formation water. The banked formation water is giving rise to a low resistivity annulus, while the incomplete flushing of oil by (in this case) low-salinity filtrate, results in a trailing **antiannulus**.

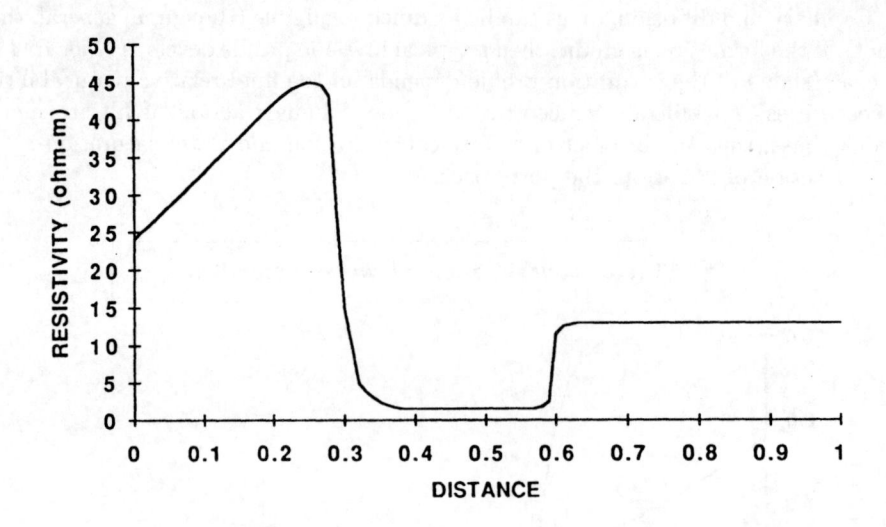

FIG. 15.14 : Resistivity profile.
This results from the water saturation and mud filtrate concentration of *Fig. 15.13*. The filtrate is assumed to be less saline than the formation water.
Courtesy of Schlumberger [24]

Depending on the filtrate and formation water salinities and the water saturation profile, three different types of resistivity profile can occur, as shown in *Figs 15.15a, 15b* and *15c*. Again, these profiles assume that capillary and diffusion forces are negligible.

When molecular diffusion and capillarity are included in the flow equations, the resulting profile shapes are dependent on the fluid loss rate. *Figure 15.16* corresponds to the case where imbibition is the dominant mechanism. The effect of capillary forces is to smear the saturation front and hence the leading front of the annulus. Molecular diffusion smears the front between the filtrate and formation waters and hence the trailing edge of the annulus. *Figure 15.16* also indicates that water saturation does not reach $(1-S_{or})$ when capillary imbibition dominates flow.[2] This has important implications for shallow reading tools and for interpretations based on deep/medium/shallow resistivity combinations.

Even simple models of invasion give rise to fluid and resistivity distributions that are significantly different from the type of model currently used in interpretation programs. An example of a conventional invasion model used in many interpretation programs is shown in *Fig. 15.17*.

Today, the approach based on shallow, medium and deep resistivity measurements to yield R_{xo}, R_t and a diameter of invasion d_i is simplistic. This explains why the log analyst sometimes has great difficulty matching a fourth measurement (for instance, R_{Ild} after using $R_{\mu SFL}$, R_{LLs} and R_{LLd}) in the resistivity model defined by the first three.

---

[2] S_{or} is the residual oil saturation.

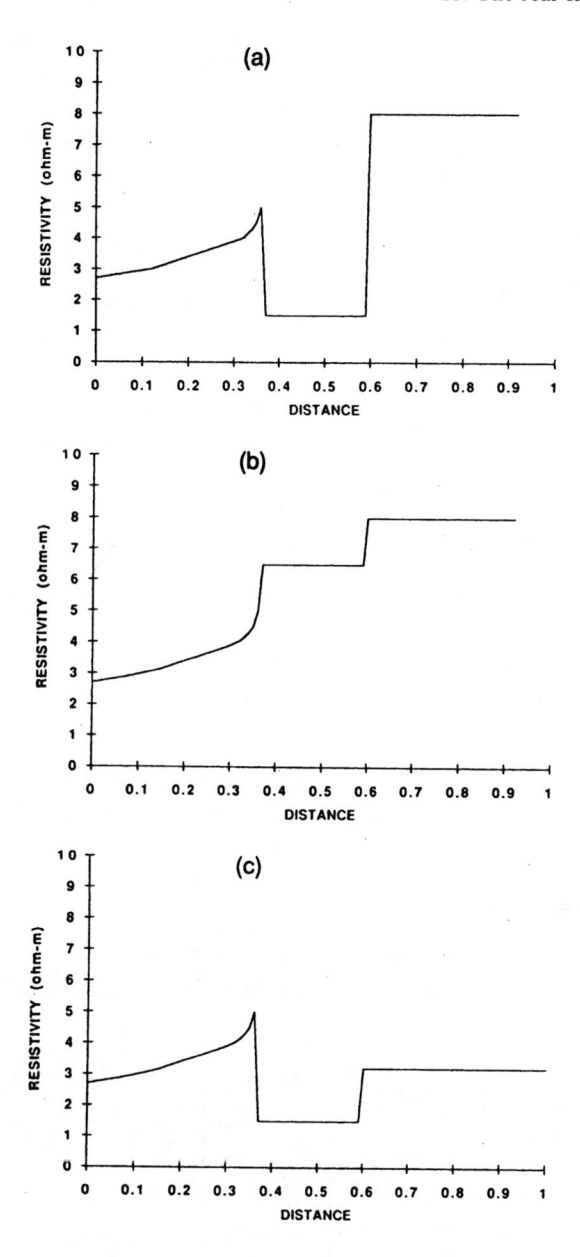

FIG. 15.15 : Possible resistivity profiles arising from invasion by water-base mud filtrate.
a. Formation water more conductive than mud filtrate, low S_{wi}.
b. Formation water less conductive than mud filtrate.
c. Formation water more conductive than mud filtrate, high S_{wi}.
Courtesy of Schlumberger [24]

210   15. The real environment

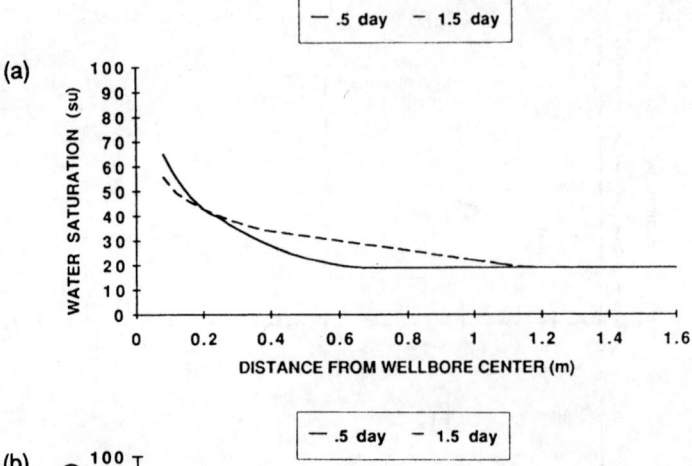

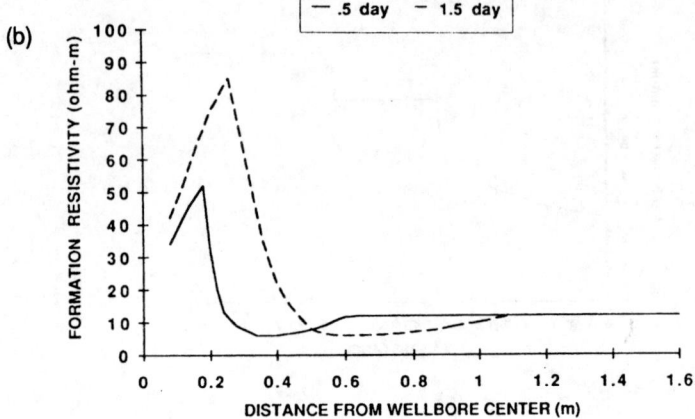

FIG. 15.16 : Profiles after 0.5 day and 1.5 day.
a. Saturation profile.   b. Corresponding resistivity profile.
Courtesy of Schlumberger [24]

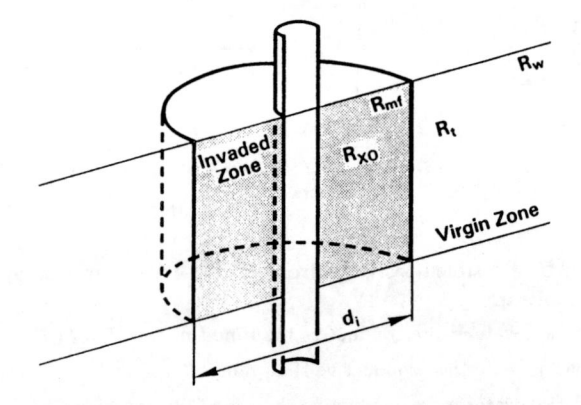

FIG. 15.17 : Classical invasion profile.

A solution to this problem is approached along the following lines:
(a) time-lapse technique: observation of invasion at different moments; comparison of data acquired during and after drilling [36]
(b) experimental simulation of the invasion
(c) modeling of invasion using more complex assumptions [28 and 39]
(d) use of a larger number of resistivity measurements with different depths of investigation
(e) use of different modes of current propagation (induction and laterolog methods).

15.5 Dielectric constant variation

Electric measurements performed at high frequency (above 400 kHz) are affected by the dielectric characteristics of the formation. They are compensated through several methods—one of them defined as the complex refractive index method, CRIM [14]. Failure to account for this effect may yield erroneous values (*Fig. 15.18*).

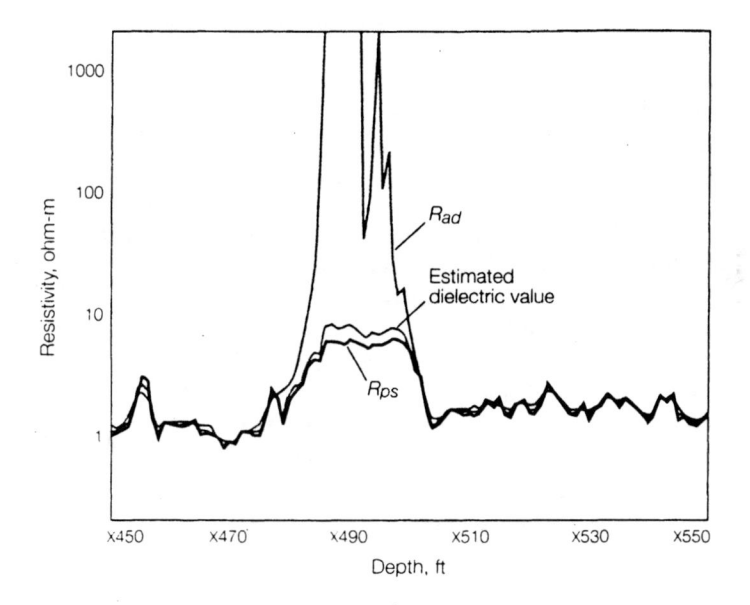

FIG. 15.18 : Anomaly caused by a pronounced dielectric effect.
This example relates to a volcanic rock in a North Sea formation. The actual value is 500 whereas the value used as an input parameter is 50.
Courtesy of Schlumberger [13]

15.6 Putting it all together

In more and more cases, it is necessary to deal simultaneously with variations of borehole shape, shoulder beds, thin beds, dielectric and anisotropy effects, dip, invasion and even crossbedding [29 to 35]. These data sets are handled with 3D modeling. Though computer[3] and expert intensive, these methods prove to be rewarding as in many cases, they enable the validation of additional hydrocarbon volume in place [38]. *Table 15.2* contains a glossary on the computer codes that enable extensive corrections through mathematical modeling.

TABLE 15.2
RESISTIVITY MODELING GLOSSARY

Acronym	Description
FEM	Finite-element method
FDM	Finite-difference method
MEM	Maximum entropy inversion method
SLDM	Spectral Lanczos decomposition method
1D (vertical)	Handles shoulder beds
1D (radial)	Handles invasion and annulus
2D	Handles invasion and shoulder beds
3D	Handles invasion, shoulders and dip
ANIS	Program for dip, layered, anisotropic media
ELMOD*	First modeling package
INFORM*	Integrated forward modeling
MERLIN*	Maximum entropy resistivity log inversion
R_tBAN	Z&S modeling package
XBED	Crossbedding code

Figure 15.19 is an example of a complex geometrical formation model.

---

[3]Computations in 2D are typically three times slower than 1D computations. 3D geometries require two order of magnitude more time than 1D computations.

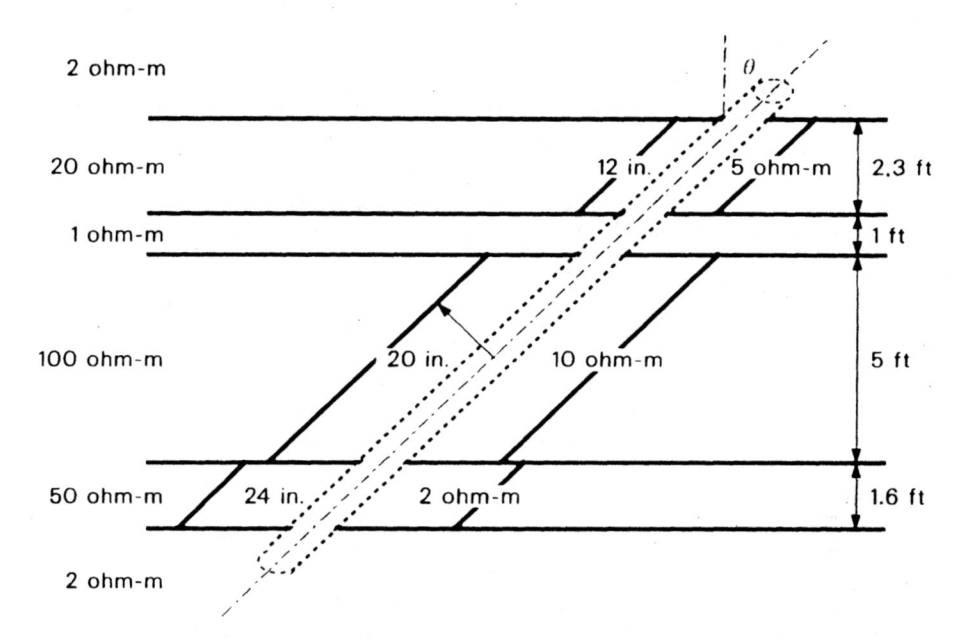

FIG. 15.19 : Six-bed invaded formation model.
Dip angle θ is 70°. Invasion radii are perpendicular to the sonde axis.
Courtesy of Schlumberger [31]

References

In this chapter, articles are first classified in categories, corresponding to the different environmental effects. They are then listed in alphabetical order.

Anisotropy

1 Ayan C., Colley, N., Cowan, G., Ezekwe, E., Wannell, M., Goode, P., Halford, F., Joseph, J., Mongini, A., Obondoko, G., Pop, J., "Measuring permeability anisotropy: the latest approach." *Oilfield Review*, Vol. 6, No. 4, 10-1994.

2 Hagiwara, T., "Response of 2 MHz resistivity devices in thinly laminated formations (anisotropy resistivity and EM log interpretation)," SPE 28426, SPE 69th annual technical conference and exhibition, New Orleans, 9-1994.

3 Klein, J. D., "Induction log anisotropy corrections," *Trans.* SPWLA 32nd annual logging symposium, Midland, 1991.

4 Leake, J., Shray, F., "Logging while drilling keeps horizontal wells on small target," *Oil and Gas Journal*, No. 38, 9-1991.

Borehole shape

5 Charlez, P., Segal, A., Heugel, O., Quenault, O., "Elaboration d'un modèle microstatique pour expliquer l'ovalisation des forages profonds dans les roches fragiles," *Proceedings du Congrès* Rocks at great depth, Pau, 9-1989. Also published by A. A. Balkema, Rotterdam, 1990.

6 Guénot, A., "Contraintes et ruptures autour des forages pétroliers," pp. 109-118, *Proceedings* Sixième congrès international des roches, Montréal.

7 Hornby, B. E., Chang, S. K., "A case study of shale and sandstone alteration using a digital sonic tool," paper H, *Trans.* SPWLA 26th annual logging symposium, 1985.

8 Jaeger, J., Cook, N. G. W., *Fundamentals of rock mechanics*, Chapman and Hall Ltd, London, 1979.

9 Segal, A., *Elaboration d'un modèle microstatistique linéaire de la rupture fragile et application à la stabilité des forages profonds*, Ecole Centrale, Paris, 1989.

Borehole size

10 Campbell, R., "Borehole geometry and log response," *The Technical Review*, Vol. 29, No. 3, 1981.

11 Liu, O. Y., "The sources of errors in slowness measurements and an evaluation of full waveform compensation techniques," SPE 16772, *Trans.* SPE 62nd annual technical conference and exhibition, Dallas, 1987.

12 Misk, A., Mowat, G., Goetz, J., Vivet, B., "Effects of hole conditions on log measurements and formation evaluation," paper 22, *Trans.* SPWLA 5th European formation evaluation symposium, Paris, 1977.

Dielectric effect

13 Bonner, S., Clark, B., Holenka, J., Voisin, B., Dusang J., Hansen, R., White, J., Walsgrove, T., "Logging while drilling: a three-year perspective," *Oilfield Review* Vol. 4, No. 3, 7-1992.

14 Wharton, R., Hazen, G., Rau, R., Best, D., "Electromagnetic propagation logging... advances in technique and interpretation," *The Technical Review*, Vol. 29, No. 1, 1981.

Dip

15 Chemali, R., Gianzero, S., Su, S. M., "The dual laterolog in common complex situations," paper N, *Trans.* SPWLA 29th annual logging symposium, 1988.

16 Fylling, A., Spurlin, J., "Induction simulation: the log analysts' perspective," paper T, *Trans.* SPWLA 11th European logging symposium, Oslo, 1988.

Mud

17 Adams, J., Blott, N., Boyeldieu, C., Chéruvier, E., Cull, R., Mobed, R., Haines Jr., P., Spurlin, J., "Advances in log interpretation in oil-base mud," pp. 22-40, *Oilfield Review*, volume 1, No. 3, 7-1989.

18 American Petroleum Institute: *API recommended practice: standard procedure for testing drilling fluids. API RP 13B*, Washington, D.C., 1980.

19 Bernard, L., Personal communication on silicate muds.

20 Geehan, T., McKee, A., "Drilling mud: monitoring and managing it," *Oilfield Review*, Vol. 1, No. 2, 7-1989.

21 Holditch, S. A., Lee, W. J., Lancaster, D. E., Davie, T. B., "Effect of mud filtrate invasion on apparent productivity in drillstem tests in low permeability gas formations," SPE 9842, pp. 299-305, *Journal of Petroleum Technology*, Vol. 35, No. 2, 2-1983.

22 McKee, A., Mowat, G., Geehan, T., "Drilling mud properties and formation evaluation," *Trans.* IPA 15th annual convention, 1986.

23 Pélissier-Combescure, J., "Formation evaluation in oil-base mud wells," paper BB, *Trans.* SPWLA 25th annual logging symposium, New Orleans, 1984.

24 Phelps, G., *Oil base mud invasion*, Schlumberger-Statoil, Stavanger, 1988.

25 Scott, P., "Oil and synthetic muds: the basics."

26 Van Oort, E., Ripley, D., Ward, I., Chapman, J. W., Williamson, R., Aston, M., "Silicate-based drilling fluids: competent, cost-effective and benign solutions to wellbore stability problems," IADC/SPE 35059, presented at the IADC/SPE drilling conference, New Orleans, 1996.

27 Williams, H., Dunlap, H. F., "Short term variations in drilling fluid parameters; their measurement and implications," pp. 3-9, *The Log Analyst*, Vol. 25, No. 5, 9-1984.

28 Woodhouse, R., Opstad, E. A., Cunningham, A. B., "Vertical migration of invaded fluids in horizontal wells," paper A, *Trans.* SPWLA 32nd annual logging symposium, Midland, 6-1991.

Resistivity modeling

29 Anderson, B., Barber, T., "Strange induction logs, a catalog of environmental effects," paper G, *Trans.* SPWLA 28th annual logging symposium, 1987.

30 Anderson, B., Barber, T., Druskin, V., Lee, P., Dussan, E., Knizhnerman, L., Davydycheva, S., "The response of multiarray induction tools in highly dipping formations with invasion and arbitrary 3D geometries," paper A, *Trans.* SPWLA 37th annual logging symposium, 1996.

31 Anderson, B., Barber, T., Gianzero, S., "The effect of crossbedding anisotropy on induction tool response," paper B, *Trans.* SPWLA 39th annual logging symposium, 1998.

32 Anderson, B., Barber, T., Singer, J., Broussard, T., "ELMOD, putting electromagnetic modeling to work to improve resistivity log interpretation," paper M, *Trans.* SPWLA 30th annual logging symposium, 1989.

33 Anderson, B., Druskin, V., Habashy, T., Lee, P., Luling, M., Barber, T., Grove, G., Lovell, J., Rosthal, R., Tabanou, J., Kennedy, D., Shen, L., "New dimensions in modeling resistivity," pp. 40-56, *Oilfield Review*, Vol. 9, No. 1, 1-1997.

34 Anderson, B., Minerbo, G., Oristaglio, M., Barber, T., Freedman, B., Shray, F. "Modeling electromagnetic tool response," *Oilfield Review*, Vol. 4, No. 3, 7-1992.

35 Barber, T., Broussard, T., Minerbo, G., Sijercic, Z., Murgatroyd, D., "Interpretation of multiarray induction logs in invaded formation at high dip angles," paper A, *Trans.* SPWLA 39th annual logging symposium, 1998.

36 Gianzero, S., Chemali, R., Su, S. M., "Induction resistivity and MWD tools in horizontal wells," paper N, *Trans.* SPWLA 30th annual logging symposium, 1989.

37 Gianzero, S., Lin, Y., Chemali, R., Dossey, W., "The effect of sonde eccentering on resistivity tools: an exact theoretical model," paper GG, *Trans.* SPWLA 26th annual logging symposium, 1985.

38 Kennedy, D., "Letter to editor," *The Log Analyst*, Vol. 37, No. 6, 11-96.

39 Singer, J. M., Barber, T. D., "The effects of transition zones on induction logs," paper L, *Trans.* SPWLA 29th annual logging symposium, 1988.

Temperature

40 Jorden, J. R., Campbell, F. L., *Well Logging I - Rock properties, borehole environment, mud and temperature logging*, SPE monographs, Vol. 9, New York, 1984.

16
Density logging

Porosity, an essential input to reserve evaluation, is obtained from different measurements. Most petrophysicists give emphasis to log-derived density porosity. But this value receives intense scrutiny if it disagrees with core-derived porosity. Density cannot be derived directly from logging. It has to be inferred from the observation of gravity changes or, more commonly, from the study of electron interactions, generally of the Compton type. The raw data supplied by the downhole detectors consists of counting rates only. The relation between raw data and density needs to be established with artificial formations. This chapter describes the evolution of the algorithms defined for the *Schlumberger* density tools as an example of the development of tool response and environmental corrections.

16.1 The first density algorithms

16.1.1 Single detector

Gamma rays emitted by a chemical source interact with electrons of the formation and the borehole [4]. If all interactions are assumed to be controlled by Compton effect, the relation between the counting rates and the bulk density, ρ_b, is

$$\rho_b = A - B \times \log N.$$

A and B are constants that depend on source strength and detector efficiency. ρ_b is related to the electronic density, ρ_e, from which the real density, ρ, can be derived. Variable N represents the counting rate of the detector. By definition, a single-detector device measures only one density related to a volume that includes formation and mudcake. Locating the detector far away from the source would minimize the effect of mudcake but only few gamma rays would reach the detector, making the measurement overly affected by random statistical error (Section 10.1.3). On the other hand, a short source-to-detector spacing would decrease the uncertainty on the measurement but would also decrease the relevant volume to be investigated, the formation volume, by observing a large volume of mudcake.

16.1.2 Dual-detector device

The one-detector losing proposition called for a single-source dual-detector design [11 and 12]. The short-spacing detection properly defines the effect of mudcake, which can then be subtracted from the long-spacing information (*Fig. 16.1*).

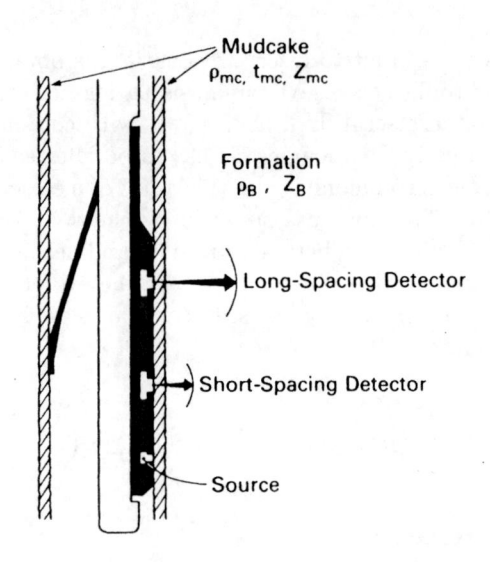

FIG. 16.1 : Idealized mudcake problem.

Courtesy of SPE [13]

The tool response of a dual-detector device is based on two equations:

$$\rho_{LS} = A_1 - B_1 \times \log N_{LS}$$

$$\rho_{SS} = A_2 - B_2 \times \log N_{SS}.$$

The difference $x = \rho_{LS} - \rho_{SS}$ allows the derivation of a correction $\Delta \rho$, which is then added to the long-spacing density:

$$\rho_b = \rho_{LS} + \Delta \rho.$$

The relation between $\Delta\rho$ and x is a curve that can be approximated by three segments of straight lines:

$$\Delta\rho = a \times x \text{ for } x < \alpha$$

$$\Delta\rho = b \times x + c \text{ for } \beta < x < \gamma$$

$$\Delta\rho = d \times x + e \text{ for } x > \gamma.$$

All constants A_1, B_1, A_2, B_2, a, b, c, d, e, α, β and γ must be determined through actual measurements performed in artificial formations, first in a set of experiments without mudcakes, then with a series of mudcakes of different densities and thicknesses. The data inputs to the tool response are indicated in *Fig. 16.2*.

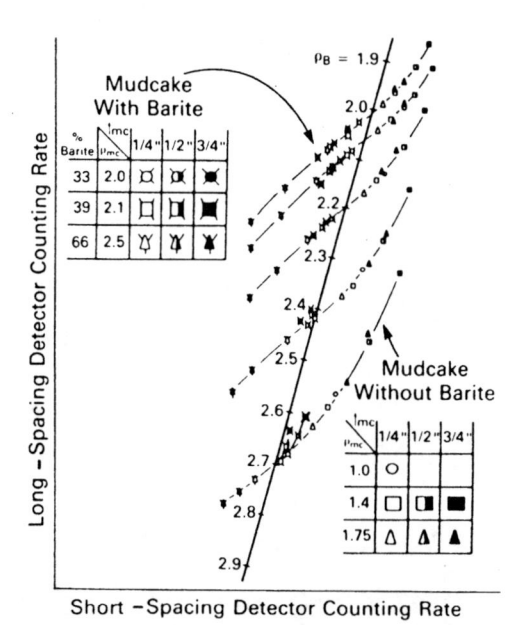

FIG. 16.2 : Spine and ribs data.
Courtesy of SPE [13]

16.1.3 Non-Compton gamma rays

Early in the life of the density tool, it was discovered that the device was affected by a parasitic phenomenon, the photoelectric effect [12]. The detectors then used were total detectors—they were intercepting gamma rays with any energy, from a few kilo electronvolt to several megaelectronvolt. In fact, the relations written in the previous paragraphs are valid only if the interactions of the detected gamma rays are all of the Compton type. To make the matter more complex, low-energy gamma rays are affected by photoelectric effect and high-energy gamma rays can be emitted from naturally radioactive formations. These two types of radiation would be detected by total detectors along with the Compton radiation.

To minimize the detection of low-energy gamma rays dominated by photoelectric effect, a cadmium shield was added on top of the long-spacing detector. This meant that no gamma rays below 200 keV could reach the detector. This solution had the severe drawback of drastically reducing the flux of gamma rays arriving at the detector. The effect of this reduction on the resulting density can be easily quantified by using the actual counting rates in the formula expressing the uncertainty of ρ_b (Section 7.5).

16.2 Spectrometric detectors

In the 1970s, the development of logging tools with spectrometric detectors provided an alternative solution to the reduction of the photoelectric effect [1]. These detectors are able to measure the energy of the incoming radiation. The incident gamma rays can be binned in different energy windows (*Fig. 16.3*).

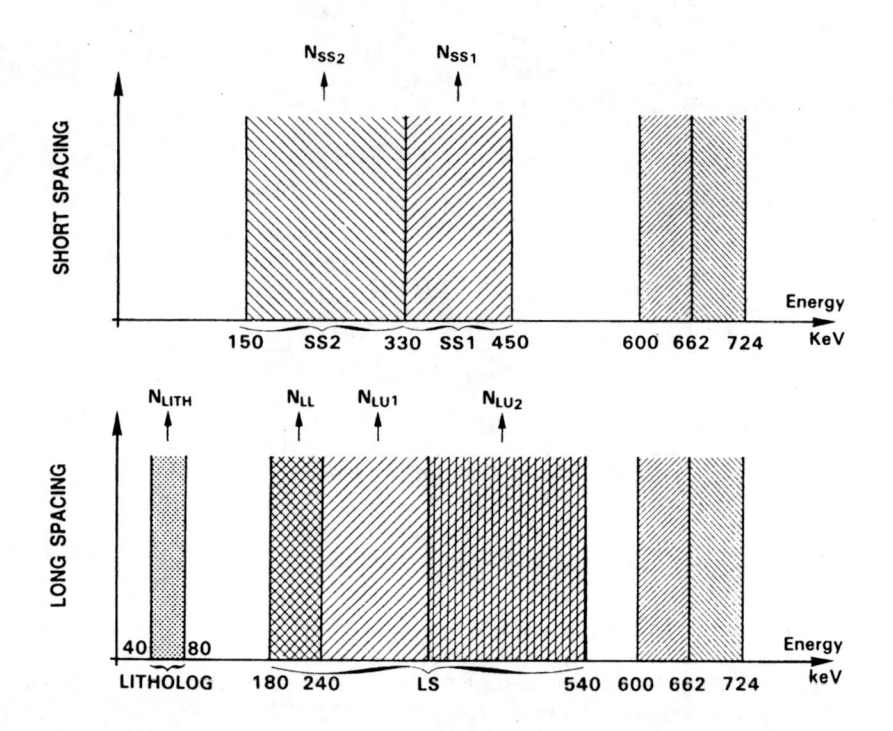

FIG. 16.3 : Spectra of the Litho-Density tool detectors.

Courtesy of Schlumberger [10]

Only gamma rays with energies above 180 keV (for the long-spacing detector) and 330 keV (for the short-spacing detector) are used for the derivation of formation density. This corresponds to the sum of N_{LL}, N_{LU1} and N_{LU2} for the long-spacing detector and to N_{SS1} for the short-spacing detector.

At this point, the tool response of the first spectrometric tools was defined in a limited number of formations. In the presence of large concentrations of barite, some **photoelectric** gamma rays could still be detected in the high-energy windows.

Therefore, there was an option to disallow the use of the 180- to 240-keV energy window of the long-spacing detector. N_{LL} counting rates were no longer used to derive the density. This choice corresponded to higher statistical variations since a large number of counts were no longer used. At the same time, as the number of artificial formations were extended from the original set, the algorithm was modified.

16.2.1 Compensating the upper windows for photoelectric information

Common sense led to the idea that the low-energy window, dominated by the photoelectric effect, could be used to correct the residual photoelectric effect in the high-energy windows. This development was named the **three-window algorithm**:[1]

$$\rho_{LS} = A_1 - B_1 \times \log(N_{LS}/N_{LS}^*) + C_1 \times \log(N_{\text{Lith}}/N_{\text{Lith}}^*)$$

$$\rho_{SS} = A_2 - B_2 \times \log(N_{SS}/N_{SS}^*) \times (C_2/\rho_{LS}).$$

The N^* counting rates are the ones observed during the shop calibration. Indeed all the constants shown above had to be redefined to the disarray of the users now hoping for a short spell of algorithm stability [3].

16.2.2 The heavy mud algorithm

Log analysts observed that in heavier muds loaded with barite, the tools were being used in conditions beyond those for which the response was defined. Experiments and least-squares analysis produced a new generation of coefficients. At this time, the second energy window ($SS2$) of the short-spacing detector was being used:

$$\rho_{LS} = A_1 - B_1 \times \log(N_{LS}/N_{LS}^*) + C_1 \times \log(N_{\text{Lith}}/N_{\text{Lith}}^*)$$

$$\rho_{SS1} = A_2 - B_2 \times \log(N_{SS1}/N_{SS1}^*) \times \log(C_1/\rho_{LS})$$

$$\rho_{SS2} = A_3 - B_3 \times \log(N_{SS2}/N_{SS2}^*) \times C_2/\rho_{LS}$$

$$\rho_b = \rho_{LS} + \Delta\rho_1 + p \times (\Delta\rho_2 - \Delta\rho_1).$$

$x = \rho_{LS} - \rho_{SS1}$	$y = \rho_{LLS} - \rho_{SS2}$	
$\Delta\rho_1 = ax$ if $x < \alpha$	$\Delta\rho_2 = 0$ if $x > 0$	$p = 0$ if $\Delta\rho_2 > 0$
$\Delta\rho_1 = bx + c$ if $\beta < x < \gamma$	$\Delta\rho_2 = a'x$ if $-\alpha' < x < 0$	$p = a_1\Delta\rho_2 + b_1$ if $-\alpha" < \Delta\rho_2 < -\beta"$
$\Delta\rho_1 = dx + e$ if $x > \gamma$	$\Delta\rho_2 = b'x + c'$ if $x < -\alpha'$	$p = 1$ if $\Delta\rho_2 < -\gamma"$

---

[1]The parameters with star subscripts relate to the readings performed in an aluminum block.

16.2.3 Experimental establishment of the response of a spectrometric tool

To establish the response of a spectrometric density measurement over a wide range of densities, a large laboratory installation is used to define all the parameters introduced by the various algorithms. It consists of

- (a) seven nonporous blocks (*Fig. 16.4*)
- (b) four artificial blocks, SiO/epoxy, CSi/SiO/epoxy, CSi/epoxy and AlMg
- (c) six blocks made of natural formations, four of limestone, one of dolomite, one of diabase.

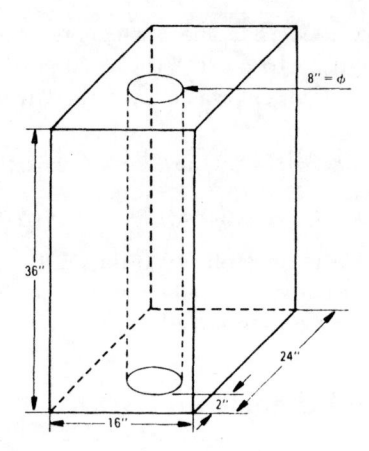

FIG. 16.4 : Calibration block.
The hole is not drilled in the center of the block. Its axis is offset. The density pad faces the part of the block with the largest amount of formation.
1 in. = 2.5 cm.

Courtesy of SPE [3]

These blocks have a 200-mm (8-in.) hole drilled through them. Density is 1.678 to 3.051 g/cm^3 and the photoelectric effect is 1.3 to 5.1 barns/electron. An accurate measurement of bulk density is required to achieve an adequate set of primary standards. Three strain gauge scales of intrinsic accuracy 3×10^{-4} are used to weigh the blocks. Each of these scales is calibrated with reference facilities located either at the manufacturer's plant or at a national laboratory. The blocks are weighed in both air and water, and Archimedes' principle is used to compute bulk density. The density of the water in the borehole or constituting the wetting liquid of the porous blocks is measured with a precision of 3×10^{-4} by a national laboratory. Bulk densities are therefore assigned to all blocks with an accuracy of ±0.001 g/cm^3. The characteristic Z/A of the blocks is derived from their known chemical composition. A knowledge of the corresponding log density can then be achieved within ±0.005 g/cm^3. *Table 16.1* summarizes the characteristics of these primary standards.

Once these primary calibration references are established, a large number of tools (48) are used and their counting rates are measured in some or all these blocks. Using all this information, a response model is generated.

The responses of the long- and short-spacing detectors are combined into the **spine**. This represents the tool response in a 20-cm (8-in.) borehole without mudcake. Table 16.2 shows the tool response in the reference blocks.

TABLE 16.1
REFERENCE BLOCKS CHARACTERISTICS

The values in brackets are uncertainties in the last significant digit.

Block	$\dfrac{\rho_{bulk}}{(g/cm^3)}$	$2 < Z/A >$	$\dfrac{\rho_e}{(g/cm^3)}$	$\dfrac{\rho_{log}}{(g/cm^3)}$	$\dfrac{P_e}{(barns/electron)}$
SiO	1.717(1)	1.016(2)	1.744(5)	1.678(5)	1.36
SiO/CSi	1.930(1)	1.015(2)	1.959(5)	1.908(5)	1.65
CSi	2.188(1)	1.103(1)	2.217(4)	2.185(4)	1.78
AlMg	2.654(2)	0.965	2.560(2)	2.552(2)	2.78
Limestone	2.713(1)	0.999	2.711(1)	2.713(1)	5.08
Dolomite	2.846(1)	0.998	2.840(1)	2.851(1)	3.14
Diabase	3.060(1)	0.989	3.026(1)	3.051(1)	4.70

Courtesy of SPE [3]

TABLE 16.2
TOOL READINGS IN REFERENCE BLOCKS

Block	ρ_b (g/cm³)	
	Reference	Tool
SiO	1.678	1.673
SiO/CSi	1.908	1.923
CSi	2.185	2.183
AlMg	2.552	2.548
Limestone	2.713	2.711
Dolomite	2.851	2.850
Diabase	3.051	3.053
Limestone 3 pu	2.665	2.669
Limestone 12 pu	2.510	2.509
Limestone 28 pu	2.235	2.233

Courtesy of SPE [3]

16.2.4 Experimental evaluation of environmental effects

16.2.4.1 Mudcake effect

To correct for mudcake and borehole effects, numerous laboratory measurements are performed in the calibration blocks and with a series of artificial mudcakes of varying compositions, densities and thicknesses. Sleeves of rubber loaded with dopants are manufactured to simulate real mudcakes in boreholes drilled with muds of widely varying properties. Thirteen sets of artificial mudcakes with densities of 1.19 to 2.32 g/cm^3 and barite contents from 0 to 50% are used (*Table 16.3*).

TABLE 16.3
LIST OF MUDCAKES

Mudcake density	Amount of barite	Thickness
(g/cm^3)	(%)	(in.)
1.193	0.0	1/8, 1/4, 1/2, 1, 1.5
1.451	5.0	1/8, 1/4, 1/2
1.453	20.0	1/8, 1/4, 1/2
1.454	10.0	1/8, 1/4, 1/2
1.457	0.0	1/8, 1/4, 1/2, 1, 1.5
1.527	0.0	1/8, 1/4, 1/2, 1, 1.5
1.910	30.5	1/8, 1/4, 1/2, 1, 1.5
1.929	53.0	1/8, 1/4, 1/2, 1, 1.5
1.958	0.0	1/8, 1/4, 1/2, 1, 1.5
1.958	5.0	1/8, 1/4, 1/2
1.960	20.0	1/8, 1/4, 1/2
1.965	10.0	1/8, 1/4, 1/2
2.327	0.0	1/8, 1/4, 1/2, 1, 1.5

Courtesy of SPE [3]

A database gathers the results of laboratory measurements performed in the artificial blocks. It includes over 900 points, one point corresponding to one tool in a particular block with one mudcake of a particular thickness and density. *Table 16.4* shows a summary of the measurements made with 13 types of mudcakes and thicknesses.

Departure curves from the spine in each of the blocks are plotted with this data set (*Fig. 16.5*). The complete database represents the laboratory simulation of densities ranging from those of diatomites to those of anhydrite with simulated mudcakes representing light muds to heavy baritic mudcakes. It allows the derivation of a set of equations for mudcake compensation.

TABLE 16.4
MEASUREMENTS WITH MUDCAKES

Mudcake density (g/cm³)	Barite content (%)	Artificial formations (density in g/cm³)						
		SiO₂ 1.678	CSi/SiO₂ 1.908	Csi 2.185	AlMg 2.552	Limestone 2.713	Dolomite 2.851	Diabase 3.051
1.193	0	8	5	10	8	13	8	5
1.451	5	3	3	3	3	3	3	3
1.453	20	3	3	3	3	3	3	3
1.454	10	3	3	3	3	3	3	3
1.457	0	8	5	10	8	15	9	5
1.527	0	8	5	10	8	13	10	5
1.910	30.5	8	5	10	9	10	9	5
1.929	53	9	10	10	10	13	9	5
1.958	0	8	5	10	8	10	8	5
1.958	5	3	3	3	3	3	3	3
1.960	20	3	3	3	3	3	3	3
1.965	10	3	3	3	3	3	3	3
2.327	0	6	5	10	8	10	8	5

Courtesy of SPE [3]

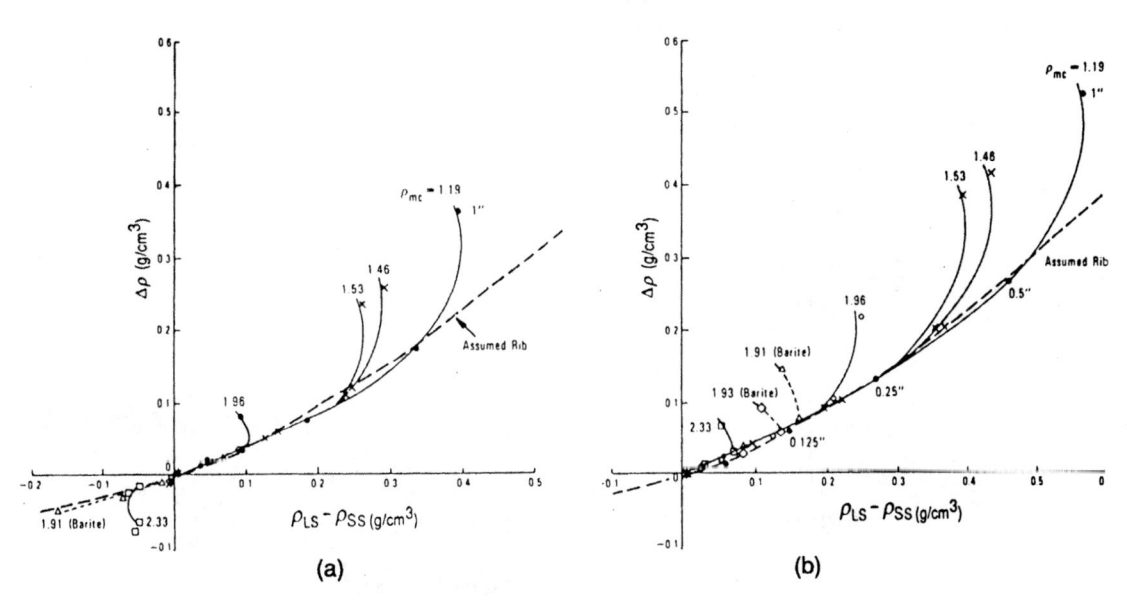

FIG. 16.5 : Examples of rib derivation in two calibration blocks.
   a. $\rho_b = 2.185$ g/cm³.   b. $\rho_b = 2.552$ g/cm³.

The difference between the long- and short-spacing readings ($\rho_{LS} - \rho_{SS}$) is shown on the x-axis. $\Delta\rho$, the correction factor, is shown on the y-axis. The curves correspond to different mudcake densities.

Courtesy of SPE [3]

16.2.4.2 Borehole effect

A second environmental effect is due to the varying borehole size. Because of the fixed radius of curvature of the surface of the measuring device, influence of mud on the measurement varies with borehole size. A parameter found to characterize this perturbation is x_{bh}:

$$x_{bh} = (\rho_{\text{measured}} - \rho_{\text{mud}})(\text{caliper} - 8),$$

in which the second term represents the difference between the diameter of the borehole measured by the caliper (expressed in inches) and the 20-cm (8-in.) reference diameter. $\rho_{\text{measured}} - \rho_{\text{mud}}$ represents the difference in density between the raw log and the mud.

An additional part of the calibration facility is designed to quantify the borehole size effect. It consists of

(a) three blocks made of the same material as the 20-cm (8-in.) borehole AlMg block ($\rho_b = 2.552$ g/cm^3) but with borehole diameters of 15, 30 and 40 cm (6, 12 and 16 in.).

(b) two blocks made of the same material as the 20-cm (8-in.) limestone formation ($\rho_b = 2.713$ g/cm^3) with boreholes of 15 and 25 cm (6 and 10 in.).

Experiments have been performed in these blocks with three different fluids as borehole mud and a correction algorithm derived as a function of x_{bh}. The measurements of the correction necessary as a function of x_{bh} are shown in *Fig. 16.6*. This correction, which is generally less than 0.02 g/cm^3, is made from the simultaneously measured borehole size and generally computed in real time.

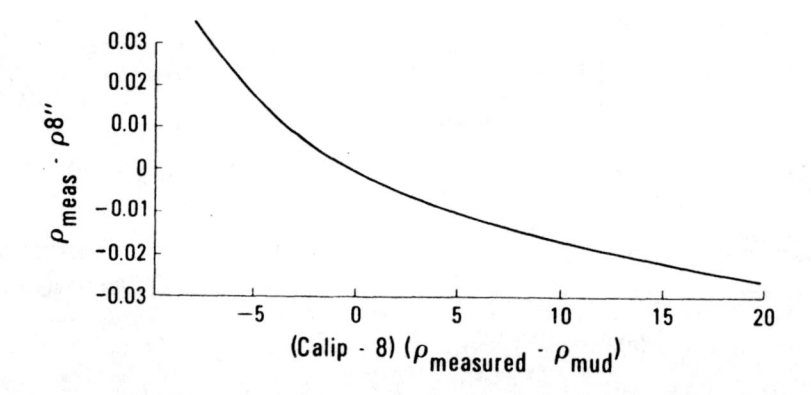

FIG. 16.6 : Borehole size effect.

Courtesy of SPE [3]

16.3 MWD density

16.3.1 Nonazimuthal density

Several MWD density tools have been developed in the late 1980s and early 1990s [15]. This technological advance was made because the lower counting rates observed in MWD nuclear logging are compensated by drilling rates of penetration much slower than common wireline logging speeds. Contact with the formation, or minimization of the mud/mudcake thickness remains a challenge. The MWD mode precludes logging with a pad. The earlier tools required the use of stabilizers to provide a tight fit with the borehole wall.

In noncircular holes, only a fraction of the MWD sensor package is in contact with the formation. Statistical methods have been developed to compensate for this condition: one of them generates "rotational density" [2].

16.3.2 Azimuthal density while drilling

The logical improvement in MWD density logging is to benefit from the rotation of the drillstring to collect data around the borehole. In a deviated well, the drillstring (including the MWD sensors) lies on the bottom of the borehole. In either configuration (MWD subs run stabilized or slick), there is good contact between the sensor and the formation. The result is an accurate density measurement if the MWD device can bin information in different quadrants [8] or sectors: the bottom quadrant (or sector) corresponds to the zone of good contact, while the other quadrants (or sectors) correspond to large standoffs.

When such a device is stabilized and there is no hole washout, the azimuthal capability helps identify formation heterogeneities around the borehole. Typically, four densities—up, down, left, right—are provided (*Fig. 16.7*).

16.4 PLATFORM EXPRESS* density

A number of drawbacks still affect the dual-detector approach: poor pad application, low statistical precision and excessive sensitivity to barite. A general limitation observed since the beginning of density logging is the negative correlation of counting rates with density, which makes imprecision greater in denser formations.

These limitations are reduced with the addition of a backscatter gamma-ray detector [5]. This detector features a positive correlation between counting rates and density. The energy spectra of the three detectors are combined to solve for five unknowns: formation density and photoelectric factor, mudcake density, thickness and photoelectric factor.

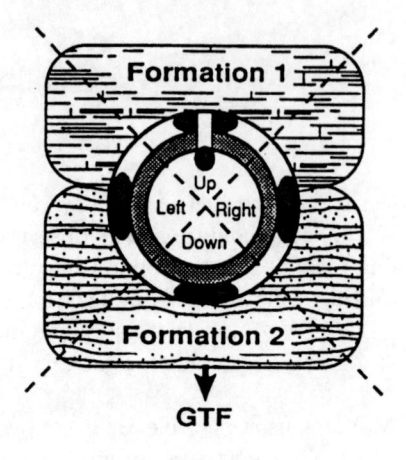

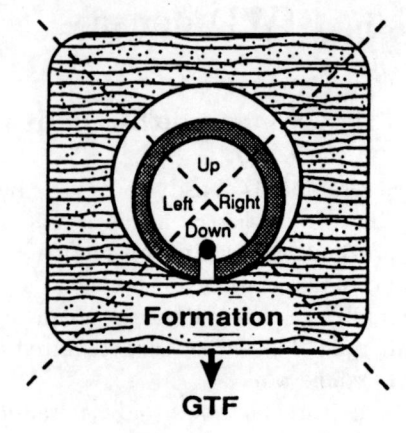

FIG. 16.7 : Azimuthal density

a. Stabilized configuration: density is measured in four quadrants and help identify heterogeneities. $\rho_{b_{up}}$ and $\rho_{b_{down}}$ relate to two different formations.

b. Slick configuration: $\rho_{b_{down}}$ represents an accurate density, even in an enlarged borehole.

Courtesy of SPWLA [8]

16.5 Later developments and summary

After the fundamental tool response and environmental database of the first *Schlumberger* spectrometric tool had been completed [14], a number of improvements involving signal processing and acquisition of additional experimental points have been made:

(a) The higher resolution of the short-spacing detection has been used to obtain sharper measurements.[2]
(b) The effects of rugosity have been investigated and a correction method developed [6].

The overall evolution of the *Schlumberger* wireline density logging tools from the 1960s to the 1990s is shown in *Table 16.5*.[3] The acronym PGT means Powered Gamma-gamma Tool. The Litho-Density Tool* is abbreviated by LDT and the PLATFORM EXPRESS* density tool by HILT. CDN* and ADN* stand for compensated density neutron and azimuthal density neutron tools.

---

[2] See Chapter 12.
[3] See [7] and [9] for the tools developed by other logging companies.

TABLE 16.5
EVOLUTION OF SCHLUMBERGER WIRELINE DENSITY TOOLS RESPONSE

Tool type	Tool response		Environmental effects		Remarks
	ρ_b (g/cm^3)	P_e	Correction	Input range	
PGT	2.0 to 2.7	None	Mudcake density ...... Mudcake thickness ..... Borehole effect	2.0 to 2.5 g/cm^3 0 to 3/4 in. 6 to 15 in.	
PGT	Same	None	Same as above plus matrix effect through cadmium window		Lower counts
LDT	1.72 to 3.1	1.3 to 6.0	Mudcake density ...... Mudcake thickness ..... Matrix/barite effect through reduced long-spacing counts Borehole effect	1.2 to 2.33 g/cm^3 0 to 1.5 in. 6 to 16 in.	Increased statistical variations
LDT	Same	Same	Same as above plus Matrix/barite effect through 3-window algorithm		
LDT	Same	Same	Same as above plus Heavy mud	ρ_m up to 3.3 g/cm^3	
LDT	Same	Same	Same as above plus Enhanced resolution Rugosity	0 to 0.5 in.	
HILT	1.0 to 3.051	1.0 to 6.5	Mud density .... $P_{e_{mud}}$ .... Mudcake thickness	1.0 to 2.34 g/cm^3 0.35 to 150 0 to 1.5 in.	
CDN	1.0 to 3.05	1.3 to 5.1	Matrix/barite effect Hole washout through rotational density		
ADN	1.7 to 3.05	1.0 to 10.0	Matrix/barite effect Hole washout through quadrants/sectors		

16. Density logging

The change of technology can be seen as a shrinking of the random (statistical) error. This is shown on *Fig. 16.8* where data from 294 wells, 147 logged with PGT tools, 147 logged with Litho-density tools is shown. The two histograms represent the distribution of data in the same homogeneous marker. While the difference in the mean density of the marker, as compared to the average value of the field, is 0.0226 g/cm^3 for data obtained with the PGT tools, it is only 0.0128 g/cm^3 for data acquired with Litho-density tools.

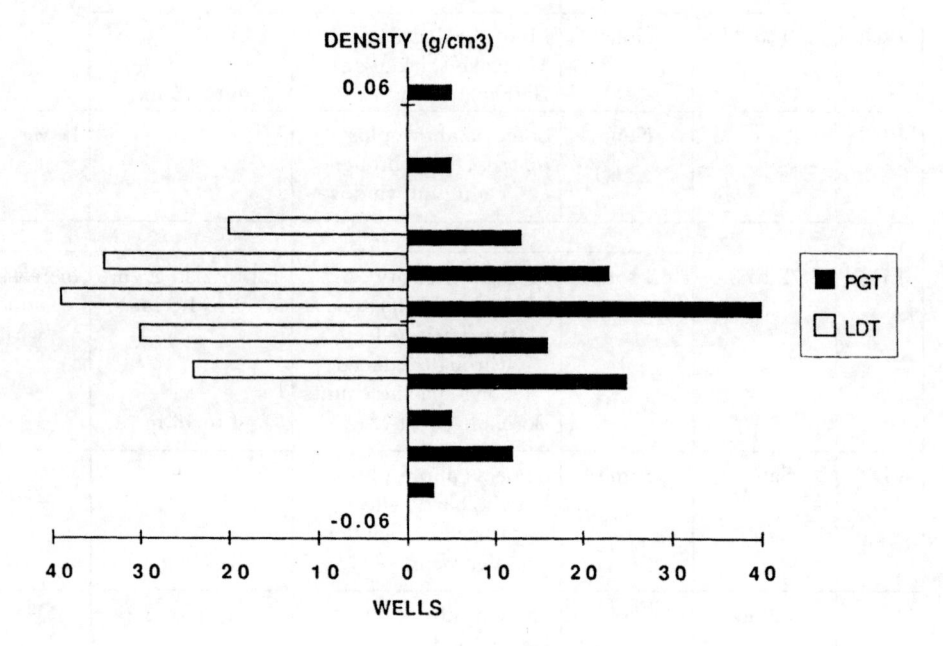

FIG. 16.8 : Reduction of random error with improved technology. The black bars correspond to the number of wells surveyed by PGT tools and whose mean density in the specified marker differs from the field average by the value indicated on the y-axis. The white bars correspond to wells surveyed by Litho-density tools. The distribution of the values supplied by Litho-density tools is sharper.

References

1 Bertozzi, W., Ellis, D. V., Wahl, J. S., "The physical foundation of formation lithology logging with gamma rays," *Geophysics*, Vol. 46, No. 10, 1981.

2 Best, D., Wraight, P., Holenka, J., "An innovative approach to correct density measurements while drilling for hole size effect," paper G, *Trans.* SPWLA 31st annual logging symposium, Lafayette, 1990.

3 Ellis, D., Flaum, C., Roulet, C., Marienbach, E., Seeman, B., "The Litho-Density tool calibration," SPE 12048, *Trans.* SPE 58th annual technical conference, San Francisco, 1983.

4 Evans, R. D., *The atomic nucleus*, McGraw-Hill, New York, 1955.

5 Eyl, K., Chapellat, H., Chevalier, P., Flaum, C., Whittaker, S., Jammes, L., Becker, A. J., Groves, J., "High-resolution density logging using a three-detector device," paper SPE 28407, *Trans.* SPE annual technical conference and exhibition, New Orleans, 9-1994.

6 Flaum, C., Holenka, J. M., Case, C. R., "Eliminating the effects of rugosity for compensated density logs by geometrical response matching," SPE 19612, *Trans.* SPE 64th annual technical conference and exhibition, San Antonio, 1989.

7 Gearhart, D. A., Mathis, G., "Development of a spectral density logging tool by use of empirical methods," paper Y, *Trans.* SPWLA 27th annual logging symposium, 1986.

8 Holenka, J., Best, D., Evans, M., Kurkoski, P., Sloan, W., "Azimuthal porosity while drilling," paper BB, *Trans.* SPWLA 36th annual logging symposium, Paris, 1995.

9 Minette, D. C., Hubner, B. G., Harris M., Fertl, W. H., "Field observations and test pit measurements of the accuracy of the Z-density gamma-gamma measurement," paper QQ, *Trans.* SPWLA 29th annual logging symposium, 1988.

10 Schlumberger, *The Litho-Density tool interpretation*, Paris, 1981.

11 Tittman, J., Wahl, J. S., "The physical foundations of formation density logging (gamma-gamma)," pp. 284-294, *Geophysics*, Vol. 30, No. 2, 1965.

12 Wahl, J. S., "Matrix effects in gamma-gamma logging," 1962.

13 Wahl, J. S., Tittman, J., Johnstone, C. W., Alger, R. P., "The dual-spacing formation density log," *Trans.* SPE 39th annual fall meeting, 1964. Also pp. 1411-1416, *Journal of Petroleum Technology*, Vol. 16, No. 12, 1964.

14 Watson, C. C., "Numerical simulation of the Litho-Density tool lithology response," SPE 12051, *Trans.* SPE 58th annual technical conference and exhibition, 1983.

15 Wraight, P., Evans, M., Marienbach, E., Rhein-Knudsen, E., Best, D., "Combination formation density and neutron porosity measurements while drilling," paper B, *Trans.* SPWLA 30th annual symposium, Denver, 1989.

17

Calibration

17.1 Definitions

The following definitions have been established by the ISO organization [3]:

Calibration (3.23[1]): The set of operations which establish, under specified conditions, the relationship between values indicated by a measuring instrument or measurement system, or values represented by a material measure or a reference material, and the corresponding values of a quantity realized by a reference standard.

What is a reference material?

Reference material (3.19): A material or substance one or more properties of which are sufficiently well established to be used for the calibration of an apparatus, the assessment of a measurement method, or for assigning values to materials.

The key words of the definition of calibration are listed with the corresponding terms used in logging:

(a) set of operations—calibration procedures
(b) specified conditions—calibration conditions
(c) relationship between values indicated by a measuring instrument and the values of a standard—calibration algorithm, calibration gain and offset
(d) reference standard—calibrator, calibration block, tank or loop.

---

[1]This number corresponds to the relevant section of the ISO document.

17.1.1 Misuse of the word calibration

The word **calibration** is used to represent other processes in a logging operation.[2]

(a) The **establishment of tool response**. Although the wording **tool response** is preferred, it is possible to use the terms **primary calibration**. The establishment of tool response has been reviewed in Chapter 13.

(b) An operational check of part or all the logging instrument in well-defined conditions (which may or may not represent realistic logging conditions). It is generally performed before and after the survey. It is now more commonly named **verification** and is described in Chapter 18.

(c) The transfer of calibration information to the surface acquisition system. This could be achieved by matching the surface electronics to well-defined downhole signals in order to scale the raw data sent by the downhole equipment. It is sometimes called **electronics calibration**. The preferred name is **surface system alignment**.

The next section summarizes the historical background that has created the confusion in the terminology related to calibrations.

17.2 Evolution of calibration methods

17.2.1 Logging calibrations in precomputer times

In the early logging era, one of the challenges was to transfer the information collected to define the tool response to a specific tool used in a specific well. The experimental setups used to establish the tool response are of large size and, most of the time, cannot be moved. Conversely, information calibration needs to be available at the wellsite. The transfer was performed through the use of successive n-ary standards, n an integer from 1 to 3.

(a) Primary standards relate to the set of accessories and to the procedure used to establish the tool response. Characteristics of these standards are as close as possible to the true formation characteristics. The objects involved are bulky and cannot be moved. They are located in special premises, typically in an engineering center.

(b) Secondary standards are found in the field, but not directly at the wellsite.[3] They are smaller than the primary standards, but still not very mobile. To this category belong the aluminum block used to calibrate the density tool and the water tank used to calibrate the neutron tool.

---

[2]In some cases, more than one definition covers the same operation. Putting rings of controlled dimensions around a caliper tool constitutes simultaneously a calibration, a verification and a surface system alignment.

[3]On large offshore rigs, it is possible to set up a special area where secondary standards can be gathered.

(c) Tertiary standards are found at the wellsite. They are small and easily transportable. The jigs for density and neutron calibration are tertiary standards.

The n-ary standards of a *Schlumberger* density tool are shown in *Fig. 17.1*.

PRIMARY CALIBRATION STANDARDS
8-in. water-filled boreholes

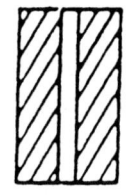

Vermont marble (ρ_b = 2.675 g/cm^3)

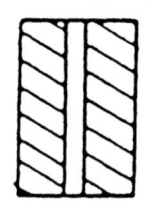

Bedford limestone (ρ_b = 2.420 g/cm^3)

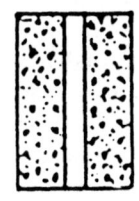

Austin limestone (ρ_b = 2.211 g/cm^3)

SECONDARY STANDARDS   TERTIARY STANDARD

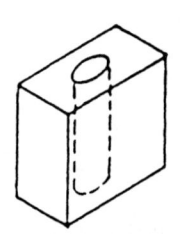

Aluminum calibration block

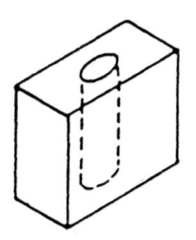

Sulphur calibration block

Field calibration jig (CALIBRATOR STRAPS ONTO TOOL HOUSING)

FIG. 17.1 : n-ary standards for the density PGT tool.

Courtesy of Schlumberger

The transfer of information was possible as the secondary standards were calibrated first in conjunction with the primary standards. In the field, secondary standards were used to calibrate tertiary standards. Finally, the tool was calibrated at the wellsite with the tertiary standard.

The manual transfer of information (keying in calibration gain and offset values into a memory) has eliminated the need of tertiary (wellsite) calibration standards. Digital transfer, which eliminates typing errors, has further improved the transfer of calibration information.

17.2.2 Surface system alignment

Before the 1980s, analog systems were used to produce logging data. Scaling of data was done by mechanically moving a galvanometer beam to intersect the location of the mechanical zero when no signal was received from the tool. Then a potentiometer was tuned to obtain a known deflection of the galvanometer when a suitable and controlled electrical signal was present. This electrical signal was also recorded as one of the steps of a calibration sequence and provided the final link between the tool response and the field recording.

Since the advent of computer technology, scaling of data has been performed in a different way. Raw data is sent uphole and recorded. For each raw channel, there corresponds a scaled channel. How are the raw and scaled channels related?

For each pair of raw/scaled channels, there are two tie points (*Table 17.1*). The low point, called zero point, corresponds to the absence of a signal from the tool. The high point corresponds to a well-defined signal.

TABLE 17.1
RAW AND SCALED VALUES

	Raw channel reading	Scaled channel reading
Low point.........	M_1	S_1
High point........	M_2	S_2
Other points......	M	S

The values M_1, M_2, S_1 and S_2 are collected during the surface system alignment. From these values, two parameters, G, the gain and O, the offset are computed:

$$G = \frac{S_2 - S_1}{M_2 - M_1}$$

$$O = S_1 - G_1 M_1 = S_2 - G M_2.$$

With G and O, scaled data, S, can be computed for every raw data point M:

$$S = GM + O.$$

The correspondence between raw and scaled data is shown graphically in *Fig. 17.2*.

17.3 Importance of calibration

Considering that tool response information can now be directly transferred in a secure way to a field surface system, why is calibration still needed [4]?

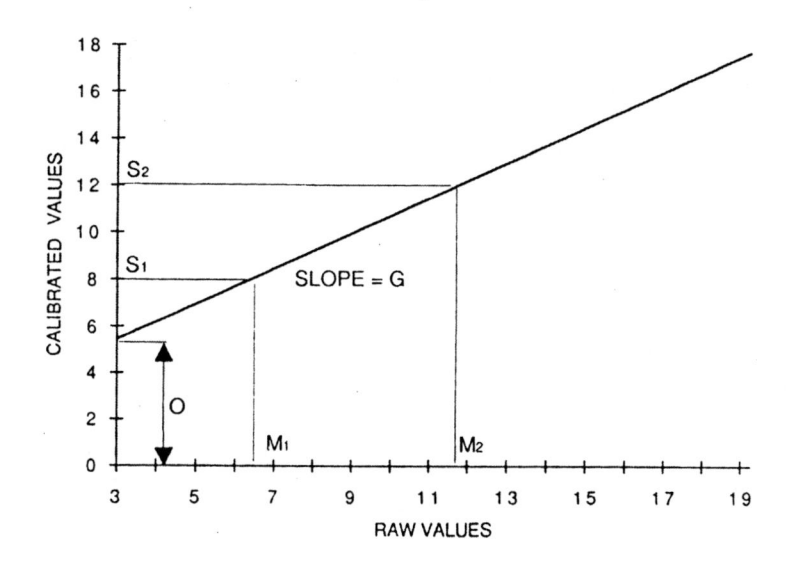

FIG. 17.2 : Principle of the surface system alignment.

17.3.1 Variation between tools

In spite of improved manufacturing processes, it is still not possible to build logging tools that are strictly identical. This statement is all the more applicable when the size of the equipment is large. MWD tools display a dispersion of characteristics even though exacting tolerances are used. These variations are compensated through the calibration process as the tools are put in reference conditions that are more constant than the tools themselves. It is not practical and time-effective to get each individual device through the tool response experiments that are applied to a limited pilot series.

17.3.2 Variation during the life of a tool

The characteristics of a specific logging tool vary throughout its life. Induction tool geometry varies with temperature, pressure and time. Consequently, its mutual inductance (compensated by the sonde error correction) also changes. The characteristics of nuclear devices detectors also evolve. The nuclear sources become weaker and the pads or stabilizers facing the formation may be thinned by wear. Frequent recalibrations are required for these reasons.

Failure to calibrate a logging measurement seriously jeopardizes the accuracy of the measurement.

17.4 Calibration requirements

Accurate calibrations rely on three requirements:
(a) quality calibrators
(b) well-designed calibration procedures
(c) controlled calibration environment.

17.4.1 Calibrators

(a) Calibrator size is important. Depending on the volume of investigation of the logging measurement, it is necessary to optimize the calibrator size so that interferences outside the calibrator are minimized.
(b) They still need the combined attributes of transportability and availability. An extremely accurate calibrator is of limited use if it is remote from operations, since tool transport could be the cause of failure and of systematic shifts from the calibration point.
(c) Just as experimental blocks are built, maintained and handled with utmost care, field calibrators must be treated as laboratory instruments. A damaged or unmaintained calibrator causes an erroneous calibration.
(d) Maximum effort should be deployed to reduce systematic error in calibrators, considering that calibrations will be undertaken by many people over the world. Tool positioning in calibrators is important. A misaligned tool will give incorrect readings. Jigs or guides can be used to give consistent readings despite the diversity of tools and operating crews. Unerasable marks, accurately indicating the position of add-on jigs, can be put on the tool during manufacturing.
(e) In some cases, the validity of the selection of the calibrator can be contested. A block of a given material may be very difficult to build in the shape of a calibrator. Field conditions should also be considered. A neutron calibrator sensitive to humidity (quite a few water molecules around a tool sensitive to hydrogen atoms) is not recommended.

17.4.2 Procedures

(a) Just as the inputs to the environmental corrections need to be defined accurately in order to take full benefit of these corrections, the conditions of the calibration need to be described with completeness and accuracy. For instance, characteristics of the fluid used in a neutron calibration tank need to be known (temperature, chemical composition, etc.).
(b) This completeness of information also concerns the hardware. In some cases, different components need to be matched to obtain accurate logs.

This match should be highlighted on the calibration record. On the contrary, if elements do not require matching, this should also be specified. For instance, for normal logging, neutron sources need not be coupled to a specific detector because of the subsequent ratio processing.

(c) Although they demand care and procedures comparable to laboratory experiments, field calibrations need to be simple. In other words, divine intervention should not be required to achieve decent standards.

(d) Shifts between successive calibrations are mostly deterministic. The chemical source of a nuclear tool has a particle emission decreasing with time. Lower counting rates should be expected. The sonde error of an induction device varies in a fairly predictable way. It is therefore essential that the information acquired during a calibration be compared to the previous information of the same type. If the trend does not correspond to that expected for the physics and design of the tool, then a tool malfunction can be considered. Extensive electronics and detector checks are then required.

(e) A good calibration requires time and cannot be rushed. Therefore, the right compromise for its frequency has to be established. It is better to do it less often but to ensure it is done correctly. The frequency should also be tuned to the rate of change of the equipment. For instance, it should be related to the half-life of a nuclear source[4] or to the rate of wear of a tool housing—or stabilizer—for contact devices. In any case, a new calibration is needed every time the tool is modified or repaired.

(f) Common sense indicates calibration points should not be selected far away from the range of values encountered in the well. For instance, a caliper tool run in a large hole, say 44.4-cm (17.5-in.) nominal bit size, should not be calibrated with 20-cm (8-in.) and 30-cm (12-in.) rings.

17.4.3 Calibration environment

The environment around the calibrator and logging tool needs to be controlled. Even if the calibrator is of reasonable size, the tool is often reading volumes located beyond the limits of the calibrator. For instance, sonde error may be improperly estimated if large conductive masses are near the calibration area. Empty space at distances greater than 10 m (30 ft) from the tool are required. This may be a serious problem offshore, where it is difficult to remove the induction logging string from large metallic masses.

The impact of the surrounding environment on calibrations is reviewed in Section 17.6.

---

[4] For instance, the counting rate reduction of a cesium source with a half-life of 30 a (yr) is 0.2% per month.

17.4.4 Control of calibration conditions

For some logging tools, parameters monitoring the conditions of the calibration are acquired in addition to calibration information. This approach has been developed for the *Schlumberger* density tool in detail.

The counting rates in the different energy windows of the long and short spacing detectors (*Fig. 17.3*) are checked.

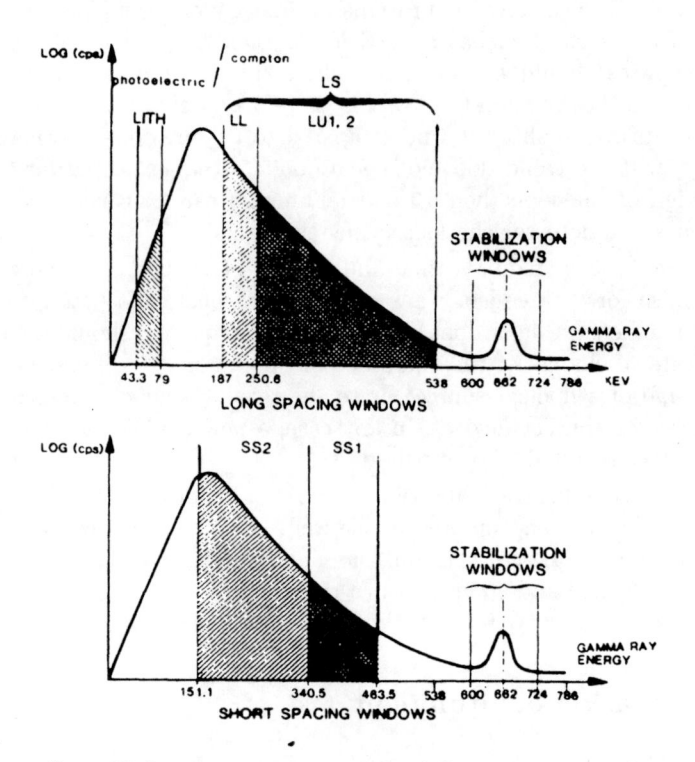

FIG. 17.3 : Counting rates of the different energy windows:
  a. N_{LL}: lower window, long spacing.
  b. N_{LU}: upper window, long spacing.
  c. N_{LITH}: very low energy window, long spacing.
  d. N_{SS2}: lower window, short spacing.
  e. N_{SS1}: upper window, short spacing.
Courtesy of Schlumberger

Five quality indicators[5] listed in *Table 17.2* are monitored:
(a) If QRLS decreases toward 0.55, the long-spacing detector is faulty.
(b) If QRLI decreases toward 0.22, the long-spacing detector is faulty.

---
[5]The subscript Al is added to the counting rates collected in an aluminum calibration block; the subscript Al + Fe is added to the counting rates acquired in the aluminum block plus iron sleeve.

(c) If QRSS starts to increase toward 0.85, the short-spacing detector is faulty.
(d) If QLIR deviates from 1.35, the tool positioning in the calibrating block is poor or the condition of the P_e sleeve is poor.
(e) Any deviation of QR away from 1 indicates a tilt of the tool in either direction inside the calibrating block.

In addition, the detector voltages are monitored with the following guidelines:

(a) An increase of 1°F is normally followed by an increase of 1 V.
(b) A change smaller than 10 V is tolerated between calibrations.
(c) A drop of 10 V is normal between the background and the block measurements.

TABLE 17.2
LITHO-DENSITY TOOL QUALITY RATIOS

Ratio	Value	Name	Tolerance
QRLS	N_{LL}/N_{LU}	Long-spacing spectrum	0.65 ± 0.05
QRLI	N_{LITH}/N_{LS}	Lithology spectrum	0.32 ± 0.06 (A or C) 0.35 ± 0.06 (D)
QRSS	N_{SS1}/N_{SS2}	Short-spacing resolution	0.72 ± 0.10
QLIR	$(N_{LITH}/N_{LS})_{Al}/(N_{LITH}/N_{LS})_{Al+Fe}$	P_e calibration	1.35 ± 0.05
QR	$(N_{LS}/N_{SS1})_{Al}/(N_{LS}/N_{SS1})_{Al+Fe}$	Density calibration	1.00 ± 0.02

Figure 17.4 gives an example of a calibration identifying a faulty tool before survey.

```
LDTC                    DETECTOR CALIBRATION SUMMARY

                MASTER    CALIBRATED
                BKGD      AL+FE     AL      UNITS
          LL    19.3      86.7      97.1    CPS
          LU    81.1      163.0     183.5   CPS
          LS    59.9      168.1     189.0   CPS
        LITH    5.8       28.6      42.0    CPS
         SS1    14.9      177.6     197.1   CPS
         SS2    10.0      254.5     280.5   CPS

                SPECTRUM QUALITY RATIOS
        QRLS    .5296
        QRSS    .7027   >  TOO LOW
        QRLI    .2227
        QLIR    1.3052
        QR      1.0129
```

FIG. 17.4 : Ratios out of specification during Litho-Density tool calibration.

17.5 Presentation of calibration

ISO (3.23.25) adds a note to the definition of calibration:
The result of a calibration may be recorded in a document, sometimes called **calibration certificate** or a **calibration report**. Such a document is often found as part of a logging print and is called a **calibration tail**.

17.5.1 Evolution of the calibration tail

Figures 17.5, 17.6 and *17.7* show the evolution of the calibration tail presentation in four stages: first, the panel generation; second, the first computer calibrations; third, a calibration at the end of the eighties; and finally the latest calibration tails. To make the comparison with standards easier, the tolerances on the different parameters are printed near these parameters.

The most difficult period for the log users has been when calibration reports included a crowd of figures with esoteric names or labels. Many inaccurate calibrations were undetected because of these complex and arcane formats.

17.5.2 Guidelines for calibration tails

(a) It is important that the terminology applied to calibration is rigorously followed. Verifications and operational checks should not be labelled as calibrations.
(b) The critical characteristics of the calibration have to be reported.
   1. What is the date?
   2. What is the reference?
   3. What are the calibration conditions and environment?
(c) The visual display of the calibration must be optimized to allow a rapid check of calibration validity. In that direction, considerable improvements have been realized thanks to computer graphics.
(d) The calibration generally contains a large number of figures related to raw data (among others, voltages, counting rates, etc.). It is imperative that the usable data be also shown.

Example

Calibration A is used for a density logging job. During the job, the wear plate is observed to be thinner than at calibration time. In a calibration B performed after the job, the counting rates of the density short spacing detector are reported to be higher by 4%. What does this drift mean in terms of density? It is recommended that the corresponding change in density be included in the tail of calibration B.

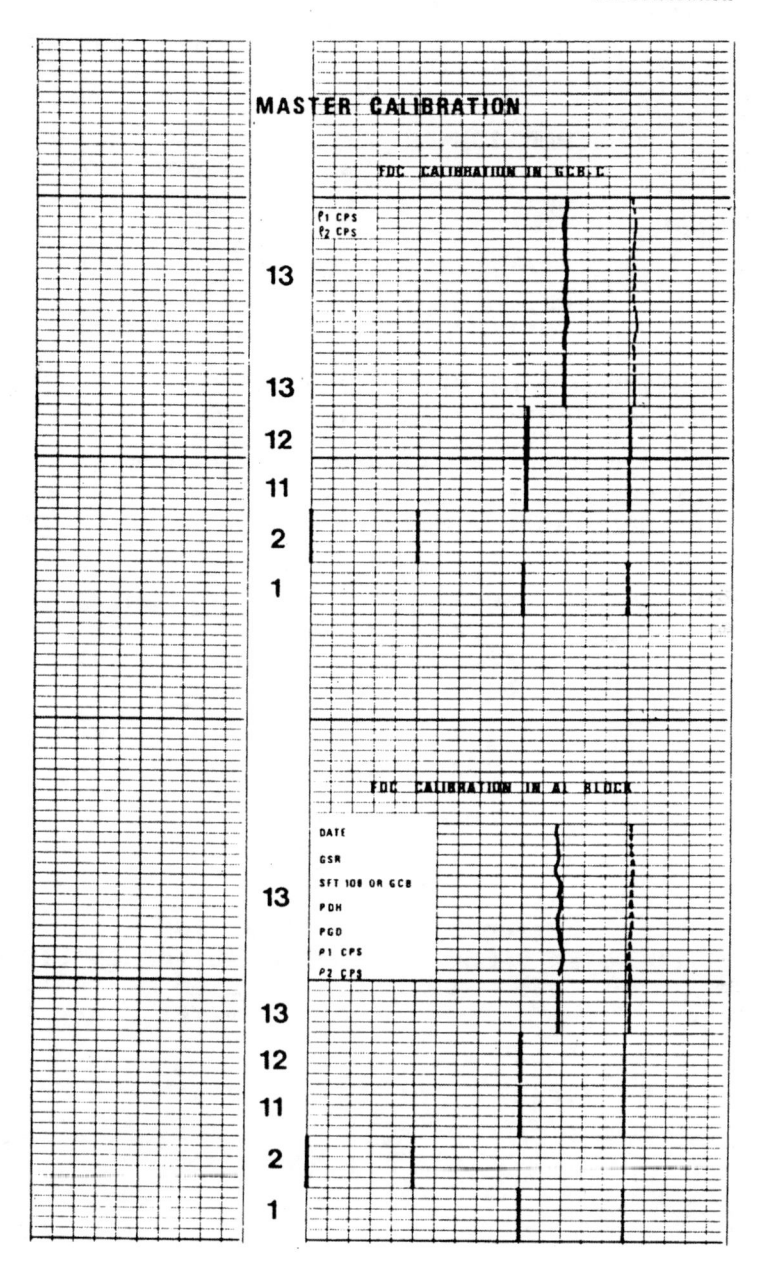

FIG. 17.5 : Evolution of the calibration presentation.
Panel and field recorder presentation. The film shows records made with the tool in the aluminum block and with a special attachment called the GCB-B. 1. Mechanical zero. 2. Recorder sensitivity. 11. Sensitivity check 1. 12. Sensitivity check 2. 13. Recording position. In this last position, statistical variations can be observed.

Courtesy of Schlumberger

SHOP SUMMARY

PERFORMED:

LDTC   DETECTOR CALIBRATION SUMMARY

```
DENSITY RESISTIVITY SONDE NUMBER        : 2714
NUCLEAR SERVICE CARTRIDGE NUMBER        : 938
POWERED DETECTOR HOUSING NUMBER         : 1854
POWERED GAMMA-GAMMA DETECTOR NUMBER     : 1869
LDT LOGGING SOURCE NUMBER               : 6778
LDT CALIBRATION MODE                    : WATE
```

MASTER CALIBRATED

	BKGD	AL+FE	AL	UNITS
LL	18.7	73.8	82.1	CPS
LU	72.9	113.4	125.4	CPS
LS	55.2	130.5	144.7	CPS
LITH	5.5	29.7	44.7	CPS
SS1	13.7	150.3	166.4	CPS
SS2	9.2	224.0	246.5	CPS

HV SETTINGS
HV LS: 1630.5 V
HV SS: 1357.8 V

SPECTRUM QUALITY RATIOS

	COMPUTED VALUE	NOMINAL VALUE	TOLERANCE
QRLS	.655	0.65	+/- 0.05
QRSS	.675	0.72	+/- 0.10
QRLI	.309	0.32	+/- 0.06
QLIR	1.357	1.35	+/- 0.06
QR	1.002	1.00	+/- 0.02

BKGD: AL: AL+FE:

CP 32.2

FIG. 17.6 : Evolution of the calibration presentation.
Computer-generated presentation with quality ratios.
Courtesy of Schlumberger

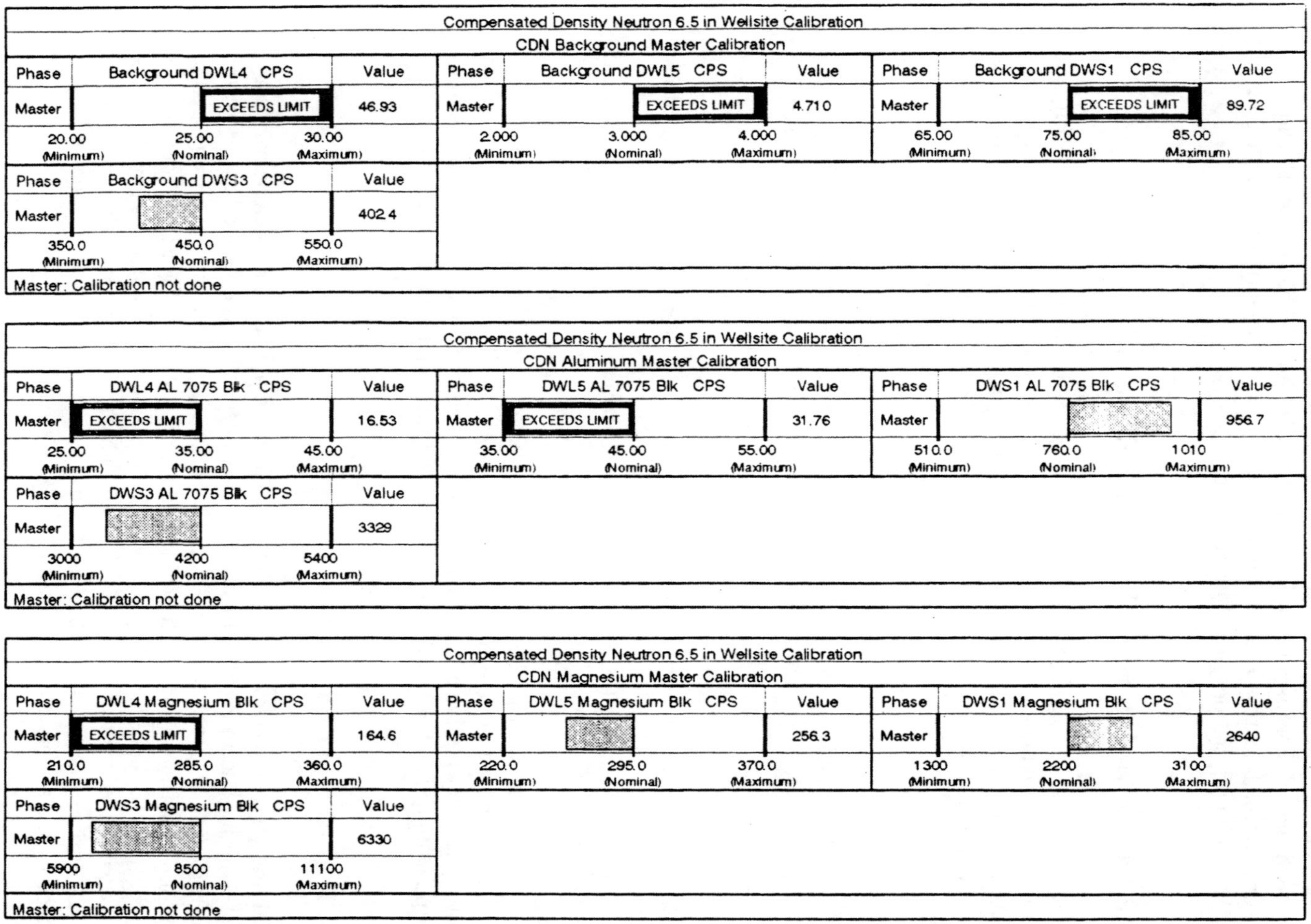

FIG. 17.7 : Evolution of the calibration presentation.
Graphical presentation: whenever a calibration step does not conform, a red flag is raised and includes the wording "Exceeds limit." In this black and white copy, nonconformity is spotted by the warning message.

Courtesy of Schlumberger

(e) It is also recommended that the external parameters influencing the calibration be displayed. For instance, it is known that the outside diameter of the housing of a thermal neutron tool has some influence on the counting rates. This value has to be printed on the calibration tail.

17.5.3 Calibration records in digital format

In many cases, calibration records are no longer limited to two values (gain and offset) per calibrated channel. Polynomial fit of raw and reference values enable more accurate measurements. As an example, quartz gauges are calibrated in a large combination of pressure and temperature conditions (*Tables 17.3* and *17.4*).

Exercice 18

1. Transfer manually the information contained in *Tables 17.3* and *17.4*. Measure the time taken for the transcription.

2. Check how many digits have been incorrectly transcribed.

It is therefore imperative to store and transfer data in a digital manner. Many logging devices are now able to store calibration data in downhole memory. This information is stored during the calibration task and retrieved when the tool is initialized at the beginning of a logging job.

In a similar way, all calibration information should be stored digitally and archived in field databases. This is the only way to be able to quality control these values and to performed corrections if needed. Quite a few logs have been salvaged because calibration coefficients have been checked to be incorrect and a calibration redone. This opportunity has been available because calibration information was at hand.

17.5.4 Meaning of tolerances

Calibration tolerances are set so that the calibrated measurement can be performed with the specified accuracy. In addition to compliance to tolerances, calibration values should follow anticipated trends. *Table 17.5* and *Fig. 17.8* represent a sequence of induction calibration records. The calibration performed on August 4 is out of line, although the corresponding values are within the absolute specifications [460-560]: the loop measurements of the medium and deep induction are different from the values observed during previous and subsequent calibrations. The log run on August 4 could be affected by a systematic shift introduced by the calibration.

TABLE 17.3
QUARTZ GAUGE CALIBRATION COEFFICIENTS

Pressure model: PCB: G999, $P = F(F_c, F_b)$, pressure matrix: 66, pressure CRC: F925; date: June 9, 2003. Temperature model, $T = F(Fb, Fc)$, temperature matrix: 66, temperature CRC: 0BE0.

F_b offset: 0.558800000000E+07 Hz; F_c offset: 0.514400000000E+07 Hz.

	F_b**0	F_b**1	F_b**2
F_c**0	+0.698203311801E+04	+0.112998846952E-01	-0.698531338632E-06
F_c**1	-0.106474337610E+01	-0.127934175768E-04	-0.943110975725E-10
F_c**2	+0.111747198295E-05	+0.483752214159E-10	+0.853194779107E-15
F_c**3	+0.263315902072E-11	0.0	0.0
F_c**4	0.0	0.0	0.0
F_c**5	0.0	0.0	0.0

	F_b**3	F_b**4	F_b**5
F_c**0	-0.891982948096E-10	-0.165479544634E-14	-0.236547679346E-19
F_c**1	+0.446562163382E-15	+0.291745572608E-19	0.0
F_c**2	0.0	0.0	0.0
F_c**3	0.0	0.0	0.0
F_c**4	0.0	0.0	0.0
F_c**5	0.0	0.0	0.0

	F_c**0	F_c**1	F_c**2
F_b**0	+0.104553592017E+03	-0.285199125965E-03	+0.660884134674E-08
F_b**1	-0.611514398184E-02	+0.193356184014E-07	+0.160171216942E-12
F_b**2	-0.329747809750E-07	+0.374663707711E-12	+0.168814770150E-18
F_b**3	-0.255693059683E-12	+0.740430357710E-17	0.0
F_b**4	-0.478116410824E-17	+0.191829305590E-21	0.0
F_b**5	-0.200247705415E-21	0.0	0.0

F_b**0 versus F_c**3: $-0.754467386034E-14$

TABLE 17.4
QUARTZ GAUGE CALIBRATION COEFFICIENTS

Clock model: PCB: G999; date: May 24, 2003.

Clock S/N : 113 (9804-0113), PCB: C319	
R offset	+0.109100000000E+01 Hz
F'_b/F'_c**0	+0.517514705943E+07
F'_b/F'_c**1	+0.191590197709E+05
F'_b/F'_c**2	+0.111495059749E+08
F'_b/F'_c**3	-0.932597037319E+10
F'_b/F'_c**4	-0.336253451462E+12
F'_b/F'_c**5	+0.161164400061E+14

TABLE 17.5
CALIBRATION RECORDS VERSUS TIME

	Master calibration summary					
Equipment..	IRT-R, IRC-234, DIS-567					
Date........	13-Jun-02	9-Jul-02	4-Aug-02	30-Aug-02	25-Sep-02	21-Oct-02
Engineer....	A. Jones	B. Smith	A. Jones	B. Smith	A. Jones	A. Jones
Sonde error						
ILM........	3.0	3.5	4.0	4.0	4.0	4.0
ILD........	7.5	7.0	7.0	7.5	7.0	7.0
Loop						
ILM........	496	497	504	497	497	497
ILD........	505	505	513	501	501	502

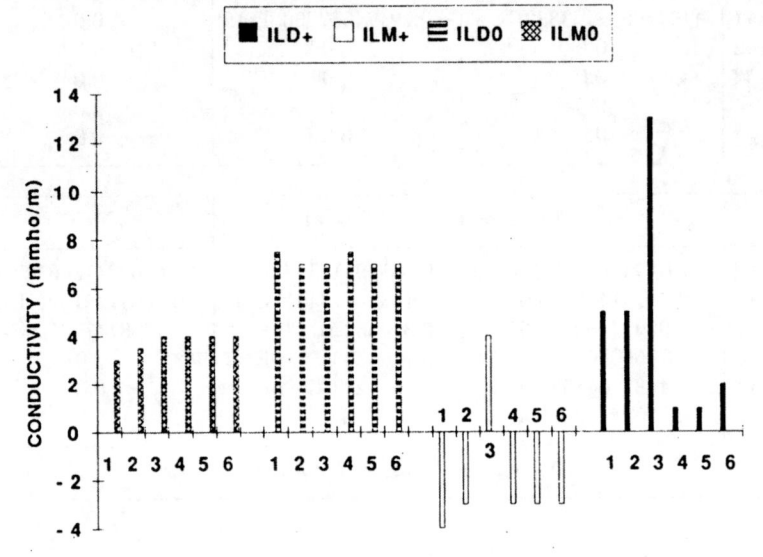

FIG. 17.8 : Variation of calibration values with time.
For the "plus" checks, the nominal value (500 mmho/m) has been subtracted from all observed values.

17.6 Error introduced by calibrations

Calibration introduces a constant error in the logging data. This error is **systematic** and cannot be estimated. It has to be avoided altogether.

Procedures need to be established to place a ceiling on potential systematic error. If tool design implies an uncertainty σ, then the ceiling on the calibration error should be a small fraction of σ. Therefore, an improved measurement not only represents a challenge to tool design but also a challenge in the creation of significantly improved calibrating procedures and equipment.

For instance, a generation of tools having a relative design uncertainty of 10% shows an upper limit of calibration error corresponding to 5%, or half the design uncertainty. If a new technology device having a 3% uncertainty is introduced, it requires a considerably better set of calibration procedures, corresponding to, say, 1.5% of the measurement, a three-fold improvement over the previous procedures.

17.6.1 Caliper calibration

The caliper is a common and important input to environmental corrections. When used with a dipmeter tool, it is a critical input to dip computation. An error of 1 cm (0.4 in.) in a 20-cm (8-in.) borehole with a 45° apparent dip results in an error on dip magnitude of 1.5°.

The caliper calibration, in appearance a trivial task, is in fact delicate. As an example, the calibration of the OBDT* Oil-Base Dipmeter Tool, developed by *Schlumberger* is performed in four steps:

(a) zero measurement without pad pressure
(b) zero measurement with maximum pad pressure
(c) zero measurement without pad pressure; this step enables quantification of pad pressure hysteresis
(d) plus measurement without pad pressure.

The acquisition software uses the caliper and pad pressure calibration values to correct the log. Variation of the caliper raw signal as a function of pad pressure is shown in *Fig. 17.9*.

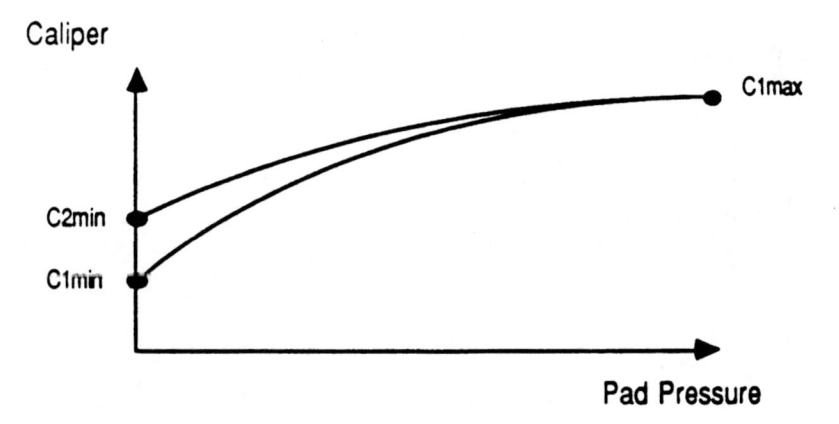

FIG. 17.9 : Pad pressure effect on caliper measurement.
Depending on pad pressure, the caliper sensor reads two voltages, C_{1min} and C_{2min}, although the caliper arms are opened in the same calibrating ring. An algorithm corrects for pad pressure hysteresis.
Courtesy of Schlumberger

17.6.2 Natural gamma ray calibration

The characteristics of a gamma ray tool change with time. The detector crystal may be slowly affected by crystal hydration. A field calibration is therefore required to normalize the response of all tools to API standards.

The gamma ray tool is calibrated with a portable jig that contains a small radioactive source or with a blanket containing some radioactive minerals. When placed a fixed distance from the axis of the tool, it produces a constant increase over the background counting rate. This increase is equivalent to a fixed figure in API units, depending on tool size and type.

To perform a correct calibration, the following points need to be addressed:

(a) When run in combination with other tools, special care needs to be exercised to perform the calibration with the same tool configuration as used during logging. This is because some logging tools inhibit the gamma ray transmission to avoid interference of gamma ray pulses with their own signals.

(b) The position of the tool during calibration must be monitored and reported. When the tool is horizontal and close to the ground or a large object, gamma rays are bounced back to the detector and counted. The influence of the distance to the floor and of the angle of the source holder as referred to the vertical is shown in *Fig. 17.10*. The tool should be at least 1.5 m (5 ft) above the ground if the tool is horizontal. It is recommended to calibrate the tool vertically whenever possible.

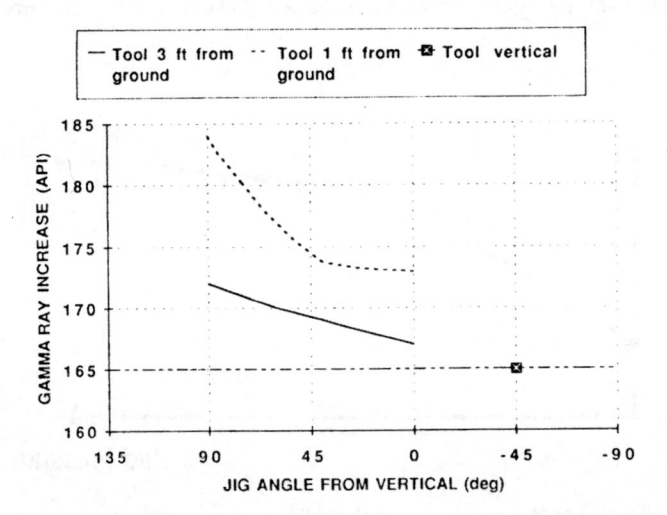

FIG. 17.10 : Influence of the position of the calibrating radioactive pill on gamma ray response.

Courtesy of Schlumberger

17.6.3 Induction tool calibration

17.6.3.1 Sonde error correction

Sonde error is the signal an induction tool would measure if it were suspended in a zero-conductivity medium. The sources of sonde error are the metallic parts used in construction of the tool, wires, coil wiring, heads, housings. The sonde design attempts to make this signal as small and as stable as possible. However, as the sonde ages, the characteristics and position of the metallic parts change and create a drift in sonde error. Temperature compensation of sonde error, achievable by hardware or software, is also an important factor.

Sonde error can be significant for an induction sonde measuring the X-component of the formation signal [1]. Most of the X-signal sonde error results from the sum of direct mutual inductances from the individual coil pairs in the induction array. Each direct mutual inductance is large, but the design of the array is such that the direct mutual inductance signals cancel one another. With pressure, the relative position of the induction coils changes slightly. It is not practical or economical to compensate these changes with a mechanical or electrical design. Therefore, it is necessary that during manufacturing, the behavior of a tool be characterized in a wide range of varying temperatures and pressures that reflect actual well conditions. From a modeling of the tool behavior and a number of controlled experiments, a set of correction coefficients is defined. Then, the acquisition software automatically adjusts the sonde error correction during logging using temperature from a downhole sensor and mud weight from a keyboard entry.

17.6.3.2 Field determination of sonde error

Accurate determination of induction sonde errors in the field has historically been a difficult and inexact art. There is no uniform method for determining sonde error correction. Some logging engineers use lifting machines while others measure sonde error with the tool on tall poles. There is nothing fundamentally wrong with either method. The major obstacle is accurately determining the signal from the surroundings. It is often assumed that the earth signal is small if the sonde is located a few feet from the ground. In fact, modeling shows that for high ground conductivities, a residual signal can still be sensed at large distances. *Table 17.6* displays the signal at 3 m (120 in.) for different ground conductivities. This signal must be subtracted from the apparent sonde error to determine the **true** sonde error correction.

An added complexity is that the signal from the surrounding environment may change with time. It depends on whether the ground near the facility is wet or dry, (it could change from season to season or even daily). Variations of 2 mmho/m of the same ground are not uncommon. Humidity may also affect a sonde recently used in salty water by creating conductive paths. The background also depends on vehicles

TABLE 17.6
BACKGROUND SIGNAL AT 3 m (120 in.)

Ground conductivity (mmho/m)	Earth signal (mmho/m)
10	0.5
50	1.0
100	2.0
200	3.0
500	5.0
1000	7.0
2000	9.0

Courtesy of Schlumberger

parked nearby. Typically, a 1m × 1m steel plate induces a 4-mmho/m signal at 3 m (10 ft) from the induction sonde. The problem of conductive surroundings is amplified offshore where large metallic masses are difficult to avoid.

17.6.4 Two-height sonde error determination

A method has been designed to eliminate the uncertainty linked to the background signal. Measurements of sonde error are performed with the sonde at two different distances from the ground. Dedicated calibration areas ensure the two positions of the sonde relative to the ground are carefully controlled. The difference of measurements is linked to the background signal and to the difference of geometries between the two setups. As these geometries are known, the background signal can be extracted (*Fig. 17.11*).

Sonde error measurements are made at 1.2 m (4 ft) and 2.4 m (8 ft). The difference is plotted on the x-axis of the correction chart and yields a 8-ft background signal on the y-axis. The background signal is subtracted from the upper measurement, which is performed with the sonde at 8 ft from the ground.

Sonde error is ultimately determined with an accuracy expressed in fractions of millimhos, typically 0.25 mmho/m to 0.5 mmho/m, while previous procedures yielded an accuracy no better than 1 mmho/m in the best conditions. This improvement could lead to significant resistivity and therefore saturation changes in high-resistivity beds logged in oil-base mud.

17.6.5 Neutron calibration

Detector sensitivity and efficiency of the thermal neutron logging tools may vary; in addition, the radioactive source strength decreases with time; a field calibration standard is required to normalize the response of all tools to the primary standards. The field calibrator commonly used is a water-filled tank [IX].

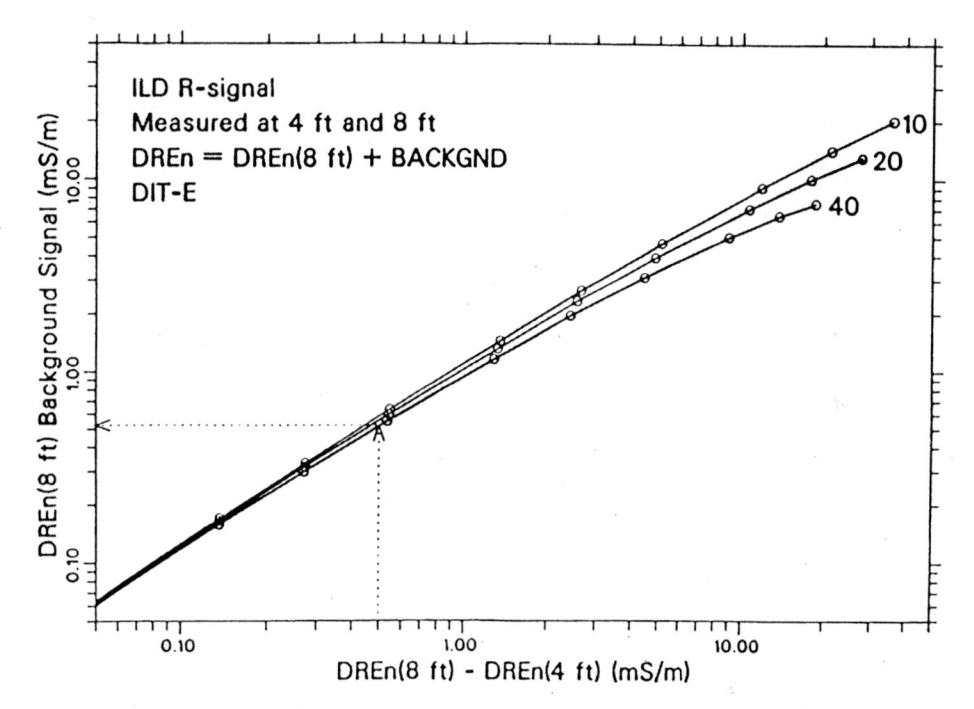

FIG. 17.11 : Correction chart of the two-height calibration method.

Courtesy of Schlumberger

Design requirements

(a) **Tool housing geometry**. Variation of housing size diameter from one housing to another and housing wear creates shifts on the calibration (*Fig. 17.12*). This effect can be removed if the housing size is measured accurately and a correction factor is computed.

(b) **Tool positioning in the shop calibrator**. Improper centering and vertical positioning in the neutron tank also cause calibration shifts (*Fig. 17.13*).

Fig. 17.14 shows how *Atlas Wireline* uses auxiliary rods to provide suitable positioning for different size sondes [VII and 2]. The small circles are neutron moderators.

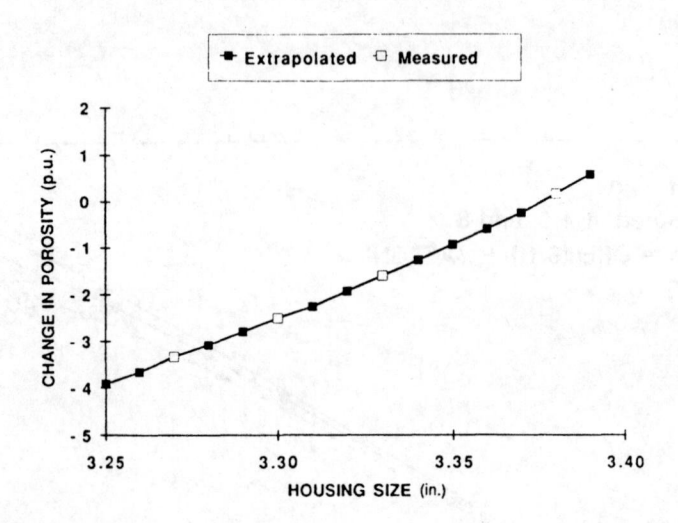

FIG. 17.12 : Effect of housing size on thermal neutron tool.

Courtesy of Schlumberger

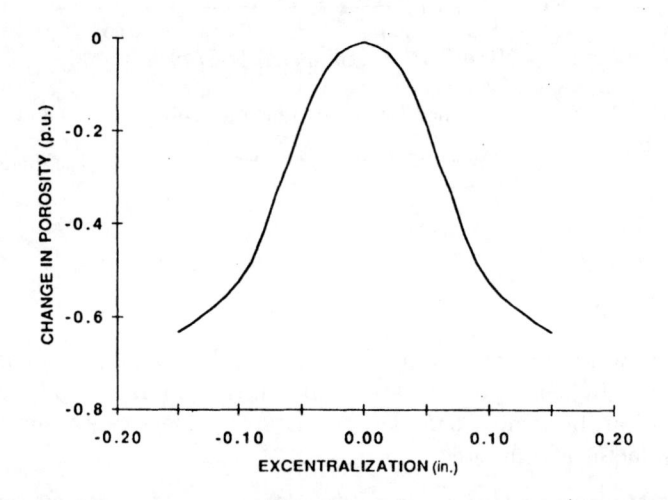

FIG. 17.13 : Effect of eccentering on the thermal neutron tool calibration.

Courtesy of Schlumberger

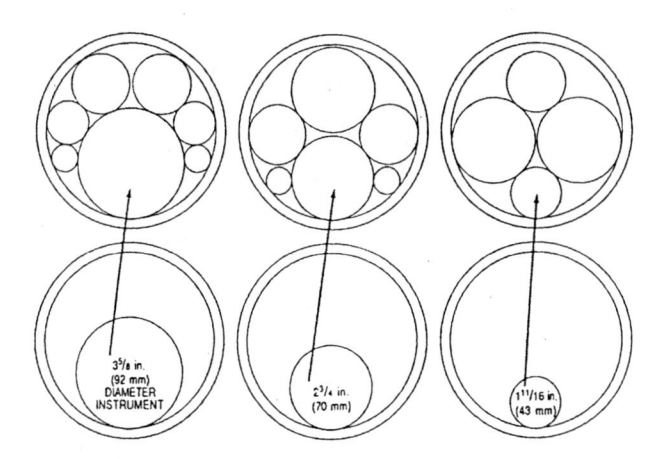

FIG. 17.14 : Setup to control the tool eccentering.
Different rods are used for different tool sizes.
Courtesy of SPWLA [VII]

(c) **Effects of calibration surroundings**. Proximity of a vertical wall or of the ground, and rain effects may also introduce systematic shifts in the calibration (*Fig. 17.15*).

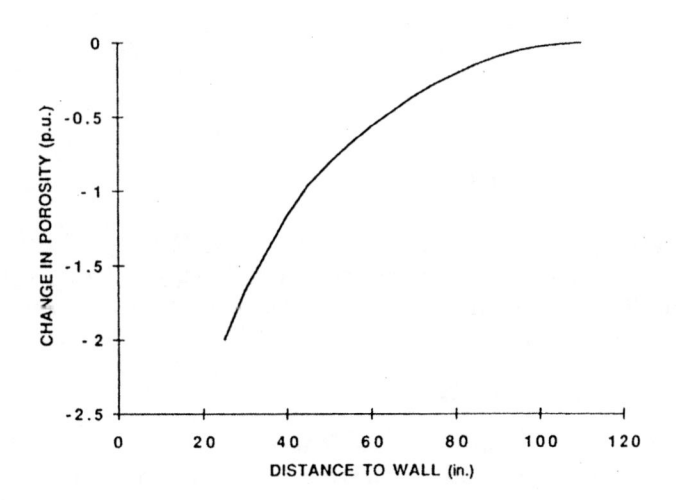

FIG. 17.15 : Variation of porosity with the distance of the calibrating tank to a wall.
Courtesy of Schlumberger

(d) **Tank fluid characteristics**: water purity, salinity, temperature and presence of antifreeze agents.
(e) **Acquisition system losses**: detector and amplifier dead time and transmission losses.

Deviations from correct procedures that introduce systematic shifts in the determination of the neutron porosity are listed in *Table 17.7*.

TABLE 17.7
EFFECTS OF POOR CALIBRATION PROCEDURES ON
THERMAL NEUTRON POROSITY

Effect	Measured porosity (pu)					
	27	28	29	30	31	32
Good calibration				▬		
Eccentered			▬			
90 cm (3 ft) from wall	▬					
Raining				▬		
Horizontal	▬					
2.5 cm (1 in.) too high				▬		
2.5 cm (1 in.) too low			▬			
70°C (160°F), no correction			▬			
4°C (40°F), no correction					▬	
85.3 mm (3.36-in.) housing, no correction					▬	
Seawater in tank				▬		
Jet fuel in tank				▬		

17.6.6 Density calibration

The master calibration of the density tool is necessary to correct for variations in the source strength, pad surface wear and detector efficiency. A correct calibration requires that pad tilt and surrounding fluid be carefully controlled.

17.6.6.1 Effect of tool setup

The alignment of the tool inside the calibration block is important. If the axes of the tool and the calibrator cavity are misaligned, different counting rates will be obtained, resulting in a shift of log density values. For instance, an error of ±1.2 cm (0.5 in.) would create an error of 0.003 g/cm^3 (*Fig. 17.16*). The potential misalignment is limited by a jig that limits the tool movement. Repeated measurements help in verifying that the density skid is properly positioned. Generally, it should be done three times, with the setup that produces the lowest counting rates being the best.

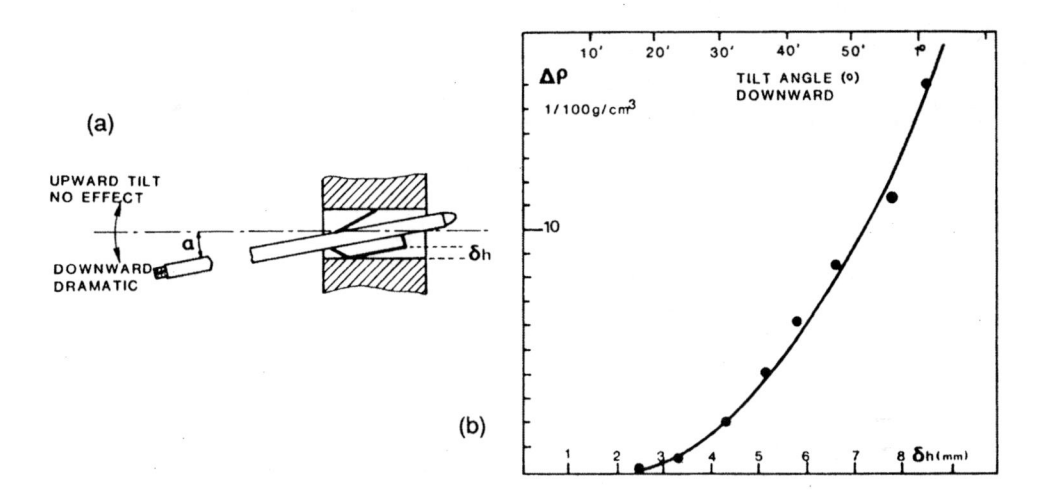

FIG. 17.16 : Effect of tilt on log readings.
a. Tool setup.
b. Chart of the effect as a function of the tilt angle.
Courtesy of Schlumberger

Vertical or horizontal calibration? That is the question. This selection has been investigated for *Schlumberger* Litho-Density tools. Repeated experiments with different tools and setups have proven that the difference does not exceed 0.003 g/cm^3.

17.6.6.2 Effect of fluid

The tool response of the density tool is established with blocks filled with water. Tools calibrated in these blocks are inserted in standard aluminum blocks shortly after calibration so that a reference point can be supplied for field operations.

Again water is used in the space between the block and the tool. It therefore makes sense to use a similar fluid in the field to perform the calibration. The principle that a field calibration should be simple could have been violated by the decision to use water. Special easy-to-use attachments have been designed for that purpose. Moreover, the use of water makes tool positioning less critical as the density contrast between water and aluminum is smaller than the one between air and aluminum. Noticeably, failure to use water introduces a systematic shift of 0.01 g/cm^3.

References

1. Barber, T., private communication, 1989.
2. *Dresser Atlas computerized logging service calibration guide*, Dresser Atlas, USA, 1986.
3. International Organization for Standardization, *Quality assurance requirements for measuring equipment*, ISO/DIS 10012-1, Geneva, Switzerland, 1992.
4. Theys, P., "Le log," *The Log Analyst*, Vol. 39, No. 6, 11-1998.

18

Monitoring of tool behavior

The response of a logging tool has been defined in the controlled conditions of an engineering center. The tool has been calibrated in the facilities of a field location. But now, the logging tool design is going to be really put to test. The tool is first transported to the wellsite, a sometimes gruesome experience on difficult roads or seas. Then, it is going to be subjected to high temperature, pressure, vibration and shock levels.

While many log users investigate thoroughly the correctness of the tool response and of the calibration, only a few question how the tool performs in field conditions. This chapter covers the monitoring methods of tool behavior between two master calibrations, and especially in the most critical phase, when the tool acquires downhole data. **Verifications** and **downhole monitoring** are respectively described.

18.1 Verification

The objective of the verification is to check that the tool is operating and whether its performance has drifted from the shop calibration. For practical reasons, this check cannot be performed over a large range of values. Typically, one or two points are checked. This control may involve signals generated

(a) internally in the surface acquisition system, in the cartridge or in the sonde
(b) by a special wellsite accessory, a verificator.

The verification controls the proper operation of part or all the logging system. The sonic verification performed in the past on analog equipment involved only the panel (or surface equipment) and mainly checked the proper functioning of analog-to-digital converters. In contrast, the caliper verification involves the complete chain from sonde to surface system and is strictly equivalent to a calibration.

Verification is often performed before and after the survey. The readings under control are always different (strictly identical readings should be considered with suspicion) and the difference is the drift, an assessment of tool stability.

Verification is more conveniently performed at surface. Signals similar to the ones recorded during logging can be produced. But at surface, the stability of the tool in realistic well conditions is not really checked.

18.1.1 Before-survey verification

The before-survey verification has two functions:

(a) **Transfer of calibration** information to the tool as it goes to the wellsite. This is achieved by using a secondary standard during the shop calibration and the before-survey verification. The secondary standard, called verificator, is of smaller size than the shop calibrator and can be moved to a wellsite location. Today, calibration information is often transferred digitally and the logging tool is not adjusted to match the verificator characteristics. "Calibrating" a logging tool with a verificator (i.e., adjusting gain and offset as the verificator is affixed to the tool) would lead to less accuracy as verificators are not "precision" instruments and have variable characteristics.

(b) **An operational check.**

In a before-survey verification, the date of calibration has to fit with logging company specifications. The tool should not be transported between the verification and the survey. A reasonable time limit between before-survey verification and the beginning of logging is 24 h.

The values of the before-survey verification should fall within the range specified by the logging company. Only a small drift, as compared with the master calibration, is tolerated.

The tool numbers listed in the verification record should match the ones shown in the log heading and the calibration tail. This requirement is lifted when some tool components are not critically matched together. This is the case of the nuclear source for neutron porosity logging.

Figure 18.1 shows an example of a compensated neutron tool before-survey verification.

18.1.2 After-survey verification

The after-survey verification checks that the tool has not drifted, in other words, that its response during the survey has not changed significantly. This is achieved by performing exactly the same tasks as for the before-survey calibration and determining the change of the values read by the tool. *Figure 18.2* is an example of after-survey verification for a compensated neutron tool.

```
CNTH                    DETECTOR CALIBRATION SUMMARY

    EQUIPMENT:

        NCCN (NEUTRON COMPENSATED CARTRIDGE #):  2692
        NSSN (NEUTRON SOURCE SERIAL #):          1288
        TNHN (THERMAL HOUSING #):                3257
        TCNB (THERMAL CALIBRATOR NEUTRON BOX #): 1341


            INPUT    PLUS       SHOP      SHOP      BEFORE    GAIN
                     REFERENCE  TANK      JIG       JIG
                     COUNTS     COUNTS    COUNTS    COUNTS

            CNTC     6031.00    6257.23   2852.82   2855.30   .964
            CFTC     2793.00    2758.08   1249.83   1264.21   1.013


            RATIO    2.159      2.269     2.283     2.259


 BACK: 30-OCT-88 10:15    JIG: 30-OCT-88 10:22    COMP:
```

FIG. 18.1 : Before-survey verification.
The conformance check consists in the verification of the **before jig counts**. They should not differ from the **shop jig counts** by more than a specified percentage. The tools listed should match with the tools reported in the shop calibration tail, except for the source.

Courtesy of Schlumberger


```
                    AFTER SURVEY TOOL CHECK SUMMARY

    PERFORMED:    03-NOV-88 13:41
    PROGRAM FILE: GTSAP   (VERSION    29.850A 00/00/00   88/02/12)


    CNTH                    TOOL   CHECK

            INPUT      BEFORE JIG    AFTER JIG

            CNTC       2855.30       2841.45

            CFTC       1264.21       1234.33


 CHANGE IN THERMAL POROSITY AT 20 PU IS    .654  PU
```

FIG. 18.2 : After-survey verification.
before jig and **after jig** counts are compared. The difference is checked against a specified tolerance.

Courtesy of Schlumberger

18.2 Need for downhole performance monitoring

Surface conditions are so different from downhole conditions that the development of downhole checks is recommended. *Figure 18.3* indicates the temperature variations to which the detectors and electronics are submitted.

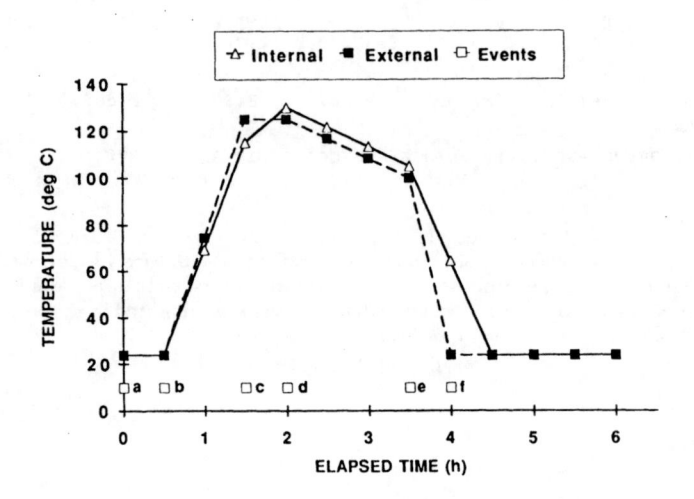

FIG. 18.3 : Evolution of logging tool temperature during a survey. The logging tool is submitted to significantly different temperatures during calibration and during logging. The events correspond to the starting times of the following operations:
a. Verification before survey.
b. Tool run in the hole.
c. Repeat section.
d. Main log.
e. Downhole verification. Tool pulled out.
f. Surface verification.

Courtesy of SPWLA, adapted from [IX]

Verifications within tolerances performed before and after survey do not guarantee a good log. The tool may drift with temperature downhole, or even fail intermittently, and come back to a normal status at surface. Because of the complexity of the equipment and the frequent lack of obvious geological markers that can be used to check tool response downhole, some independent verification of logging tool behavior is needed. More curves are becoming available that are not used directly for formation evaluation, but for checks of tool performance. The $\Delta\rho$ curve (*Fig. 18.4*) is the prototype of a curve of this type. Interpreters are familiar with the information provided by $\Delta\rho$ even though it is seldom used quantitatively for formation evaluation.

Quality curves have been developed for many modern tools. The first type of checks investigates the validity of the measurement individually. In this category, the curves associated with the *Schlumberger* thermal neutron, Litho-Density, spectral gamma ray and Phasor*/AIT* Array Induction Imager and *Halliburton Logging Services* Compensated Natural Spectrometric Gamma Ray tools are described in the following sections. The second type of check is analyzing the coherence of the collected information. Array induction, MWD density-neutron[1] and sonic fall in this category.

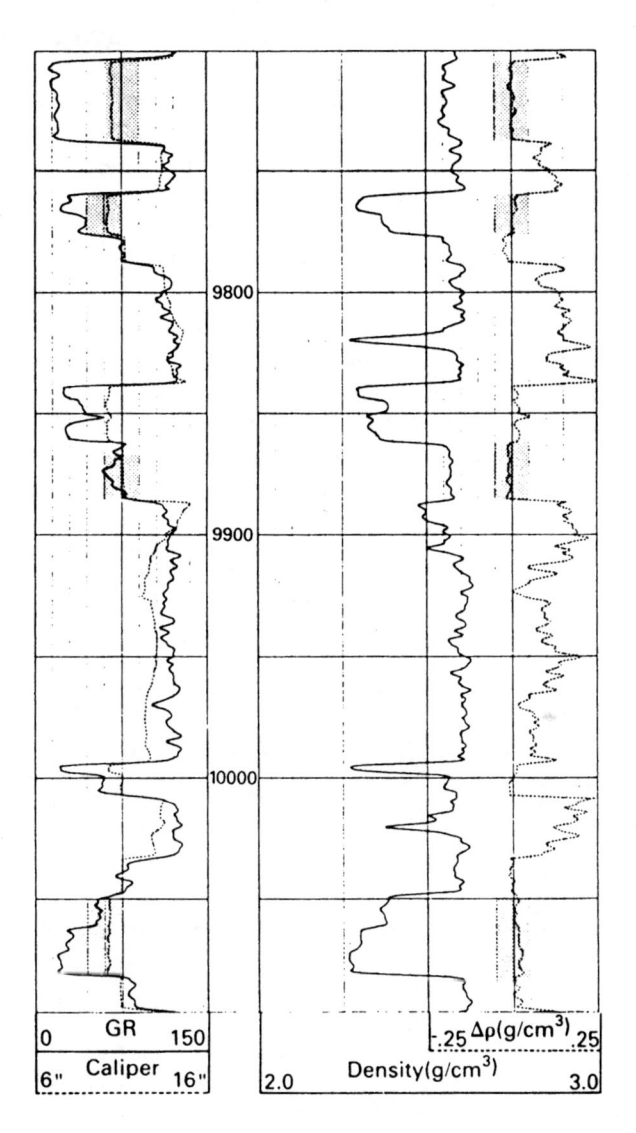

FIG. 18.4 : Presentation of the density correction.
The $\Delta\rho$ curve identifies smooth section of borehole. A rule of thumb indicates that, above 0.15 g/cm^3, the confidence in the ρ_b curve is significantly reduced.
Courtesy of Elsevier [V]

---

[1]MWD quality curves may be indexed to time instead of depth.

18.3 Individual downhole checks

18.3.1 Thermal neutron quality control diagrams

The downhole electronics of the thermal neutron porosity tool designed by *Schlumberger* can be affected by two failures that would not be recognized during a verification sequence: double pulsing of the detector electronics and intermittent spiking. These two occurrences are detected by the observation of a parameter, the instantaneous near-detector recalibration factor [IX and 3].

In proper operating conditions, this parameter (TALP) is equal to unity. Slight shifts may be observed if the logging environment changes or if the tool is logged with a source different from the one used in shop calibration. This shift would not affect the validity of the log whose final output, neutron porosity, derived from the ratio of counting rates, is independent of the neutron source.

Counter overflow or double pulsing, on the contrary, would result in a much stronger shift of TALP. The failure diagnosis is better performed on a plot (*Fig. 18.5*).

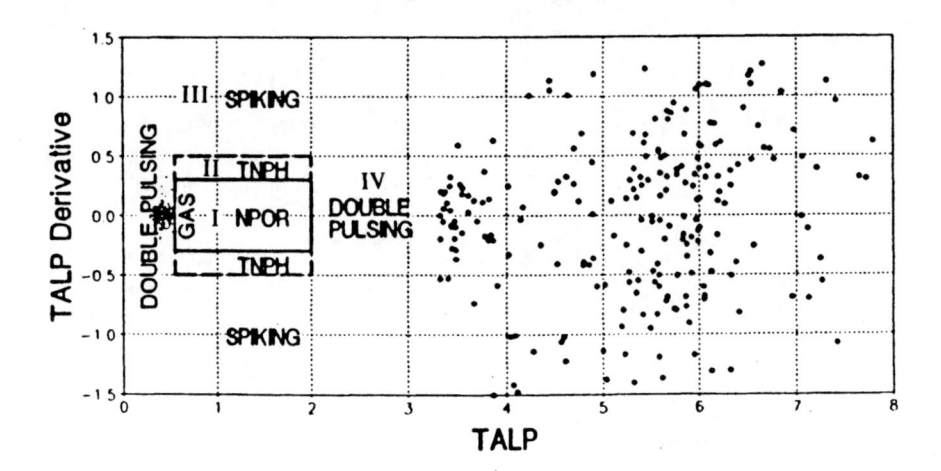

FIG. 18.5 : TALP diagram for thermal neutron porosity.
TALP itself is shown on the x-axis while the derivative of TALP versus depth is recorded on the y-axis. The plot is divided into four zones:

    a. Zone I: Standard and enhanced resolution porosity processing are possible.

    b. Zone II: Only standard porosity can be derived.

    c. Zone III: Spiking is detected.

    d. Zone IV: Double pulsing is detected.

Courtesy of SPWLA [IX]

18.3.2 Litho-Density tool quality control curves

Tool quality indicators are monitored during calibration (Section 17.4.4). As the tool is run in the hole, additional parameters are monitored (*Table 18.1*). The indications below are used to diagnose anomalies in the tool behavior.

TABLE 18.1
LITHO-DENSITY TOOL QUALITY RATIOS

Ratio	Value	Name	Tolerance
QRLS	$(N_{LL}/N_{LU}) \times [1 - 0.06 \times (2.6 - \rho_b)]$	LS spectrum	0.65 ± 0.05
QRSS	$(N_{SS1}/N_{SS2}) \times [1 + 0.20 \times (2.6 - \rho_b)]$	SS resolution	0.72 ± 0.10
QLS	$QRLS - (N_{LL}/N_{LU})_{Al}$	LS spectrum	0 ± 0.025
QSS	$QRSS - (N_{SS1}/N_{SS2})_{Al}$	SS resolution	0 ± 0.025
FFLS		LS form factor	
FFSS		SS form factor	
LSHV		LS high voltage	
SSHV		SS high voltage	

(a) **Detector performance.** QRLS and QRSS ratios are weakly sensitive to formation properties. They reflect the downhole performance of the detectors.

(b) **Crystal resolution.** The QLS and QSS ratios monitor the downhole resolution. If QLS < 0, the resolution of the long-spacing detector is degrading. If QSS > 0, the resolution of the short-spacing detector is degrading.

(c) **Crystal hydration.** When the crystal is already hydrated during calibration, the QLS curve drifts positively. The QSS curve drifts negatively. The high voltages decrease with temperature, by 50 to 150 V. High temperature dehydrates the crystal and improves resolution.

(d) **Barite effect.** Counting rates in the LL and SS2 energy windows are sensitive to photoelectric absorption. Therefore, the QLS and QSS curves correlate with the P_e curve. QLS decreases in high P_e zones. QSS increases in high P_e zones.

Figure 18.6 shows an example of Litho-Density quality curves while logging.

18.3.3 Natural gamma ray spectrometry control curves

The derivation of thorium, potassium and uranium weight percents from a spectral gamma ray tool requires rigorous control of the alignment of the photomultiplier energy windows. Logging companies have developed diagnostic methods allowing a positive check of the proper functioning of the device.

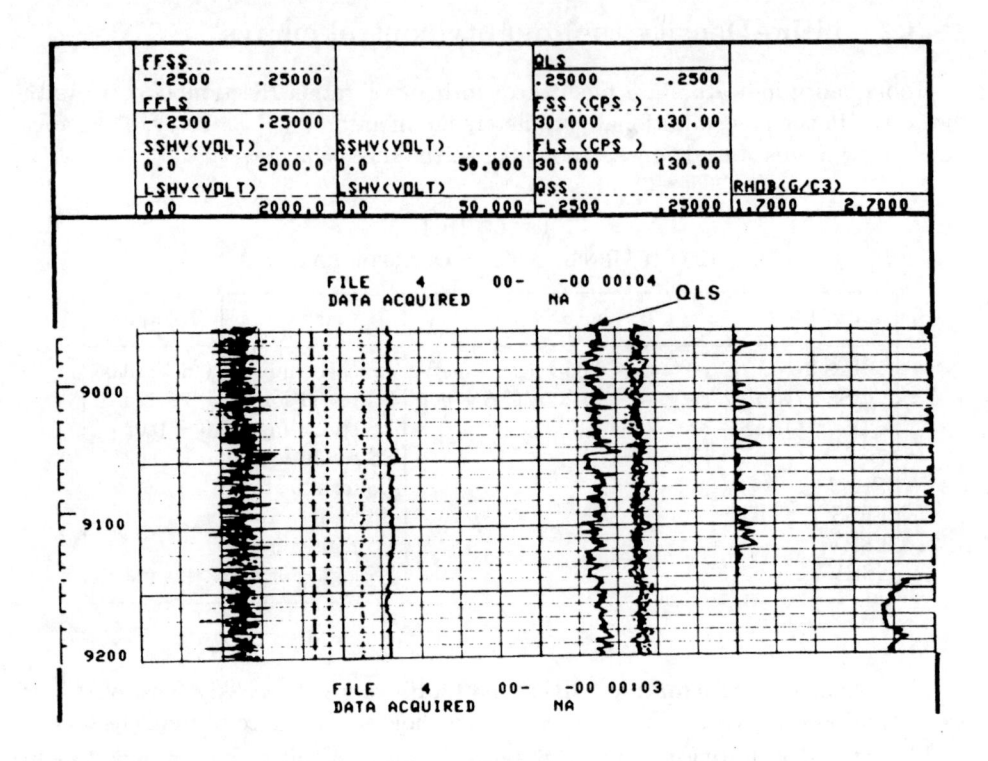

FIG. 18.6 : Example of Litho-Density quality curves monitoring.

Courtesy of Schlumberger

Halliburton Logging Services tool

Halliburton Logging Services Precision Logging System provides quality curves for the Compensated Spectral Natural Gamma Ray log (*Fig. 18.7*):

(a) The "Fit-err" curve is the normalized fitting error for observed counting rates relative to calculated concentrations. It indicates how well the computer program applied a weighted least-squares fit to the measured data.

(b) The "Americium" curve is the smoothed counting rate accumulated in an energy channel close to 60 keV, the energy of the stabilization source. It monitors the detector and stabilization system operation.

(c) "Noise" is the counting rate accumulated in an energy window below the photoelectric window. It monitors the amount of noise in the photomultiplier and in other system electronics.

(d) "Uranium S. dev" represents the estimated uncertainty for the uranium concentration expressed in parts per million. Similarly, "Thorium S. dev" and "Potassium S. dev" are the uncertainties for the thorium and potassium concentrations, respectively expressed in parts per million and weight percents.

(e) The "Source factor" is the fractional contribution of the potassium counting rate to the total counting rate. It is used to correct the openhole lithology ratio for source dependence.

(f) The "Square root spectra" is the square root of the counting rates per channel of the Lo and Hi spectra—respectively 0 to 350 keV and 0 to 3000 keV summed over a 10-m (30-ft) depth interval and normalized. This information verifies the gain stability and helps recognize noise.

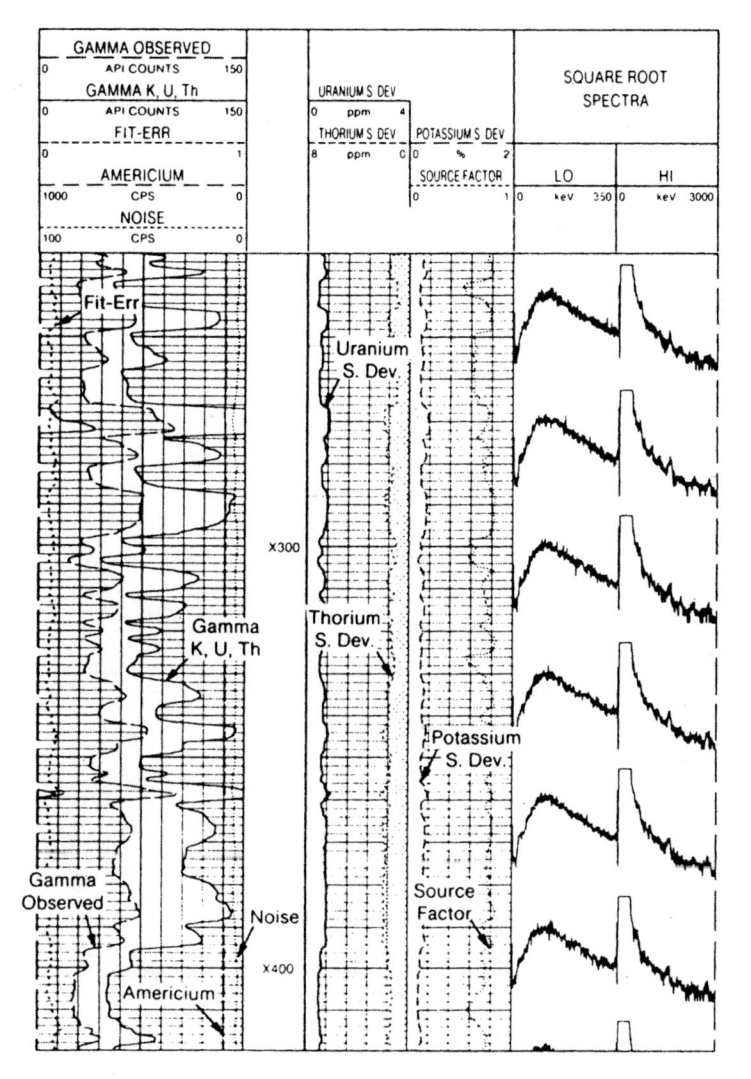

FIG. 18.7 : Compensated Spectral Natural Gamma Ray log provided by the Precision Logging System.

Courtesy of Halliburton Logging Services [1]

Schlumberger tool

The *Schlumberger* log provides four monitoring curves:

(a) LQCL and LQCU. These two curves monitor changes in lower and upper stabilization window counting rates, relative to those observed during the calibration. These two values should be close to zero with no systematic trend above 1.0.

(b) CHIS. It monitors the departure of the five-window counting rates from those observed in the primary standard formation. CHIS should be less than 6 for 95% of the log.

(c) DHVF monitors the detector voltage. Instantaneous variations should not be larger than 2 V and temperature-related variations should not be larger than 2 V/°F.

Figure 18.8 is an example of CHIS histograms.

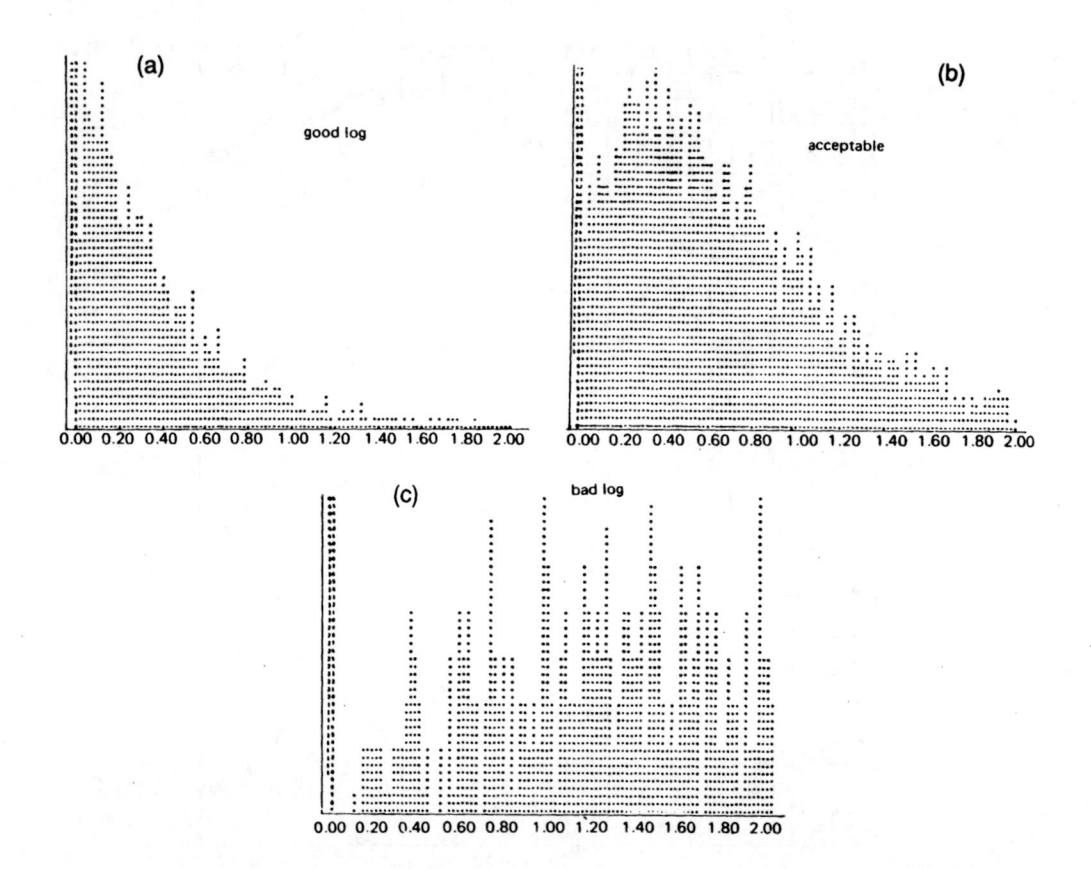

FIG. 18.8 : CHIS histograms.
a. Good. b. Acceptable. c. Log affected by tool failure.
Courtesy of Schlumberger

18.3.4 Induction control curves

Dual induction tool

Log quality flags for the induction appear during logging whenever the difference between auto calibration at successive 15-cm (6-in.) increments exceeds tolerance levels. Flags should appear at the bottom of the log before the tool actually moves in the hole. *Figure 18.9* shows the flags before pick-up.

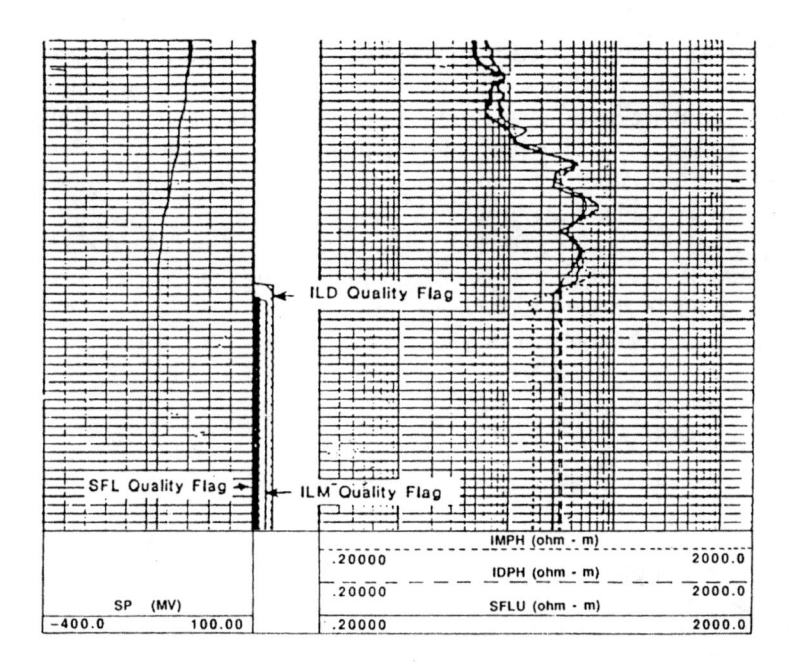

FIG. 18.9 : Induction log quality flags.
The *ILM* and *ILD* quality flags are raised before tool pickup.
Courtesy of Schlumberger Educational Services [2]

18.4 Consistency checks

With the advent of multisensor (detector banks, receivers, transmitters, buttons) tools, a new approach to data consistency can be implemented.

Multisensor tools yield multiple measurements of similar formation characteristics with some degree of independence. If the tool response characteristics are known to a reasonable degree of completeness and accuracy, this independence can be used to generate a quality indicator.

After perturbing factors such as the environment and formation heterogeneity are minimized, a statistical comparison of the information from the different sensors is made. This is analogous to comparing logs recorded with different tools of the same type over the same logging interval or recording the same parameter with different passes. However, for a multiple sensor tool, the comparison data is acquired simultaneously.

18.4.1 Array induction tool

This check has been designed from the data acquired with *Schlumberger* AIT [6], which contains eight induction arrays with spacing ranging from few inches to several feet. From a combination of arrays, transmitter frequencies and use of in-phase and quadrature signals, 28 conductivity measurements can be inferred. These different measurements still produce coherent and predictable patterns, while deviation from some basic patterns indicates a problem.

This approach has been successfully used to detect wellbore fluid contamination, displaced standoff, incorrect sonde error, standoff mounted incorrectly, array malfunction (broken ground), incorrect entry of mud salinity causing an inaccurate borehole correction and incorrect tool centralization (*Fig. 18.10*).

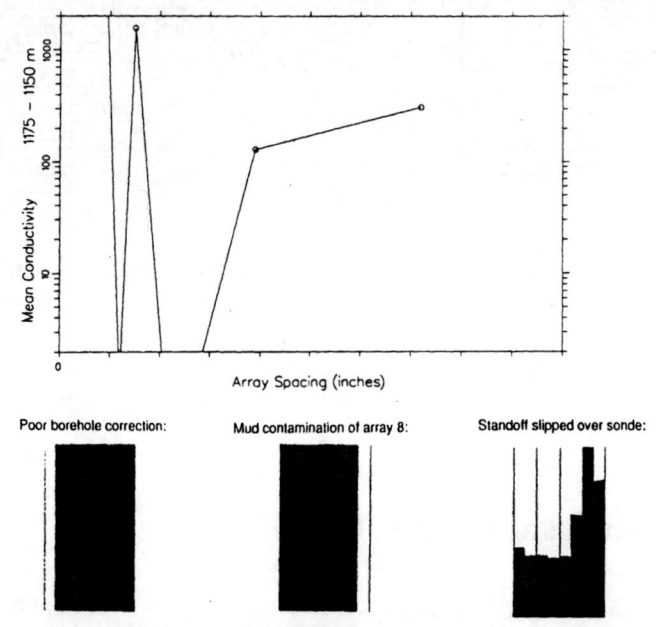

FIG. 18.10 : Multiple sensor consistency check.
    a. Metal components of a standoff displaced onto the array section of the sonde produce error signals in several arrays.
    b. Presentation of the array monitor.
Courtesy of Schlumberger [6]

18.4.2 Multidetector nuclear tool

MWD nuclear tools are equipped with several detectors to azimuthally scan the formation. The counting rates of the different tubes should be similar. *Figure 18.11* shows an example of a tool with a partial failure.

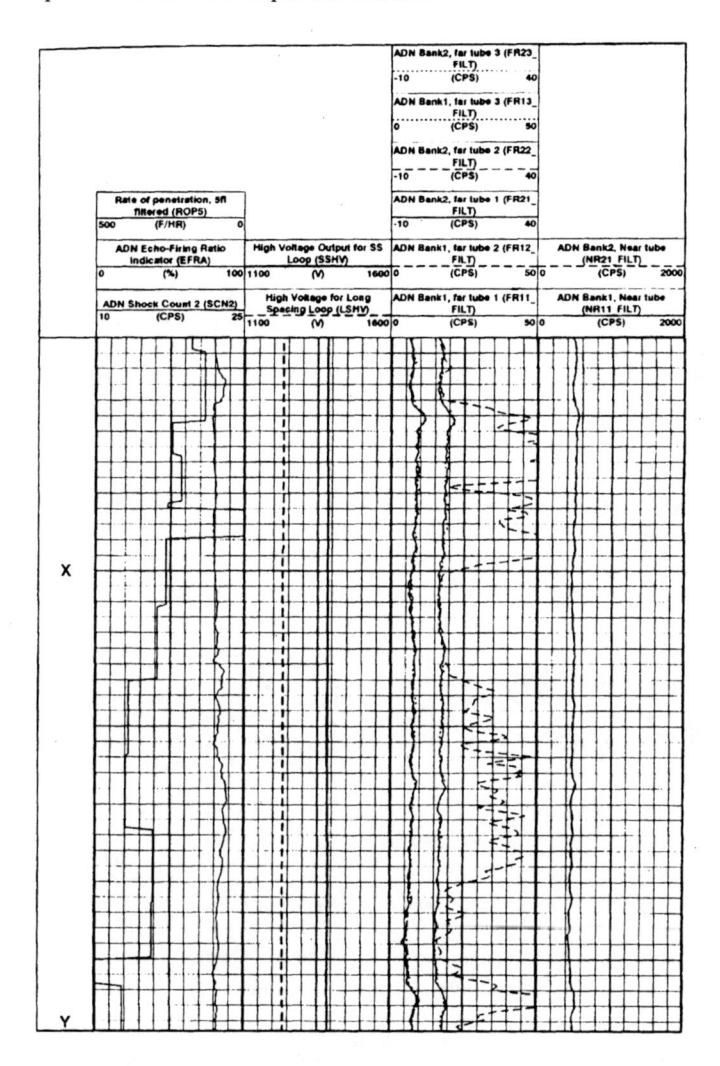

FIG. 18.11 : Partial failure of a multidetector neutron tool.
In a correct log, all neutron bank counting rates (FRxx) should track. In this specific example, curves FR12 and FR22 overshoot the other FRxx curves. Tube 2 is malfunctioning. A correct, although less precise log, can be computed after removing the counting rate contribution of tube 2.
Courtesy of SPWLA [8]

18.4.3 Coherence plots in the frequency domain

Measured data sampled in the depth domain can be transformed in the frequency domain. The coherence [4 and 5] and phase of the signals are analyzed to determine the limiting frequency beyond which all data relates to noise (*Fig. 18.12*).[2]

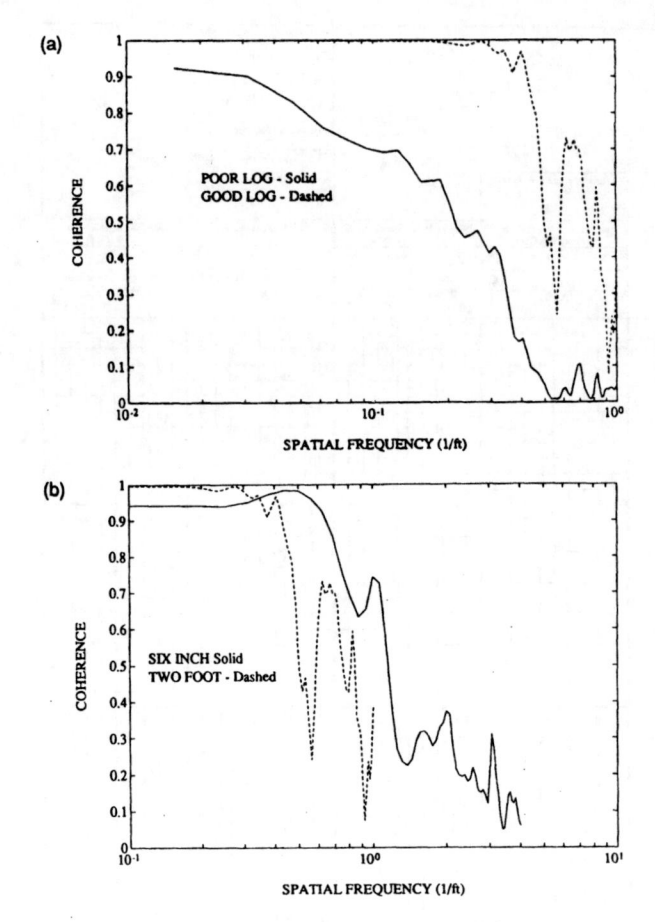

FIG. 18.12 : Coherence analysis. *Courtesy of Schlumberger.*

a. Coherence of two sonic logs recorded in the same well. The solid curve was logged at 30 m/min (6000 ft/h) with a lot of cycle skipping. Poor coherence is shown. The dashed curve consists of several passes at 12 m/min (2400 ft/h) with little or no skipping. The coherence decreases substantially at a 0.6-m (2-ft) wavelength, which corresponds to the 2-ft span of the sonic tool.

b. Analysis of the same interval logged with the 15-cm (6-in.) *Schlumberger* digital sonic tool. The data is recorded with a 3-cm (1.2-in.) sampling rate and is shown in solid with the log of *Fig. 18.12a*. The coherence is extended and remains in phase below 1 ft (0.3 m).

---

[2] The inverse of this frequency is the effective vertical resolution.

18.5 Integrated control monitoring

Quality control curves associated with a selection of petrophysical parameters have been described. For the single measurement ρ_b, there are eight quality control curves. For the three elemental curves obtained from a natural spectral gamma ray device, a similar number of controls is necessary. The multitude of checks that have to be implemented precludes an efficient use of them when a large combination of sensors is run simultaneously.

The task becomes formidable if it is requested in real time. The use of the controls as detailed above is then confined to a postmortem analysis as the quality curves are displayed in a playback mode.

The objective of a recent software development by *Schlumberger*, called LQMS for log quality monitoring system, is to provide the logging engineer with the necessary support to monitor control curves in **real time**. Any anomaly is analyzed in conjunction with information from other sources. The result is submitted to the engineer who can then decide whether to rerun the log, call for another tool or proceed with logging.

The "stripe" presentation (*Fig. 18.13*) summarizes the findings of the surface system [7]. It has the advantage of displaying flags on the main log, along the curves that can be used for formation evaluation. The user can refer to the complete listing if one of the flags is raised in a zone of interest.

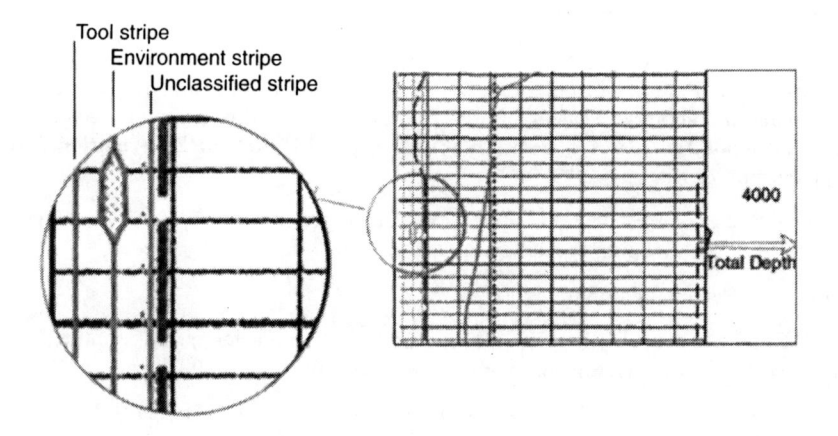

FIG. 18.13 : Stripe presentation.
Track 1 and depth track are shown on the right of the figure. Three diagrams at the left of Track 1 indicate possible anomalies in tool function, environmental problems and other factors. Anomalies are present when the diagram takes the shape of a sausage. They are also listed with a detailed explanation in a summary at the tail of the log.

Courtesy of Schlumberger [7]

18.6 Conclusions and recommendations

A display of tool health indicators is generally not part of the standard package delivered by the logging company. Many log users are intimidated by this information that has not the same appeal as density-neutron porosity crossovers or increasing resistivities. Still the quality control curves are available upon request. They constitute a valuable document that confirms the validity of the log and should be an integral component of a log package.

As it is not convenient to peruse through two prints simultaneously, the development of quality flags that summarize control information and are displayed on the main log should be part of the log product definition.

There are many instances released by oil and service companies only with great reluctance, when a reasonable looking log is used and later is invalidated by a detailed analysis of the quality curves record. Log users should definitely not take the shortcut of leaving quality curves information aside.

References

1 *Compensated Spectral Natural Gamma Ray*, Welex, A Halliburton Company, Houston, 1988.

2 Dewan, J. T., *Log quality control course*, Schlumberger Educational Services, Houston, 1987.

3 Galford, J.E., Flaum, C., Gilchrist, W. A. Jr., Duckett, S. W., "Enhanced resolution processing of compensated neutron logs," SPE 15541, SPE 61st annual technical conference and exhibition, New Orleans, 1986. Also, pp. 131-137, *SPE Formation Evaluation Journal*, Vol. 4, No. 2, 6-1989.

4 Kerford, S. J., Georgi, D. T., "Application of time series analysis to wireline logs," paper M, *Trans.* CWLS logging annual symposium, 9-1987.

5 Flaum, C., Personal communication, 1990.

6 Head, E. L., Seydoux, J., Perkins, J., "A new technique for log quality control," paper K, *Trans.* SPWLA 15th European logging annual symposium, 5-1993.

7 Burnett, N., Jeffries, J., Mach, J., Robson, M., Pajot, D., Harrigan, J., Lebsack, T., Mullen, D., Rat, F., Theys, P., "Quality," *Oilfield Review*, Vol. 5, No. 4, 10-1993.

8 Theys, P., "Le log," *The Log Analyst*, Vol. 39, No. 5, 9-1998.

19
Measurement of depth

> Depth is one of the most fundamental of petrophysical parameters. Its accurate measurement and the evaluation of its uncertainty under different conditions are vital to petroleum engineering decisions.
>
> *Steve Kimminau*

Although depth is generally not highlighted as an item in the logging company invoice, it is essential to log validity. Logs would have little value if all depth numbers were scratched off the film or print or if depth indices were removed from databases. While most log measurements are documented with calibration tails, parameter tables, etc., the characteristics of the depth measurement are poorly traced.

19.1 Importance of depth

The **logging depth** is the distance along the hole between a reference point (rotary table, bottom sea level, mean sea level, ground level, etc.) and a given point downhole, located on the logging tool. Logging depth has different applications for which different accuracy requirements are needed:

(a) **Absolute depth** is an essential input to the determination of the location of the reservoir, along with the directional data that allows a conversion to **true vertical depth**.[1,2]

---

[1] Absolute depth is also important for the logging company as, generally, the charges to the oil/gas company are a function of the deepest measured depth reached.

[2] Directional survey data is reviewed in Chapter 20.

Absolute depth comes into play when several wells are involved. Accurate knowledge of depth is required to answer questions about hydraulic continuity, hydrocarbon volumetrics, location and characteristics of faults, etc.

To evaluate the impact of an error on absolute depth, a simple field geometry can be assumed. It includes five wells, four at the extremities of a square of side a, one in the center of the square. The well being surveyed is the one at the center. The oil height is H. The error on the depth of the top and bottom of the reservoir on the surrounding wells is assumed to be nil. The error δh on the depth would be transmitted to the volume[3] as $(1/3)a^2 \times \delta h$.

Numerical application

$a = 500$ m. $H = 100$ m. $\delta h = 3$ m. $\delta V = 0.25 \ 10^6$ m^3 or around 1% of the total rock volume.

(b) **Relative depth** involves two depth horizons (for instance, the top and bottom of the reservoir) and is used to evaluate reservoir thickness. If the absolute depths of these horizons are affected by the same systematic error, the thickness is still computed correctly. Relative depth may be affected by depth shifts (e.g., cranking) performed in the interval between the two horizons,

(c) **Correlated depth** is the depth obtained by correlating one logging curve with another one, called the reference, assumed to be correctly positioned in depth. This is achieved through the use of depth-matching programs explained in Section 19.5. Failure to perform depth-matching could lead to serious errors on the evaluation of the formation petrophysical parameters because the inputs to this evaluation would relate to formations at different depths.

(d) **Differential depth** is obtained by monitoring the depth difference observed at the same point (e.g., a radioactive bullet) in a well, but at different times. The most important application is the monitoring of subsidence. An accuracy of a few tenths of an inch is needed for this purpose. This requirement is met through special methods [6].

19.1.1 Standards of depth control

There are four specifications related to depth:

(a) Absolute depth accuracy. Publicized values are of the order of 1 m (3 ft) at 3000 m (10,000 ft). A value of 0.3 m (1 ft) is often advertised. Field case studies have shown that one part per thousand is a more realistic value [9].

(b) Depth match between logs of the same type, run over overlapping intervals. The standard is expressed as follows: *Depth should be within* **x** m *(***x'** *ft) of the previous log in the same well.*

---

[3]Corresponding to a pyramid of side a and height δh.

(c) Depth match between curves acquired with different tool strings: *Logs in the same suite are within* **y** m *(***y'** ft*) of each other.*
(d) Depth match between curves acquired with a single tool string: *Curves recorded on the same tool string are correctly memorized and within* **z** m *(***z'** ft*) of each other.*

19.2 How is depth measured?

19.2.1 Wireline depth measurement

The classical way to determine depth is to measure the amount of cable going in or out of the hole at surface. Some amount of correction is then implemented.

The measurement of depth seems trivial. However, it is a demanding operation that requires the establishment of stringent procedures:

(a) marked cable
(b) properly calibrated and functioning tension device on surface, but, also, preferably downhole
(c) properly calibrated and functioning measuring wheel
(d) appropriate computer interface and software
(e) proper establishment of a reference depth
(f) control and stability of the setup.

The following factors affect the accuracy of the depth measurement performed with a wireline cable:

(a) human factors
  - setting zero; resetting after activation of heave compensator
  - reading stretch charts; using correct algoritm and software
  - cranking
(b) surface setup stability
(c) cable
  - elastic stretch
  - inelastic stretch
  - thermal stretch
  - pressure effects
  - twisting and rotating
(d) tension distribution and measurement
  - friction
  - tool sticking
  - cable keyseating
  - borehole tortuosity
(e) measuring wheel accuracy.

Table 19.1 includes some typical values of the potential error contributed by these different factors for a heptacable [5]. A depth of 3050 m (10,000 ft) is assumed.

TABLE 19.1
FACTORS AFFECTING DEPTH: WIRELINE

Factor	Potential error (m)	Potential error (ft)	Remarks
Elastic stretch	2.4	8	
Inelastic stretch	1.8	6	First run only
Temperature	0.6	2	300°F at BHT
Cable twisting	0.6 to 1.5	2 to 5	1 ft for each 50 turns
Measuring wheel	1 to 3	3 to 10	Current generation
Tidal effects	1.5	5	Floating vessel
Sheave movement	0.6 to 1	2 to 3	
Viscous drag	0.6 to 1	2 to 3	
Tool sticking	up to 12	up to 40	
Yo-yo	0.6	2	

The behavior of the wireline cable is further studied in the next subsections. The length of the logging cable is affected by

(a) **Static effects:** the behavior of the cable when the logging string is not moving.
(b) **Dynamic effects:** the behavior of the cable when the tool is moving. Dynamic effects are superimposed on static effects.

19.2.1.1 Static behavior

The length of a static cable is a function of many variables [7], including

(a) cable mechanical tension
(b) cable temperature
(c) hydrostatic effects on the cable
(d) cable age.

Obviously, the first three parameters vary with depth. It is therefore necessary to analyze the behavior of a segment dl of the cable. The following relation can be written:

$$dl = dl_0(1 + S)$$

where dl_0 is the length of the cable under reference conditions and S is the stretch. The stretch coefficient incorporates all the effects that increase or decrease cable length.

The **hydrostatic pressure** on the cable can be related to true vertical depth and mud weight, but its effect on the cable is difficult to assess. It should be further studied.

The **age** of the cable has a significant impact on its length. The layers of a new cable are not completely stabilized unless the cable is submitted to artificial aging just after manufacturing, and, in any case, before field usage. Considerable variation of cable length can be observed if such a procedure is not followed. The length of an aged cable would stay reasonably constant under similar conditions of tension and temperature.

In other words, the static behavior of a cable can be reasonably represented as a function of two variables, tension and temperature:

$$S = (E_0 + \epsilon\theta)T + (\alpha_0 + \beta T)\theta,$$

in which T is the tension and θ is the temperature.

Some typical values of the coefficients E_0, α_0, β and ϵ for two of the most common logging cables (seven conductor and monoconductor) are in *Table 19.2*.

TABLE 19.2
CABLE CHARACTERISTICS

Cable type	E_0 (ft/ft/lbf)	α_0 (ft/ft/°F)	β (ft/lbf/°F)	ϵ (ft/ft/°F/lbf)
Heptacable	$7.5\ 10^{-7}$	$-3\ 10^{-6}$	$-0.06\ 10^{-9}$	$1\ 10^{-9}$
Monocable	$20.0\ 10^{-7}$	$6\ 10^{-6}$	$0.50\ 10^{-9}$	$2\ 10^{-9}$

Numerical example of the computation of the correction

A seven-conductor cable is subjected to the respective average tension and temperature of 6000 lbf[4] and 120°F.

The derived stretch coefficient is $4.14\ 10^{-3}$. This means that at 3050 m (10,000 ft), failure to apply a correction would imply an absolute depth error of 12.5 m (41 ft).

In practice, charts (*Fig. 19.1*) are used in the field to correct depth. Computer generated stretch correction capability at the wellsite would provide a more accurate and efficient procedure [9].

19.2.1.2 Inputs needed for the static correction

For static correction, ϵ and β are neglected because the corrections they introduce are of the second order. Then E_0, α_0, T and θ are the only inputs to the stretch correction. For better accuracy, T is assumed to be uniform along the cable length:

$$T = \frac{T_{\text{surface}} + T_{\text{bottom}}}{2}.$$

---

[4]To obtain a tension of 6000 lbf at 3050 m (10,000 ft), high deviation is required.

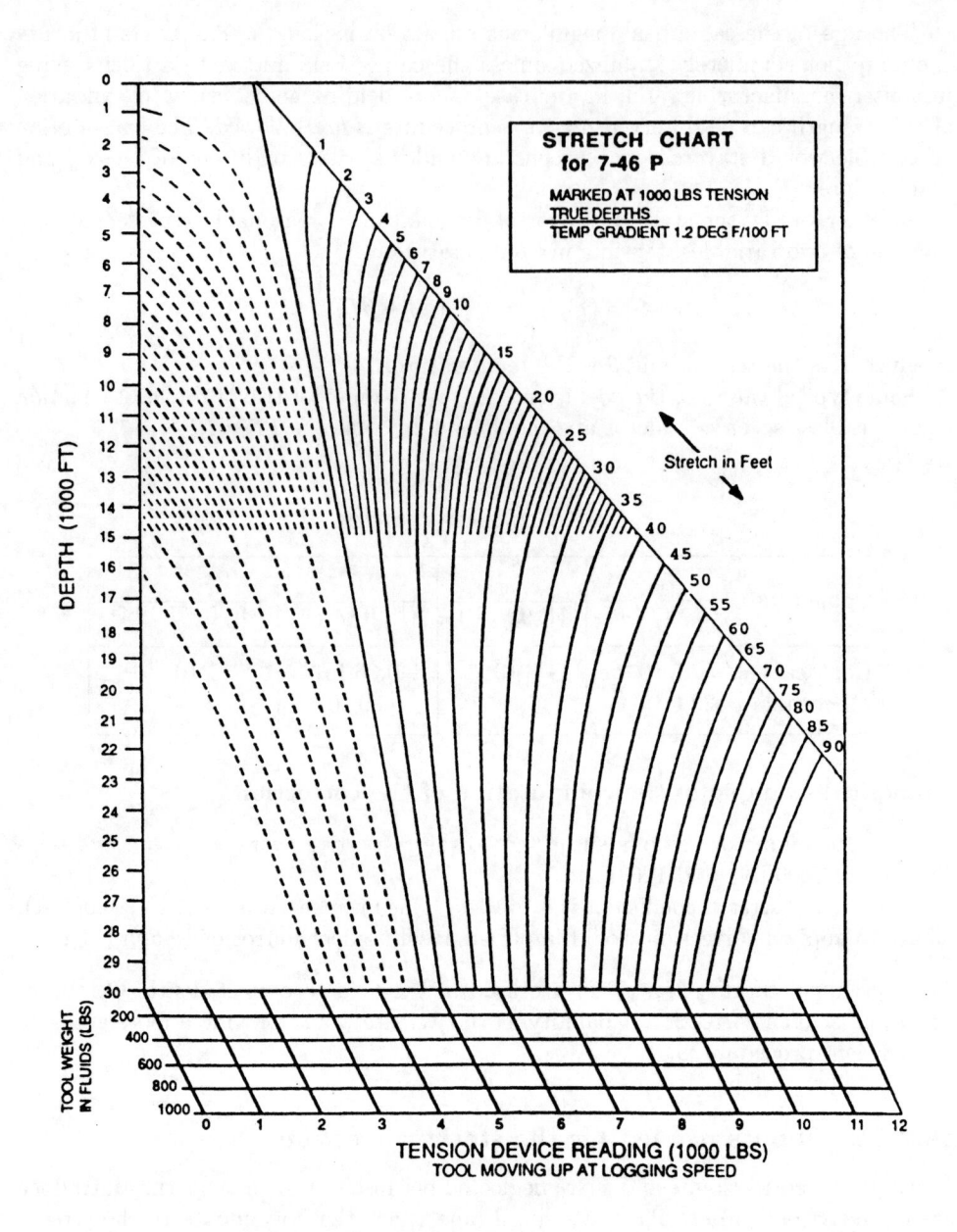

FIG. 19.1 : Correction for stretch.
Temperature and tension stretches are simultaneously accounted for. The temperature gradient is assumed constant.
Courtesy of Schlumberger

T_{surface} is measured at surface. Care must be exercised to ensure the location of the tension device is well defined (located above the upper sheave or at the logging unit). If the direction of force on the tension device is not parallel to the cable and a multiplier, a function of the angle between the cable and the tension device axis, is used. T_{bottom} is measured by a sensor located at the head of the logging string.

Parameters E_0, α_0, T and θ have the respective uncertainties σ_{E_0}, σ_{α_0}, σ_T and σ_θ. The uncertainty on the stretch coefficient, σ_S, is such that

$$\sigma_S^2 = (\frac{\partial S}{\partial \alpha_0})^2 \times \sigma_{\alpha_0}^2 + (\frac{\partial S}{\partial E_0})^2 \times \sigma_{E_0}^2 + (\frac{\partial S}{\partial T})^2 \times \sigma_T^2 + (\frac{\partial S}{\partial \theta})^2 \times \sigma_\theta^2,$$

$$\frac{\partial S}{\partial \alpha_0} = \theta, \ \frac{\partial S}{\partial E_0} = T, \ \frac{\partial S}{\partial T} = E_0 \text{ and } \frac{\partial S}{\partial \theta} = \alpha_0.$$

Then : $\sigma_S^2 = \theta^2 \times \sigma_{\alpha_0}^2 + T^2 \sigma_{E_0}^2 + E_0^2 \times \sigma_T^2 + \alpha_0^2 \times \sigma_\theta^2.$

Numerical application

$\frac{\sigma_{\alpha_0}}{\alpha_0} = 5\%$, $\frac{\sigma_{E_0}}{E_0} = 5\%$, $\sigma_\theta = 10°F$, $\sigma_T = 100$ lbf, $\theta = 150°F$, $T = 7000$ lbf

$$S = 2.4 \ 10^{-4}$$

This represents a propagated uncertainty of 73 cm (2.4 ft) at 3050 m (10,000 ft).

19.2.1.3 Dynamic behavior

(a) **Yo-yo**

Harmonic motion of the tool around the equilibrium cable length is called yo-yo. Yo-yo results from two causes [1]: First, a variation in the drag coefficient between the tool—pad and sonde—and the formation; second, variations in hole diameter. Yo-yo can account for depth variations of ±0.6 m (2 ft) in 3050 m (10000 ft). Yo-yo can be monitored by downhole accelerometers. High sampling frequency is required to record the fast variations of tool acceleration.

(b) **Stick and slip**

Stick and slip can be far more serious than other effects. It happens when strong frictional forces occur between the tool and the formation. The tool stops until sufficient overpull is applied to free it. Just after the release, the tool speed is high due to cable elasticity. The speed can reach two to five times the nominal speed. In these conditions, the quality of the measurements requiring sufficient duration is seriously jeopardized.

In difficult hole conditions, shifts up to 12 m (40 ft) could be seen at 3050 m (10,000 ft). To minimize this effect, a good strategy is to have a **reference** measurement on a slick logging string and then perform depth matching on the subsequent measurements.

The speed of the tool in yo-yo and stick-and slip situations is shown on *Fig. 19.2*. The effect of irregular tool movement on formation images is shown on *Fig. 19.3*.

282   19. Measurement of depth

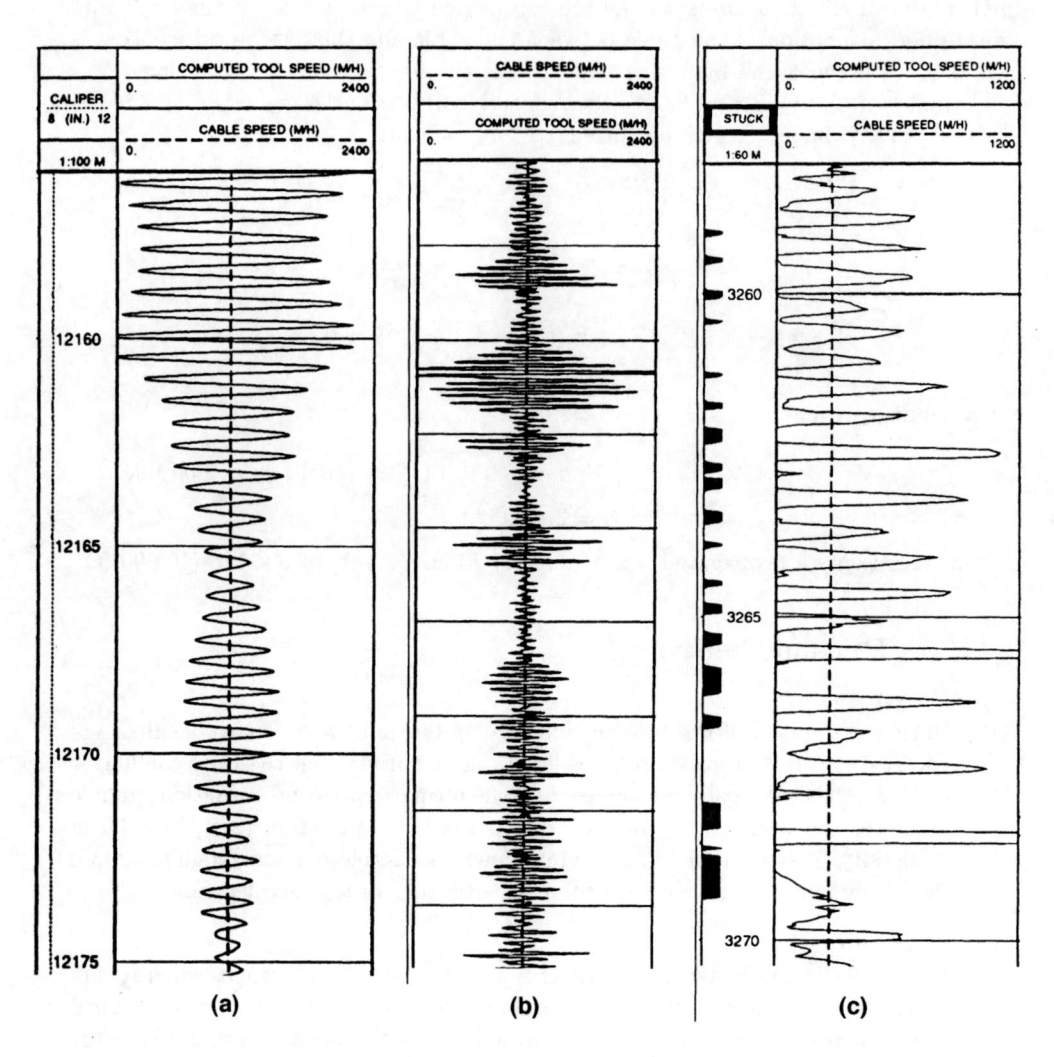

FIG. 19.2 : Tool speed in yo-yo and stick-and-slip conditions.
a. Yo-yo resulting from friction.
b. Yo-yo initiated by variation of borehole diameter.
c. Stick and slip.

Courtesy of SPWLA [1]

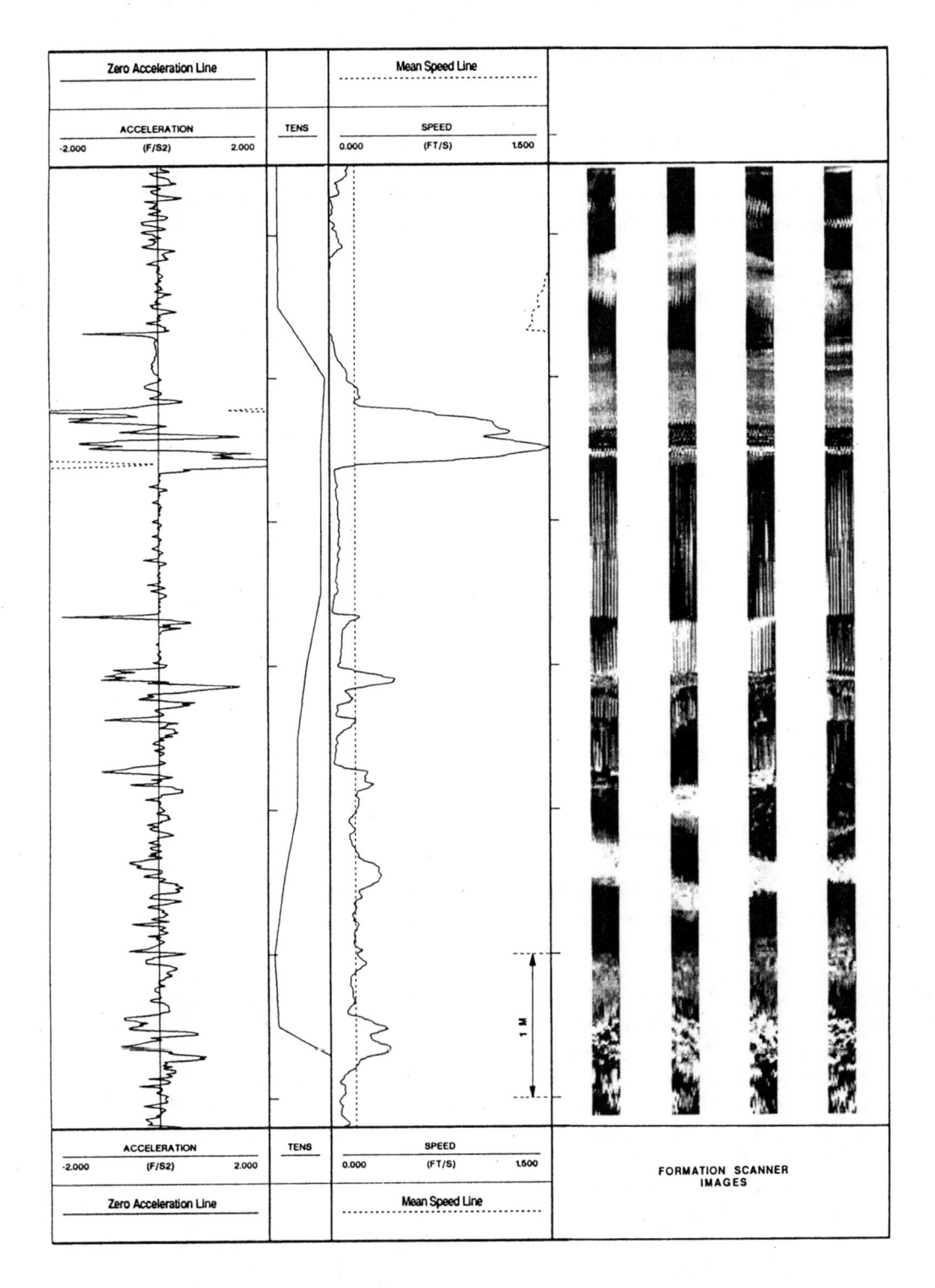

FIG. 19.3 : Effect of irregular tool movement on high-resolution data.
The Formation MicroScanner images are distorted when acceleration varies.
Courtesy of Schlumberger

19.2.2 Driller's depth

Considering the potential difficulties affecting wireline depth measurements, it may be tempting to tie logging depth to the depth derived from the drillpipe tally. Driller's depth is obtained by measuring each drillpipe or drillcollar at surface with a steel tape and adding these measurements. The factors listed in *Table 19.3* **are not** taken into account since driller's depth is not corrected for stretch or for any other effect.

Measurement at surface can be done with the pipes lying on the catwalk or with the pipes strapped in the derrick. The two values may disagree if the pipes have been submitted to too much weight on bit or too large pulls.

TABLE 19.3
FACTORS AFFECTING DEPTH: DRILLPIPES

Factor	Potential error for a depth of 10,000 ft	
	(m)	(ft)
Drillpipe stretch	5 to 10	16 to 33
Thermal expansion	3 to 4	10 to 13
Pressure effects	1 to 2	3 to 6
Ballooning effects	2	6
Other effects............	1	3

19.2.3 MWD depth

MWD logging requires a continuous depth index. Between drillpipe connections, a continuous depth is provided by a depth encoder that is linked to the displacement of the travelling block (*Fig. 19.4*). A tension device is coupled to the encoder so that it is possible to recognize if the block moves with pipes or without; for instance, when the drillpipes are in the slips for connection. When the measured load from the tension device is above a preset threshold, the depth is incremented by the encoder. When it is below the threshold, the depth is not incremented. In addition, the depth encoder is calibrated so that the number of pulses is translated into lengths above a reference (drillfloor or kelly bushing). At connection the MWD engineer is requested to tie the encoder depth with the drillpipe tally (even if the sequence of drillpipes is not correct and therefore the total depth is erroneous).

For this reason, MWD depth agrees with driller's depth and does not match wireline depth (the former depths are not stretch corrected, the latter is). Typically, at 3000 m (10,000 ft), a difference of more than 3 m (10 ft) should be expected.

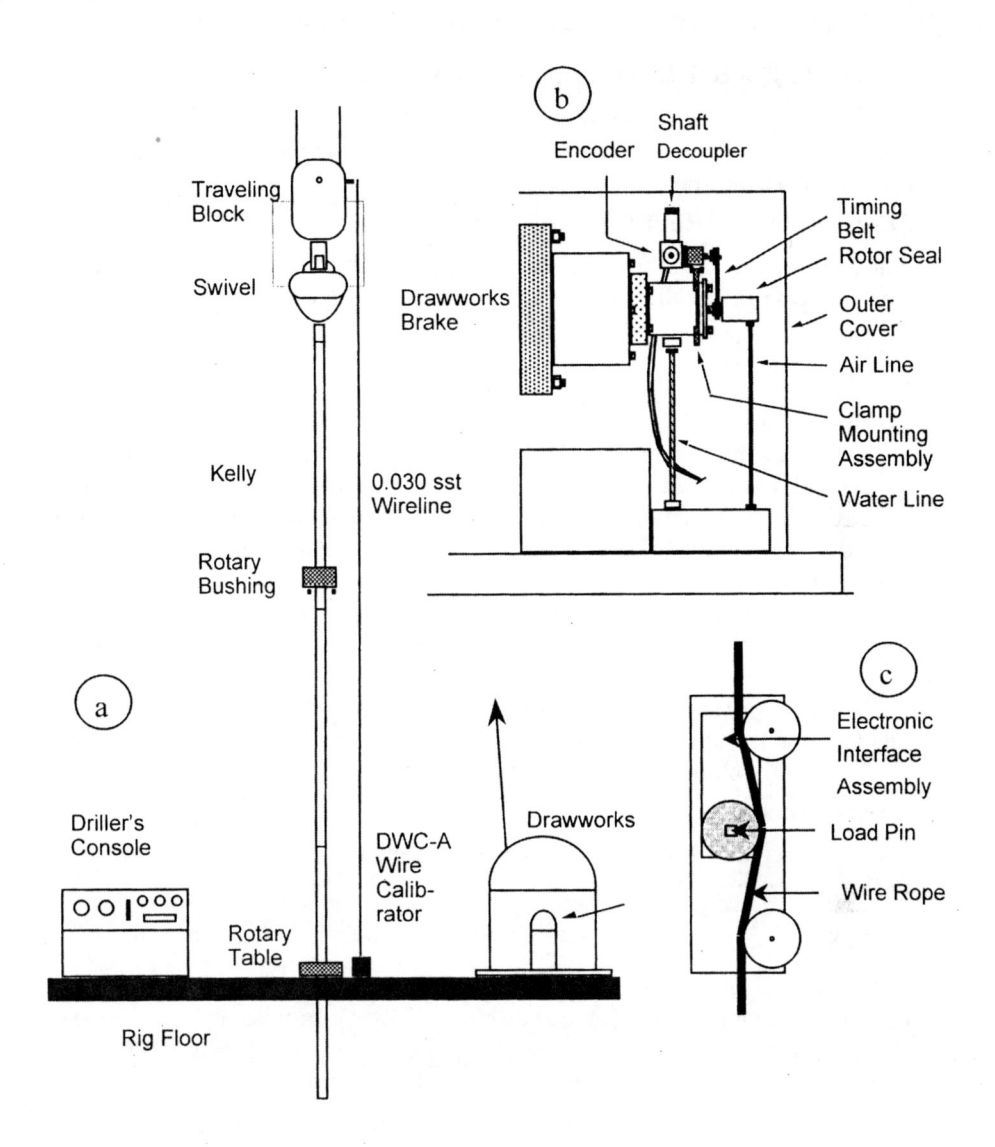

FIG. 19.4 : MWD depth system.
a. Overall system.
b. Depth encoder blowout.
c. Tension device.
    Courtesy of Schlumberger

19.3 Causes and effects of depth mismatch on logs and interpretation

There are two categories of reasons why logs are not matched in depth: discrepancies within and between tools.

(a) **Discrepancies within the tool**

The measurements originate from the same logging tool or from a combination of tools run simultaneously; e.g., density and neutron porosities, induction curves and images.

Causes

- Obstruction of the movement of the tool while logging in the borehole will result in some variation in the tension at the surface. Except in the most extreme circumstances when the tool or BHA is irretrievably stuck, the cable or drillpipe at surface will continue to move at a constant velocity. Surface movement in these cases is not reproducing tool movement and therefore results in compressed or expanded sections of data being shown on print or recorded on tape. As these problems occur in real time, the curves are affected at different depths (measurements are made at different depths at each moment) and therefore can appear shifted in depth when compared.
- Differing vertical resolutions result in nonaligned bed boundaries.
- Rotation of the tool in the borehole causes the sensors to scan different formation volumes.

(b) **Discrepancies between tools**

These discrepancies appear between measurements made on consecutive runs in the well.

Causes

This group of problems includes the problem of tension variations while logging and the more important variations resulting from mismatches in consecutive logging runs. Also, nonconformity to depth-control procedures and depth-system calibration could cause between-tool discrepancies.

19.3.1 Effect of depth matching errors

(a) Many environmental corrections depend on input from different tools of the same or consecutive logging runs. A good example is the caliper used in the borehole corrections on many tools. Depth differences between the caliper and the tool being corrected result in erroneous corrections.

(b) Interpretations of reservoir characteristics depend on the input of several measurements. Again, depth mismatches lead to incorrect results ([8] and Section 19.3.2).
(c) A depth mismatch may in certain cases have a direct impact on the validity of the measurement. This occurrence is reviewed in Section 19.3.3.

19.3.2 Apparent gas effect caused by depth mismatch

Irregular tool movement could have catastrophic consequences on interpretation. *Figures 19.5* and *19.6* depict a density-neutron combination. *Figure 19.5* shows the memorized and unmemorized curves when tool movement is regular. *Figure 19.6* shows similar curves, but this time with a tension pull (the tool stopped momentarily). In this second example, the memorized curves are such that an apparent gas effect appears.

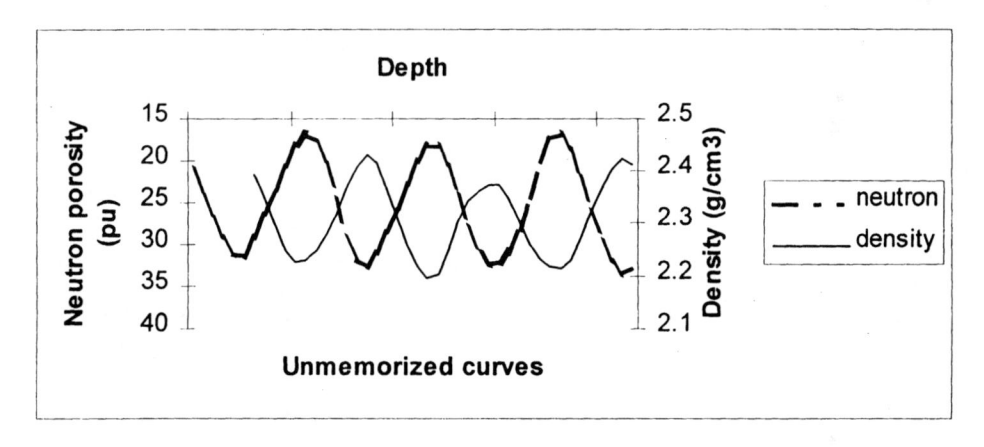

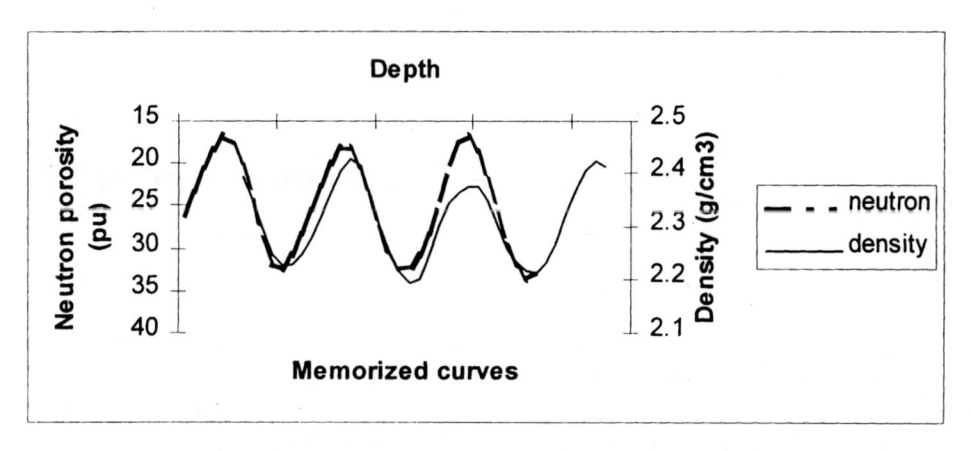

FIG. 19.5 : Effect of memorization on log readings.
Readings when tool movement is regular.

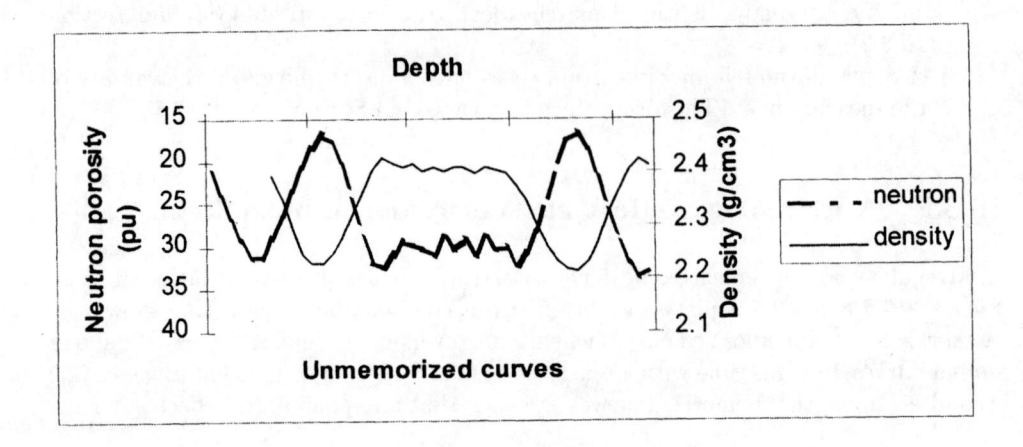

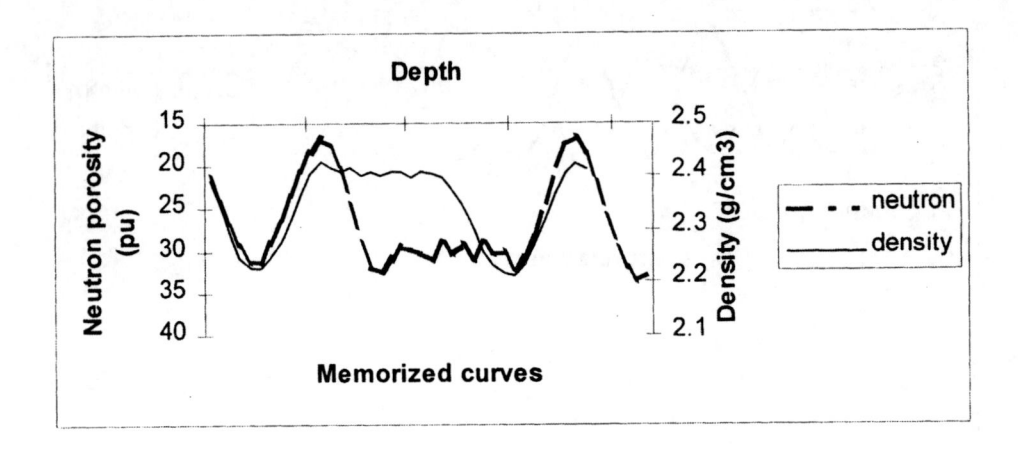

FIG. 19.6 : Effect of memorization on log readings.
Readings when tool movement is not regular. An apparent gas effect appears when the tool stops.

Total CFP has designed a program called Destickage [3] that removes the curves once the tension is above a certain threshold. In a similar way, Schlumberger has developed a presentation entitled Stuck Tool Indicator that changes the color or the coding of the curve once a tension threshold has been crossed (*Fig. 19.7*).

The minimum requirement on identification of irregular tool movement is that the tension should always be recorded during logging.

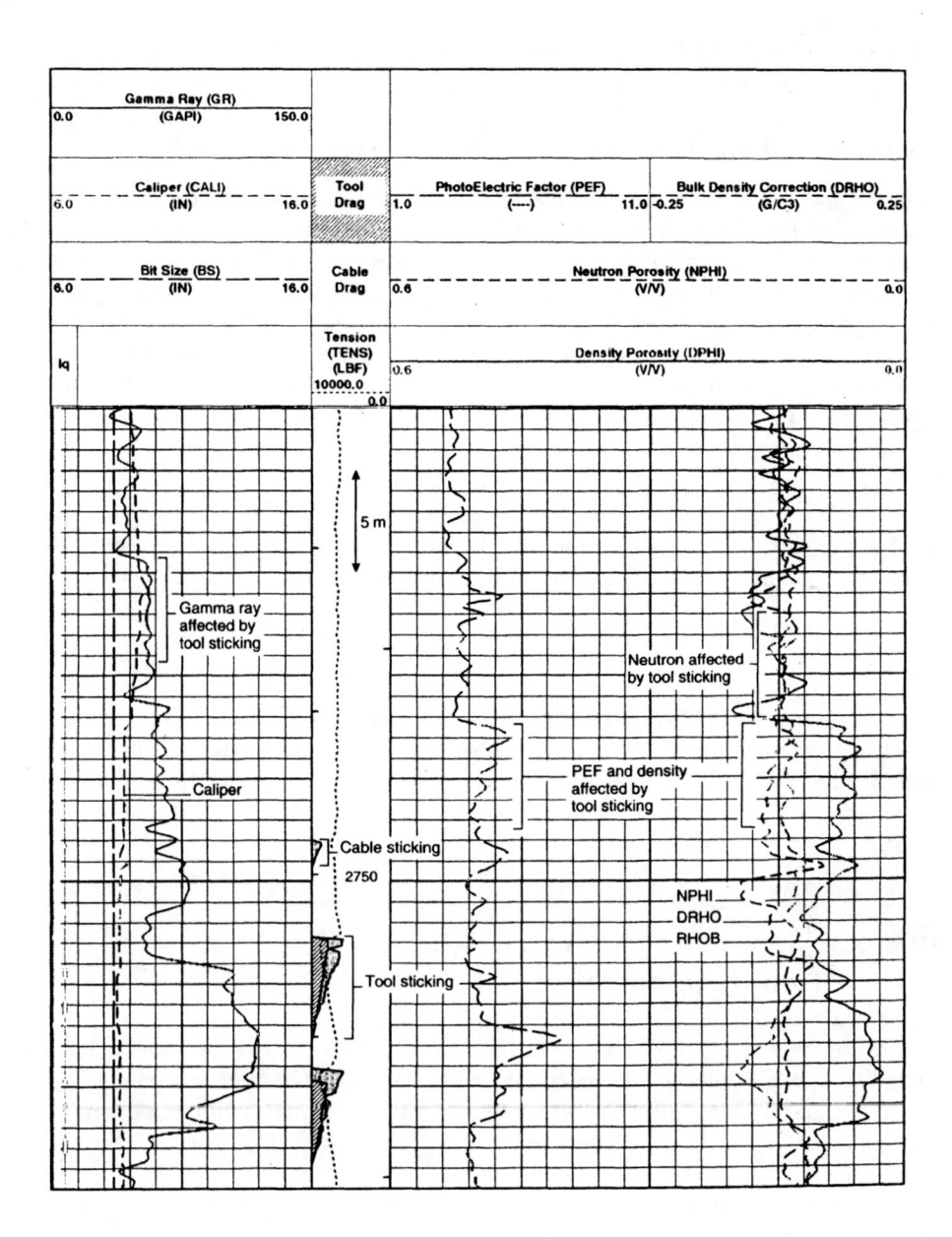

FIG. 19.7 : Stuck Tool Indicator.
The curves become gray when the derivative of tension versus depth exceeds a set threshold. The gamma ray, P_e, density et neutron porosity curves do not change color at the same depth because their memorization distance to the reference point—the bottom of the tool string—is different.
Courtesy of Schlumberger

19.3.3 Mismatch introduced by depth-derived compensation

Depth-derived borehole compensated sonic data

Long-spacing sonic measurements are closer to seismic transit times than short-spacing ones because they are not affected by the zone damaged during drilling. The presence of washouts and hole irregularities requires that the sonic, even long-spaced, be compensated for borehole effects. The classical borehole compensation geometry is not practical for long-spacing because it implies very long sondes. The depth-derived borehole compensation calls for the memorization of the signal before it is combined with other measurements. It is essential that the measured transit times are put on depth prior to combination and addition [4]. $\Delta t_{\log}$, read from the log, is

$$\Delta t_{\log} = \frac{\Delta t \times \delta}{\delta_0},$$

in which Δt is the actual formation transit time, δ the actual span between receivers and δ_0 the nominal span (2 ft in the case of the long-spacing sonic). The error is then

$$\sigma^2_{\Delta t_{\log}} = \sigma^2_\delta \left(\frac{\Delta t}{\delta_0}\right)^2 + \sigma^2_{\Delta t}\left(\frac{\delta}{\delta_0}\right)^2.$$

As a numerical application, consider the incidence of a 15-cm (6-in.) depth mismatch on the raw transit time measurements. Assuming no error in the measured transit times, a 0.6-m (2-ft) nominal span and a formation transit time of 305 μs/m (100 μs/ft), a 7.6-cm (0.25-ft) error in the span δ gives an error of 38 μs/m (12.5 μs/ft) in $\Delta t_{\log}$.

19.3.4 Apparent depth mismatch from high apparent dip

An apparent depth mismatch may be observed even though the cable is not submitted to dynamic effects. This occurs when the logging tool crosses a boundary with a high apparent dip because the sensors have different depths of investigation. When the tool moves up, the deep-reading sensors[5] see the boundary sooner than the sensors with a shallow penetration. This is illustrated with a simple induction short normal combination tool run in a well with a high deviation and crossing a flat boundary separating a resistive formation and a conductive one (*Fig. 19.8*).

---

[5]This is true except for the induction device, which is lagging compared to shallower investigation sensors. The induction current loops need to be closed in order to have a signal change and the tool string must be completely surrounded by the upper formation before it sees it.

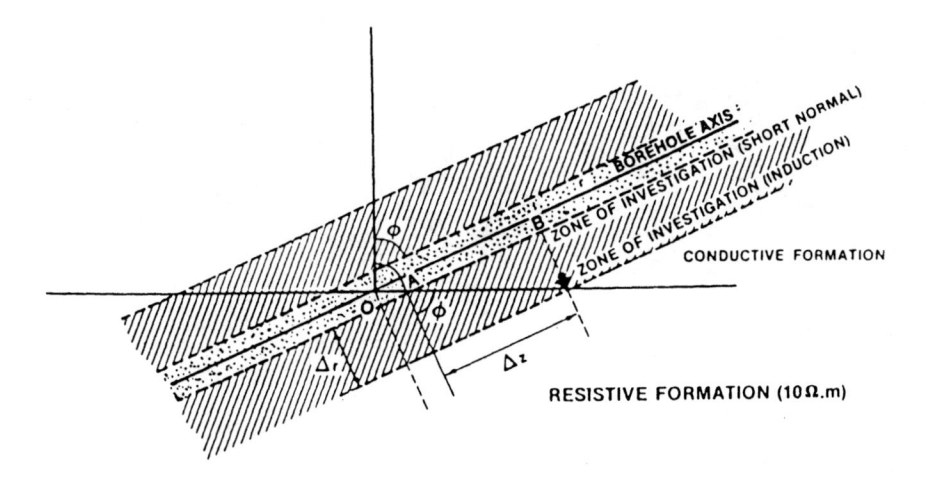

FIG. 19.8 : Model of formation boundary with high apparent dip.
The resistive formation is seen by the short normal electrodes up to point A and by the induction loops up to point B. The angle of deviation of the well, ψ, is related to the depth lag between the curves, Δz, and the difference of depth of investigation, Δr. The actual measured depth of the top of the formation is neither at point A or B but farther down. Its exact position can be derived from the value of ψ.

Courtesy of Schlumberger [VI], section on horizontal wells

Figure 19.9 shows a depth mismatch caused by high apparent dip. It can be quantified considering the lag between the curves—2.7 m (9 ft). The short normal has a depth of investigation of 10 cm (4 in.) while the induction has one of 80 cm (30 in.). The apparent dip (ψ) can be deducted as the angle having a tangent value equal to $\frac{\Delta z}{\Delta r}$ with Δz being the lag and Δr the difference of depth of investigation. In this example the derived angle is 76.5°; while a directional survey gave 78.8°. The interesting corollary of this depth mismatch is that the actual bed boundary is at neither the induction nor the short normal inflection, but a short distance from them. In this case, it is 42 cm (1.4 ft) deeper than the short normal break.

Similar depth mismatches affect all the tools every time the logging string crosses an apparent dip. Sensors with the shallowest depth of investigation display a signal contrast at a depth closest to the real formation boundary depth. They are preferred to deeper investigation tools for evaluation of reservoir tops and bottoms. For instance, high-resolution or shallow-investigation resistivity curves are to be selected instead of induction curves.

It is noteworthy that the depth mismatches caused by varied depth of investigation should be recognized and not erased through a depth-matching processing. They contain valuable information about the geometry of the formations; they should not be thrown away.

There are two ways to compensate for depth mismatches: perform a speed correction or perform depth-matching.

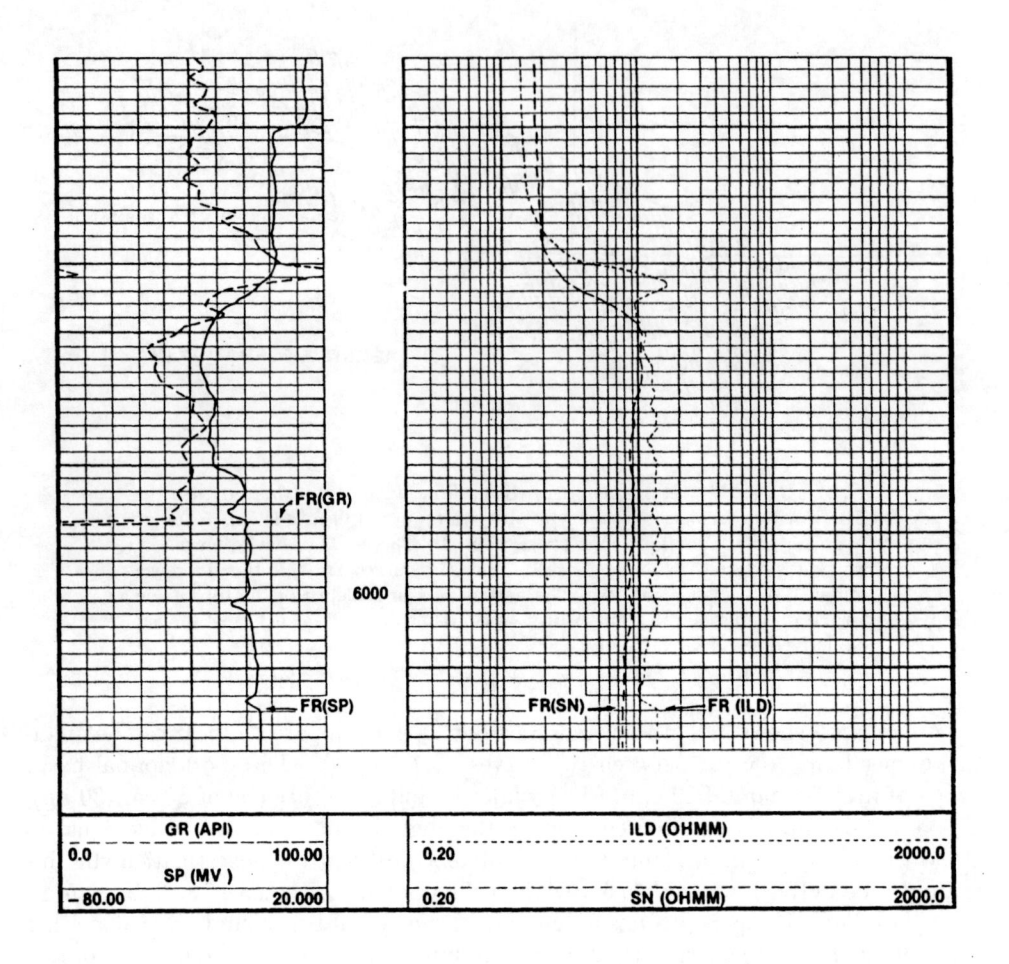

FIG. 19.9 : Example of depth mismatch caused by high apparent dip.
Courtesy of Schlumberger [VI]

19.4 Speed correction

The current depth measurement systems perform at surface. The depth measurement is based on the fundamental assumption that the displacement of cable or pipe at surface is followed by an equal displacement of the sensor downhole.

The information collected on downhole motion proves that this assumption is not always met. This issue can be compensated for by the speed correction.

19.4.1 Speed electrode

The first speed correction was implemented on the dipmeter tool. On one of the pads, a second electrode was added, at a well defined distance h from the first one. Nominal logging speed is v_0. The responses of these electrodes to the same thin event were correlated and shown to be separated by a distance H. The actual tool speed, v, was calculated as:

$$\frac{v}{v_0} = \frac{h}{H}.$$

The correction was applied only if this ratio was larger than 0.5 and smaller than 2.

19.4.2 Downhole measurement of acceleration

With continuous measurement of acceleration and of cable movement and some filtering, it is possible to assess tool velocity and depth. This method is a postjob processing applied to formation imagers. It is now applied to tools such as PLATFORM EXPRESS in real time [1].

19.5 Depth matching

The problems of depth mismatch fall in two groups, those with solutions and those without. In many cases, like those of the tool becoming stuck, there is often no way to retrieve the corrected data. If the magnitude of the problem is not too extreme, there are two solutions:

(a) **Respect of the depth control operational procedures**: See Section 19.6.
(b) Careful application of a **depth shifting software** that can handle the various complexities mentioned above without providing a completely misleading answer. Depth matching processing supposes that some kind of interpretation of the curves is performed. The analyst has to take the decision of correlating or anticorrelating the features observed on the curves.

To depth match with a reasonable chance of success, it is necessary to select the curve—called reference—that is believed to be indexed to the correct depths. The measurements that are the least affected by undesirable depth effects originate from mandrel tools with no, or weak, excentralizer so that friction against the formation is minimized. Also, for a wireline log, the depth of an upgoing record would have more validity than the one of a downgoing record. *Figure 19.10* shows an example of depth matching. In this example, it results in a substantial change in the estimation of hydrocarbons in place.

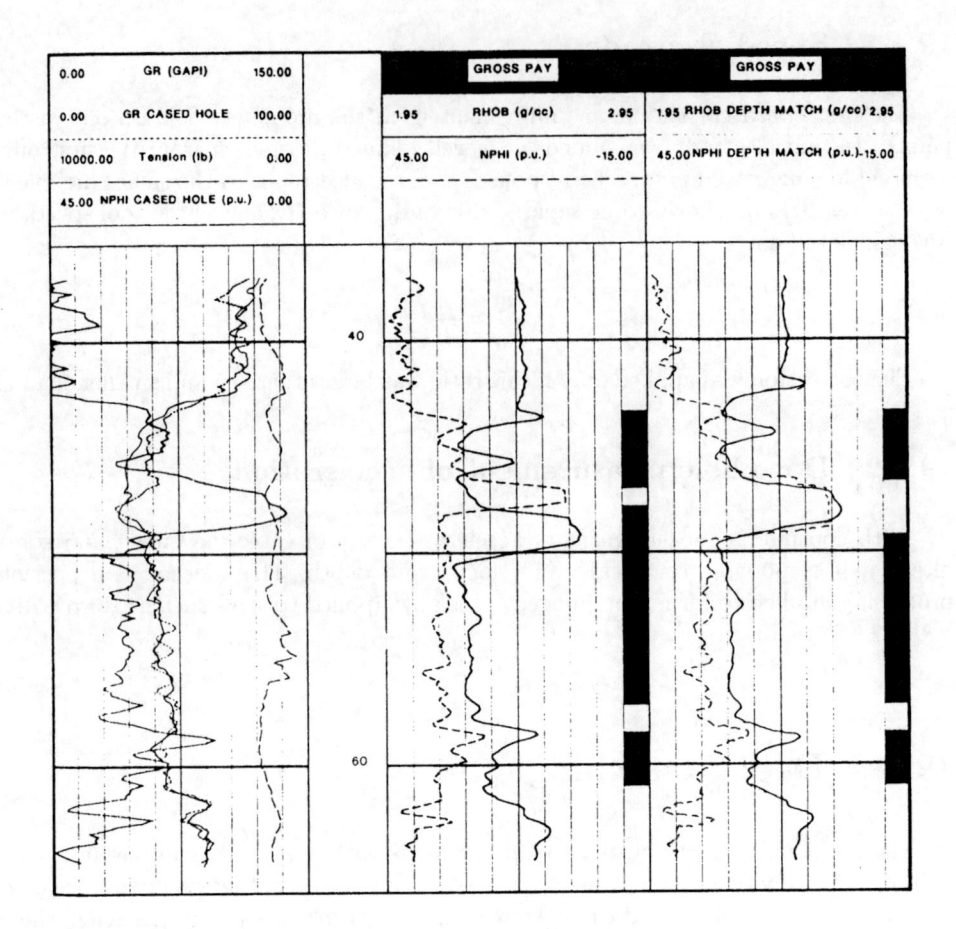

FIG. 19.10 : Depth matching error and correction.
An original neutron porosity density log is shown in track 2. While the two curves are in depth at the top and bottom of the interval, discrepancies appear between xx44 ft and xx50 ft. Correctly depth matched curves are shown in the right track. The net pay thickness derived from the uncorrected log data is overestimated by 13% [XVIII].

19.6 Guidelines of depth control

Acquiring accurate depth is a challenge. Rigorous procedures set by the logging company depth policy are required. In addition, the circumstances of the depth acquisition of a specific logging job must be documented. This applies in particular to the depth datum (zero or reference) and to the reporting of correlation with more shallow depth horizons.

19.6.1 Depth policy

Considering that depth consistency is more critical than absolute accuracy, operating procedures related to depth have to be applied universally. They are gathered in a depth policy, examples of which are given in *Tables 19.4* and *19.5* (*Schlumberger*).

TABLE 19.4

SCHLUMBERGER DEPTH POLICY

The IDW is the Integrated Dual Wheel spooler depicted in *Fig. 10.5*. FIT and TRIM are preventive maintenance procedures, respectively the fast inspection of tools and the tool review and inspection monthly.

---

1. The primary depth reference is the IDW.

2. Both depth systems must be allowed.

3. Magnetic marks are the backup of the IDW measurement. They must be monitored while running in and logging up.

4. The cable must be marked whenever possible, but in no case should the period between marking be more than six months.

5. FIT maintenance of the IDW is mandatory after each trip to the well. TRIM is required every three months.

Courtesy of Schlumberger

TABLE 19.5

ANADRILL DEPTH POLICY

The first statement implies that MWD depth differs from wireline depth.

---

1. Anadrill MWD/LWD depth is driller's depth.

2. The precision depth assembly is the standard depth system.

3. Depth shifts should not be made in any zone of interest.

4. All editing of raw depth/time files must be documented.

Courtesy of Schlumberger

19.6.2 Documenting depth acquisition

The depth measurement requires documentation that is at least as complete as other measurements. The following information needs to be reported on the log or fed to the database [10]:

(a) equipment type and number (measuring wheel, depth encoder, calibrator)
(b) calibration date and characteristics
(c) depth reference used (Section 22.2.2)[6]
  If a tie-in with a more shallow section is performed, it should be reported. It is recommended that a minimum overlap length is used for this tie-in. An adequate interval is 65 m (200 ft). Generally, nuclear tools are run first in open hole, then in cased hole over the same interval during another run.
(d) corrections applied (especially about stretch)
(e) amount and location of "cranking." See paragraph (b) in the next section.
(f) closure upon return to surface or depth reference
(g) remarks, especially deviations from standard procedures
(h) memorization distances (on the print/film, *Fig. 19.11*) or encapsulated in a tool sketch).

SENSOR MEASURE POINT TO TOOL ZERO

MTEM	12.93	METER	HTEN	-.28	METER
GR	8.89	METER	MRES	12.93	METER
CNTC	5.99	METER	CFTC	6.15	METER
LL	.79	METER	LITH	.79	METER
LU	.79	METER	LS	.79	METER
SS1	.64	METER	PARI	.64	METER
DTCL	.79	METER	SS2	.64	METER
DTPL	.79	METER	DTCS	.64	METER
LLLC	.79	METER	DTPS	.64	METER
LULC	.79	METER	LLUC	.79	METER
SLLC	.64	METER	LUUC	.79	METER
SULC	.64	METER	SLUC	.64	METER
CALI	.81	METER	SUUC	.64	METER
AX	9.96	METER	TENS	-.28	METER
AZ	9.96	METER	AY	9.96	METER
FY	9.96	METER	FX	9.96	METER
TNRA	6.30	METER	FZ	9.96	METER

FIG. 19.11 : Memorization distances for the density-neutron tool.

19.6.3 Other recommendations

(a) A new cable should be marked twice (the second time after 10 jobs) to confirm proper aging.

---

[6]Note that the ground level is a more durable reference than the drillfloor.

(b) Cranking is a special procedure whereby the logging engineer changes the depth while the logging tool or cable does not move. It should not be more than 15 cm (6 in.) in any 30-m (100-ft) interval. It should not be performed at all in zones of interest.

19.7 Future developments

Major improvements of downhole sensors and surface acquisition systems are underway. Depth measurements have not been following the same fast trend. The following concepts are now being investigated by logging companies:

(a) control of the measurement sequence by time, not depth
(b) measurement of the time taken by an electric pulse sent from surface and echoed downhole
(c) optical methods for use with fiber optic cables
(d) improved measuring wheel or linear depth interval measurement systems
(e) improved tension measurement and modeling
(f) measurements of pressure/fluid density versus depth.

References

1 Belougne, V., Faivre, O., Jammes, L., Whittaker, S., "Real time speed correction of logging data," paper F, *Trans.* SPWLA 37th annual logging symposium, New Orleans, 6-1996.

2 Bolt, H., *Wireline depth determination*, 1998.

3 Imbert, P., personal communication, 1987.

4 Jøranson, H., "A comparative study of full waveform acoustic tools," *Trans.* SPWLA 30th annual logging symposium, Denver, 1989.

5 Kirkman, M., Seim, P., "Depth measurements with wireline and MWD logs," Formation evaluation while drilling symposium, Bergen, 6-1989. Also presented at the SPWLA MWD seminar, London, 23-4-1990.

6 Menghini, M. L., "Compaction monitoring in the Ekofisk area chalk fields," pp. 735–739, *Journal of Petroleum Technology*, Vol. 41, No. 7, 7-1989.

7 Moulin, P., private communication, 1990.

8 Ramberger, R., Wagner, K., "Dynamic depth shift of the count rates in compensated neutron logs," paper I, *Trans.* SPWLA 10th European formation evaluation symposium, 1986.

9 Sollie, F., Rodgers, S., "Towards better measurements of logging depth," paper D, *Trans.* SPWLA 35th annual logging symposium, Tulsa, 6-1994.

10 Theys, P., "Le log," *The Log Analyst*, Vol. 38, No. 6, 11-1997.

20
Directional surveys

20.1 Introduction

In the past, when planning vertical wells, it was assumed that a depth measurement would enable the complete determination of the location of the logging sensors. With the growth of directional or horizontal wells, coordinates in a plan perpendicular to the vertical are required. This information originates from directional surveys and is used as an input to the computation of **true vertical depth**:

(a) **Azimuth** is the orientation of the wellbore relative to the North.[1]
(b) **Inclination** is the measurement of the deviation of the wellbore from vertical.
(c) **Toolface**[2] (used in drilling) is the measurement of the angle between the scribe line on a bent portion of the BHA relative to the high side of the wellbore.
(d) **Relative bearing** (used in wireline logging) is the measurement of the angle between the upper side of the tool and a reference point in the tool (e.g., electrode 1).

This chapter deals mainly with directional surveys acquired while drilling or just after drilling. These surveys are used for decision-making (steering) or as definitive information to characterize the trajectory of a well. In some cases, the information applies to formation evaluation wireline surveys; directional logs are acquired in combination with formation scanners to derive the orientation of the image.

The magnetic and gyroscope surveying methods are reviewed after a list of the potential errors that can be encountered.

---

[1] The need to define an appropriate North reference is discussed in Section 22.2.3.
[2] From vertical to about 8° inclination, the magnetic tool face is used (a plumb-bob would not define an accurate direction). Above 8°, gravity tool face is used.

20.1.1 Potential errors in directional surveys

Directional surveys have their share of classical errors [2]:

(a) random errors
   1. unpredictable environmental variations
   2. round-off errors
   3. orthogonality errors when sensor is rotating
   4. mud pump-induced fluctuation in mud pressure

(b) systematic errors
   1. errors present in only one survey
      - reference errors in connection with free gyro surveys
      - residual errors in magnetic declination corrections
      - drill collar sag for MWD measurements
   2. errors systematic for all surveys in a given area
      - magnetic measurements without magnetic declination corrections
      - errors in the geodetic reference network

(c) blunders
   1. use of wrong initialization parameters
   2. use of wrong calibration constants
   3. instrument used beyond operational specifications
   4. memory or processor error
   5. inferior data management practices (e.g., manual transfer of data).

The correct characterization of errors is essential in directional surveying as it affects the propagated uncertainty considerably [10]. Assuming that most errors are random leads to compensation of errors from one surveying station to another and a smaller propagated positioning uncertainty [5]. Assuming that all errors are systematic may lead to pessimistic positioning uncertainty. This approach still seems closer to reality [6 and 9]. Data management issues are discussed further in Chapter 23.

Gross errors on directional surveys are found at a high rate [1]. Azimuth errors larger than 15° over long well sections have been detected. Many of these errors could have been avoided if adequate quality control procedures had been in use.

In spite of the inherent difficulty to quantify accuracy, it has been done for directional surveys [11]: surveying systems have been checked in a high-angle directional hole that came out through the side of the Grand Canyon. The exit point of the hole was surveyed by an independent method (infrared range finder) with an accuracy of ±1.0 ft. The surveys performed in the hole drilled down to 1650 m (5000 ft) in measured depth placed the bottom of the hole within 2.4 m (7.2 ft) of its true location. Unfortunately, a mistake in the choice of reference grid did not allow the correct prediction of the hole exit point at the time of drilling and it is only after large corrections had been applied that the quoted accuracy was obtained.

This example proves that well managed directional surveys have the potential to be accurate, but that actual surveys are often affected by poorly defined procedures.

20.2 Magnetic surveys

20.2.1 General

Magnetic surveys measure the magnetic field from an array of magnetometers, combined with an array of accelerometers (*Fig. 20.1*). From these measurements, it is possible to infer the spatial angle between an axis of reference linked to the logging string and a direction of reference aligned with the Earth's magnetic field.

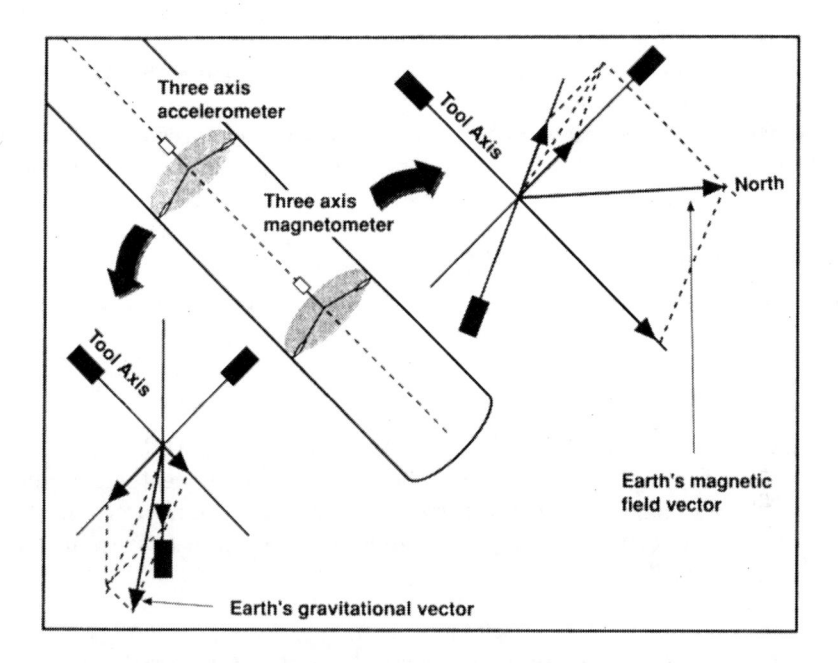

FIG. 20.1 : Model of the magnetic directional sensor package.
The three-axis accelerometer measures the Earth's gravitational vector relative to the tool axis and a point along the circumference of the tool (scribe line). The three-axis magnetometer measures the Earth's magnetic field relative to the tool axes.
Courtesy of IMS.

Difficulties arise from the variations of the Earth's field, from the parasitic contributions from other magnetic fields (magnetic interference) and from the misalignment of the sensor package with the axis of the borehole (sag effect).

20.2.2 Earth's magnetic field

The total Earth's magnetic field has two components:

(a) the principal field induced by the movement of the Earth's nucleus
(b) the transitory field generated outside the Earth and that is subject to
   1. secular variations (15 nT)
   2. diurnal solar variation (35 nT)
   3. 11-yr cycle
   4. magnetic storms (in the order of several hundred nT).

As observed, the total Earth's magnetic field is a function of location and time. Its evolution versus time is modeled and declination maps updated, along with maps of rate of change. A study of the models derived from magnetic field observatories has shown that much erroneous or uncontrolled information is present in their database [3]. A critical review of the models is required especially in areas where the input data of Earth's magnetic field is sparse.

In critical cases, it is necessary to perform a direct measurement of the Earth's magnetic field close to the wellsite. This is done with a magnetic observatory system [7].

20.2.2.1 Magnetic field strength

Only the horizontal component of the Earth's magnetic field is measured. It is equal to

Horizontal component = Total magnetic field × cos(dip angle).

The dip angle is the angle between the tangent to the Earth's surface and the magnetic field vector. The computation of the horizontal component lets us anticipate that the measured quantity is two to three times smaller in some areas, while errors and corrections stay of the same magnitude. Relative errors in these areas are therefore far more significant (*Table 20.1*).

TABLE 20.1
TYPICAL CHARACTERISTICS OF THE EARTH'S MAGNETIC FIELD

Areas	Total magnetic field strength (nT)	Magnetic dip angle (deg)	Horizontal component (nT)
Gulf of Mexico	50,450	60	25225
East Canada	54,000	70	18469
Beaufort Sea	58,500	84	6115
North Sea	50,000	70	17101

20.2.2.2 Magnetic interference

Magnetic interferences may be created by

(a) the drillstring
   1. magnetic drillpipes
   2. hot spots on collars or magnetometer sub
   3. batteries
(b) external factors
   1. a fish left in the hole
   2. nearby casing
   3. formations containing magnetic minerals (pyrite, hematite)
   4. muds (hematite).

Significant azimuth errors result from the lack of action in presence of magnetic interferences. Two complementary routes are possible, the use of non magnetic drill collars, and the use of a correction software.

Non magnetic drill collars (NMDC)

Drill collars made of special alloys (Monel) can be positioned in the BHA. The length of NMDC collars is derived from charts [4]. The length depends on the latitude of the well (NMDC required length is larger the farther away from the Equator), on the hole angle (longer with larger angles), on the angle with the North-South direction (longer in the East-West direction). As much as 30 m (90 ft) of NMDC collars have been used.

Additional measurements and correction software

To compensate for any residual interference effect after the use of NMDC, it is necessary to perform several directional surveys at the same depth, but with different toolfaces. From these measurements, a correction to the magnetic field is derived.

20.2.3 Misalignment errors

There are three types of misalignments associated with directional surveys: sensor (within the sensor package), instrument (sensor package with the principal axis of the drill collar) and collar (collar axis relative to the wellbore axis). The first type is compensated through calibration (though a residual error may be present). The second type is treated as a systematic error whose upper limit is derived from the geometry of the sensor package and the collar. The third type, also called sag effect, is a function of wellbore geometry and the bottomhole assembly. It can be quantified for good hole conditions, but it is rather unpredictable with washouts, breakouts and keyseats.

20.3 Gyroscope surveys

20.3.1 Gyroscope technologies

Gyroscopes belong to the following categories:

(a) mechanical (rotating mass)
- rate
- integrating

(b) interferometric fiber optic
- open loop
- closed loop
- resonator

(c) ring laser
- active
- passive

(d) vibratory
- hemispherical resonator
- vibrating beam .

(e) electrostatic.

Most gyroscopes used in the oil industry are mechanical. They are classified according to the number of flywheels (one to three), of degrees of freedom of movement and the level of restriction of movement (gimbaled or strapped). Unrestrained systems form rate integrating gyroscopes; restrained systems (the axis of the gyroscope is controlled by a torquing motor) form rate gyroscopes. Three-axes systems have a distinct advantage in horizontal holes where they have reduced survey errors and improved accuracy.

Ring laser and interferometric fiber optical gyroscopes are based on the Sagnac effect [8] originally discovered in 1913 and proven in an experiment by Michelson and Gale in 1925—by sending ordinary light through water pipes, they were able to detect the shift produced by the rotation of the Earth. Both gyroscopes use a rotating beam of light to detect rotation. Vibratory gyroscopes sense the Coriolis effect.

20.3.2 Sources of errors in gyroscope measurements

Many types of errors affect gyroscopes. Fortunately, only a few have significant importance in the context of directional surveying.

The principal errors affecting mechanical gyroscopes originate from mass imbalance around the spin axis of the gyroscope, fixed restraints (aerodynamic friction and friction in bearings) and thermal drift. Solid state gyroscopes have four sources of error: bias drift caused by the electronics, angular random walk, angle white noise and scale factor error—especially critical with strapdown systems used for inertial navigation.

Whatever the choice of gyroscope technology is selected, it is associated to the presence of accelerometers in the surveying device. These accelerometers add to the overall error budget. Accelerometer bias, scale factor and overall noise levels contribute to this budget. In addition to these are the errors linked to the misalignment of the sensor package with the wellbore axis.

Gyroscope survey tools can also be classified with the operating methodology as a criterion. Attitude and heading reference system (AHRS) is a multishot style of measurement where an attitude of the survey tool is measured at different depths and an algorithm is then used to show the continuous trajectory of the borehole. The balanced tangential method is generally used in the oil industry. Errors of centralization of the tool within the borehole may be significant.

The inertial navigation system (INS) calculates the tool position in space from the accelerometers and gyroscopes. Centralization errors are not a factor but accelerometer bias is critical.

20.4 Importance of accurate directional surveys

In addition to helping the petrophysicist locate the piece of formation interpreted, accurate directional survey have two important uses.

20.4.1 Collision avoidance

In many cases, multiple boreholes are drilled from a single surface location. Adjacent wells may be producing and a collision could result in an extremely dangerous situation. Accurate surveys of previously drilled wells are used to plan the new well. In the process, each well trajectory is associated with an ellipse of uncertainty computed from an estimation of the errors encountered on each survey. Error budgets vary depending on sensor type, well orientation and other factors.

20.4.2 Undulations in well trajectories

Experience in drilling and producing horizontal wells has lead to the conclusion that angular variations (undulations) in the well trajectory are detrimental. They favor the creation of water and gas chokes along the producing string. Undulations can be minimized by an early acquisition of surveying data that allows a continuous and fast reaction to changes in well trajectory. *Figure 20.2* shows an example of production limited by gas and water chokes.

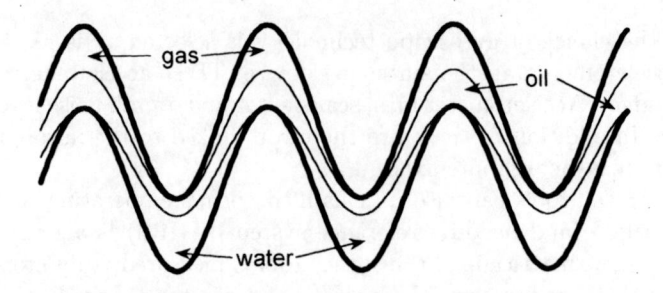

FIG. 20.2 : Limitation of oil production in an undulating well.
The flow of oil is reduced by the presence of gas and water in the production string. The vertical scale of the figure is exaggerated.

References

1. Ekseth, R., Nakken, E. I., Jensen, L. K., "Wellbore position uncertainties calculation, a tool for risk based decision making," Offshore mediterranean conference, Ravenna, Italy, 1997.

2. Ekseth, R., *Uncertainties in connection with the determination of wellbore positions*, Trondheim, 1998.

3. Emery, X., "Analysis of the IGRF model," 1997.

4. Grinrod, S. J., Wolff, C. J. M., "Calculation of NMDC length required for various latitudes, developed from field measurements of drill string magnetization," SPE 11382, IADC/SPE drilling conference, New Orleans, Louisiana, 2-1983.

5. Harvey, R. P., Walstrom, J. E., Eddy, H. D., "A mathematical analysis of errors in directional survey calculations," *Journal of Petroleum Technology*, Vol. 23, No. 11, 11-1971.

6. Lange, J. I., Twilhaar, G. D. N., Pelgrom, J. J., "Accurate surveying: an operator's point of view," SPE 17213, 1988.

7. "Nemo set to unravel driller's magnetic North mysteries," *Offshore Engineer*, 1997.

8. Sagnac, G. "L'éther lumineux démontré par l'effet du vent relatif d'éther dans un interféromètre en rotation uniforme," p. 708, *Comptes Rendus de l'Académie des Sciences*, Vol. 157, 1913.

9. Truex, J. N., "Directional surveys problems, East Wilmington oil field, California," *American Association of Petroleum Geologists Bulletin*, Vol. 55, No. 4, 1971.

10. Wolff, C. J. M., Dewardt, J. P., "Borehole position uncertainty—analysis of measuring methods and derivation of systematic error model," pp. 2339-50, *Journal of Petroleum Technology*, Vol. 33, No. 12, 12-1981.

11. Wright, J. W., "Directional drilling azimuth reference systems," SPE 17212, IADC/SPE drilling conference, Dallas, 1988.

Part III

Data quality control

21

Data quality plan

> It is better to aim at perfection and miss it than aim at imperfection and get it.
>
> *Thomas J. Watson, IBM*

> The lack of agreed requirements is the biggest single quality issue.
>
> *Bret Thompson*

> Failure to plan is plan to fail.

Historically, the quality of logging has been judged on the occurrence or lack of failures and the measurement of rig time. The performance of a logging company is based on its ability to avoid or minimize failures and to reduce the time the rig is involved in logging operations. The rush to minimize loss of rig time may compromise the quality of data acquired.

The concept of logging quality must be widened to **total quality**, which includes

(a) service efficiency (minimization of failures, optimization of rig time)
(b) data quality (intrinsic quality of data)
(c) data relevance (ability of data to describe the formation or to enable the customer to take the correct decisions).

21.1 Definition of quality

Quality, without a qualifier, is a misnomer [13 and 14]. Quality cannot be defined only in qualitative terms. It has to be **quantified**. This need is understood in quality assurance programs from which definitions of the word quality are extracted.

(a) The totality of features and characteristics of a product or service that bear on its ability to satisfy a given need (*BS 4778, the British standards*).

(b) **Conformance with specified requirements** (*NS 5801, the Norwegian standards*, [10]).

The second definition is simple and actionable. It splits the journey toward quality in two steps:

(a) Requirements must be defined, understood and clarified. This process is formalized with suppliers and customers agreeing on specifications.

(b) Products and services conforming to the preset requirements are delivered.

Unfortunately, in practice, the first process is poorly completed. The data user is often left under two basic misunderstandings:

(a) True values, exactly reflecting the characteristics of the formation or downhole conditions, are acquired by logging companies.

(b) As a corollary, all acquired values of the same formation characteristic or downhole condition are the same.

This misinformation leads to unending questions about why a log value does not match another one.[1] It also leads to disappointment as the logging company is just not able, by the constraints of physics, to deliver those true values that are so much wanted by the oil company. Finally, it leads some organizations to use data that has not been completely through the complete measurement chain—an example being the direct use of data that has not been corrected.

21.2 Data quality plan

The problems evoked in the previous section can be avoided if a data quality plan is followed. It has three phases:

(a) job planning and preparation

(b) execution and conformance to the plan

(c) control of the execution; differences between execution and plan; feedback and collection of what has been learned.

---

[1] Read more about comparison of data in Chapter 27.

The importance of these three phases are shown on *Fig. 21.1*.

Data Quality

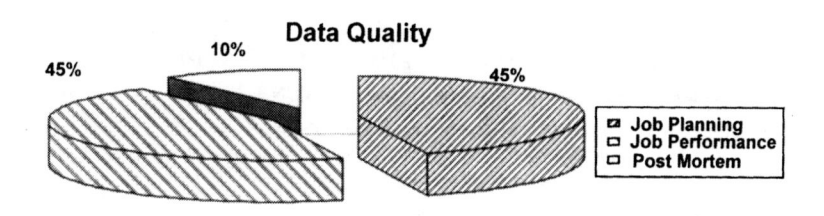

45% — Job Planning
10% — Job Performance
45% — Post Mortem

FIG. 21.1 : The components of data quality.

21.2.1 Job planning and preparation

If an oil and gas company has no well defined objectives in terms of data acquisition, there is not much chance the data acquired will satisfy the needs. The customer is required to go beyond the "shopping list" approach—one triple combo/two sidewall core guns for a wireline job, 200 permeabilities/8 thin sections/2 water analyses for measurements in a core laboratory [6].

1. The first step in the overall process is to define the objectives that motivate the collection of data.

2. The second step is to search for options:

    (a) logging measurements
    (b) logging companies
    (c) availability of equipment and trained personnel.

Cost is a consideration, but the following has to be kept in mind:

   (a) Incorrect data that leads to the wrong decision is expensive.
   (b) So is data that is not going to be used at all.
   (c) Data that has not been acquired and that would have eased the solution of a problem represents also an indirect cost.

Logging measurements options are considered under the perspective of their specifications. Mechanical and environment specifications are verified so that the tool is likely to survive in the wellbore. Detailed metrological specifications enable the user to find out if the measurement is going to fit the purpose previously set. For instance, the split between accuracy and precision is useful.

While accuracy cannot be improved easily, precision can be changed by using time-dependent parameters—measurement duration and signal processing. Examples of specifications for two MWD logs are given in *Table 21.1*.

TABLE 21.1
SPECIFICATIONS OF COMMON LWD TOOLS

a. Accuracy and repeatability are distinct for ADN.
b. Three percentages of the geometrical factor are considered for each ARC5 resistivity measurement. This gives a much more concrete idea of the diameter of investigation of this device.

ADN measurement specifications				
Measurement	Range	Accuracy	Statistical repeatability	Log vertical resolution
Compensated epithermal neutron porosity	0 to 100 pu [limestone]	±0.5 pu below 10 pu. 5% between 10 & 50 pu	@30 pu: 1.5 pu @60 m/h (200ft/h)	61 cm (24 in.) 30 cm (12 in.) with α-processing
Formation bulk density from spectral $\gamma - \gamma$	1.7 to 3.05 g/cm^3	±0.015 g/cm^3 in good borehole	±0.006 g/cm^3 @ 15 m/h (50 ft/h) @2.4 g/cm^3	30 cm (12 in.) 15 cm (6 in.) with α-processing
Photoelectric factor P_e	1.0 to 10.0 units	±5% units	±0.1 @15 m/h (50 ft/h) in limestone	5 cm (2 in.)

ARC5 resistivity measurement specifications					
Spacing (in.)	10	16	22	28	34
Range (Ω/m) Phase resistivity	All spacings: 0.2 - 200				
Attenuation resistivity	0.2-20	0.2-30	0.2-50	0.2-50	0.2-50
Accuracy Phase resistivity < 60 Ω/m > 60 Ω/m	All spacings ±3% All spacings ±0.5 mmho/m				
Attenuation resistivity < 25 Ω/m > 25 Ω/m	All spacings ±5% All spacings ±2 mmho/m				
Geometrical factor Phase resistivity	Diameter (in.)				
10%	14	19	24	28	33
50%	28	36	44	52	60
90%	66	73	81	88	95
Attenuation resistivity					
10%	44	50	57	63	69
50%	80	88	96	103	111
90%	150	157	163	170	176
Vertical resolution Phase resistivity	All spacings 0.7 ft at 0.2 Ω/m 2 ft at 200 Ω/m				
Attenuation resistivity	All spacings 1 ft at 0.2 Ω/m 8 ft at 200 Ω/m				

Courtesy of Anadrill [3]

This listing is given as an example and may become quickly outdated. Specific information can be obtained from the logging company. The specifications for most other logging tools are available in [3, 8 and 12].

The review of the logging companies includes a search on how well their measurements conform to their specifications; in other words, the difference between marketing hype and real delivery. Also, a measurement is the result of many different processes, starting far away from the well location and well before a logging job. The user at some point needs to be informed on how extensive and accurate is the tool response, if corrections are available, if procedures to calibrate the tool are well defined **and** adhered to, etc. This "general study" may be performed once and for all for a given measurement. Industry groups (e.g., Log Characterization Consortium) have the charter to perform this work that can be shared by all member companies. This approach avoids duplication of efforts.

Logging companies may display differences in delivery according to the geographical location. Accordingly, it is a good idea to visit the shop managing the field operation, to inspect the calibration and maintenance areas, to query how tools are prepared for special jobs. An inspection of the archiving system is often useful.

3. The third job planning step is propagation of errors.

   In most cases, logging companies specifications cannot be directly applied; specifications relate to measurements, while the oil and gas companies require uncertainties on processed results, such as residual oil saturation that is computed from several measurements. Information on measurements specifications enable the propagation of uncertainties to decision-making parameters, as developed in Chapter 7 and reviewed in Chapter 28.

   Formation modeling may be used to assess the magnitude of the environment corrections and verify that their magnitude is acceptable for the final data user.

4. The fourth step of planning includes a clear understanding and agreement of references, including those on **depth** and **location coordinates**. **Units** systems must be clearly identified.

5. Finally, job preparation includes contingency plans that enable quick decisions if the master plan requires modification.

Many oil companies have extended their job planning strategy. Planning can go to extreme details when time-lapse data acquisition is contemplated. A specific example [1 and 2], includes the following steps and techniques:

(a) positioning a glass-reinforced fiberglass window in the casing for future resistivity logging
(b) providing latch points over the reservoir section for precise and repeatable depth-positioning of a borehole gravity meter
(c) computing the propagated uncertainties of the depth and gravity measurements to the residual oil saturation.

21.2.2 Job execution

Once the requirements have been well spelled out, job execution is generally easier. Reality may not duplicate what was planned completely, but deviation from the plan may not have any substantial consequence if the job objectives are fully understood by the data user and the data collector. Creativity and initiative are still required as many wells display some kind of surprise—geology, environment, etc. It is precisely because of this expectation of the unexpected that wells are logged: to spot underground anomalies.

In line with the overall well objectives, real-time decisions at the wellsite are crucial. They are better taken if they were anticipated during job planning. Data quality control should already be initiated in this phase. As data flows from downhole, it has to be checked as, in case of failure or degraded quality, it may be necessary to relog, call for another tool, etc.

21.2.3 Review after the job

At the conclusion of the acquisition job, the work is not over yet. It is necessary to evaluate how the data collection fits with the original plan. Any issue requires analysis, not to find a scapegoat, but to learn from the problem—and to find an eventual solution. Equipment and data quality failures should be reviewed as a team by the data user and the data collector.

21.3 Log quality control

Definition

LQC is a set of methods that identifies and analyzes data deviations from established standards and allows the design of a remedy.

What?

The first electric log, run in *Pechelbronn*, France, in 1927, was a sequence of points versus depth. The data was likely to have a high level of validity and relevance as it confirmed the other pieces of information available. The **log** had no calibration, no repeat section. All of the print was dedicated to what is called today the main log.

Nowadays, most logs include **quality evidences**, pieces of information that confirm the validity of the formation-related data, such as repeat section, relogged intervals, quality control curves and calibration tails. Most of these quality evidences do not add any information about the formation, but they increase the level of confidence that can be attributed to the main information. The more sophisticated and precise the logging tools, the more quality evidences are needed, to the point that they may occupy more space on the print than the main log itself.

Log quality control is therefore mostly based on the observation and checks of the quality evidences.

Who?

(a) The oil companies express their requirements that become future logging tool specifications.
(b) Engineering designs the tool and establishes standards.
(c) Manufacturing builds the tool so that it is capable of meeting the standards.
(d) The sales organization informs the field about the actual standards and limitations of the logging tool.
(e) The field technical staff organizes a proper setup for the tool (calibration areas, temperature cycling, etc.). It maintains and repairs the tool so that it stays within specifications.
(f) The oil/gas companies control the well environment where the tool is used. They are the main users of the log data.
(g) The engineer produces the log.
(h) The wellsite client witnesses assist the operating logging engineer for real-time controls.
(i) The field manager of the logging company reviews and controls the acquired data.
(j) Log analysts interpret the logs, keeping in mind the performance and limitations of the tool and the log quality control reports originating from the wellsite or the field location.

When?

Time is particularly critical. Log quality control methods are not meant only to establish rigid **postmortem** controls. They should be able to provide **real-time** results so that the log value is immediately assessed. Once the logging company equipment has been rigged down, irreversible operations are initiated. The borehole is likely to be cased a short time after logging and the casing precludes the performance of a large number of logging measurements.

Where?

Is it ludicrous to think that real-time log quality control is possible? At the wellsite, external information is not yet available. Cores are not fully analyzed for weeks or months. The full picture of the field is not yet complete. In addition, log quality control surveys reveal that only a small percentage (30% to 40%) of nonconformities is detected at the wellsite. During these wellsite checks, only a third of the total number of nonconformities recognized by experts are detected. Nevertheless, every effort should be made to perform the best possible control at the wellsite.

Going one step further in time and one level higher in quality awareness, log quality control can be thought of as defect prevention. During job preparation, the logging engineer can anticipate ways of getting the highest probability to get a 100% correct log the first time. Logging conditions and environments offer such a wide spectrum that it is impossible to predict them in a theoretical manner. The analysis of the problems encountered while logging wells in the same field is still the best way to control the risks for future logging jobs, so that what is a postmortem for a job becomes part of job planning for the next one.

TABLE 21.2
AUDIT PERFORMED BY OIL COMPANIES

This audit was performed on 360 logs. Only 119 logs had been controlled at the wellsite. Only 36% of the total number of nonconformities has been detected at the wellsite.

	Nonconformities	Per log
Total	630	1.75
Job-related	526	1.46
Environment-related	104	0.29
Detected at wellsite	76	0.64

Courtesy of SPWLA [14]

21.3.1 Log quality control requirements

(a) **Need for standards**

Rigid and quantitative standards are required. Once they are fixed, data is tested against the specifications for success or failure. No gray area should be permitted or subjectivity may intervene in the system and reduce its efficiency. Every time the data fails a test, a nonconformity[2] is reported. Quality experts seek a zero nonconformity [5]. Should logging companies also aim at the same objective? Perfection seems difficult to reach in borehole conditions—thousands of pounds of pressure and several hundreds of degrees Celsius. Also, the laws of physics preclude the performance of perfect measurements, which means than any value observed on a log would differ from the true value that the log user would like to obtain.

---

[2]**Nonconformity** [10] is the departure or absence of one or more quality characteristics or quality system elements from specified requirements. This word is preferred to **defect**, which often implies a moral judgment. Nonconformity may originate from human error, but also from the special conditions characterizing a well or an operating environment.

(b) **Need for procedures and training**
   The establishment of standards does not resolve all the log quality issues. Operating engineers need to be trained to use the standards and must be in an environment conducive to rigorous and objective reporting.
(c) **Need for audits**
   A log quality audit is a planned and systematic review of the wellsite personnel's ability to verify conformance to requirements.
   Log quality control at the wellsite is limited by a conflict of interest: the same person performs the job, or supervises it in the case of the field manager, and rates it. Subjectivity in log quality control cannot be removed unless the first persons who rate the log are aware of audits at a later stage.
(d) **Need for reporting and information handling**
   The results of the control must be put in a format that enhances transmission and analysis. Software that uses the form as an input is advantageous. Such software should preferably be ported on a personal computer. Database operations are effective in defining areas that require correction and ordering them by priority.

21.3.2 Log quality control reporting

Many users and providers of logs have developed log quality control forms [7, 9 and 11]. They display the differences in control requirements made necessary by the wide variety of conditions prevailing in different geographical areas. Specifics aside, the ideal form has these attributes:

(a) Be used by as many persons as possible. Certainly, many people have good ideas to develop log quality control forms, but this explosion of creativity has to be gently controlled. A common form, even if not very imaginative, has the advantage of allowing easy consolidation.
(b) Strike a compromise between simplicity and comprehensiveness. Different nonconformities have to be gathered in the same category. This may be done at the expense of completeness but it allows consolidation of a manageable number of classes. For instance, the check of a pressure gauge in hydrostatic conditions before and after a repeat formation tester pretest could be called **pressure stability check**. In practice, it is clustered with the category **response in known conditions**. A reasonable number of categories is 20.
(c) As much as possible, perform some kind of normalization if this does not increase the complexity of the form. Normalization means referring the number of actual nonconformities to the maximum number possible. A modern service, for instance a formation imaging tool, involves calibration tail films, repeat sections, tool control film, etc. A simpler service, for instance the positioning depth control of a perforating gun, implies a short record that may not conform on only two items, depth-matching and information completeness. With a non-normalized system the first service may be associated with up to a dozen

nonconformities in the worst case, while the second service could be linked to two nonconformities at most. Two nonconformities on the first service might be acceptable, while they are not for the second one.

(d) Split between logging company and environment related nonconformities. It is recommended to build a data bank containing information on the acquisition of inferior data even if the logging company is not responsible for the low quality.

Example

Keeping track of the footage of logging intervals where hole conditions have precluded acquisition of accurate formation data may help assess the options: a change in drilling practice, a change of mud, modification of logging hardware.

(e) Use ticks or marks. Some control forms use ticks or Yes/No questions while other forms use a numerical approach. The second approach has the advantage of giving a rating that can be used more directly, but it may have a negative psychological impact. The operating engineer judging his/her own work would be reluctant to report nonconformities associated with large penalties.

Important note

The rating of a log is directly linked to the specifications associated with the tool run. Considering two tools with different specifications but measuring the same formation, one may not conform while the other one does—even though they give exactly the same reading.

Example

Tool 1 has a vertical resolution of 60 cm (2 ft) while tool 2 has a resolution of 15 cm (6 in.). They are both run over a sequence of geological beds with different thicknesses (*Fig. 21.2*).

Tool 1 observes bed A—1 m (3.3 ft) thick—but not bed B—0.3 m (1 ft) thick. Tool 2, when in proper operating condition, recognizes beds A and B. The logs produced by the two tools are different but they are both considered as conforming. Now, tool 2 is assumed to be affected by a failure that deteriorates its vertical resolution—suppose that the short spacing detector of the density tool is out of order; consequently, enhanced resolution processing is no longer possible. The log recorded by tool 2 is similar to the one recorded by tool 1. But, in this case, the log obtained with tool 2 does not conform since a superior tool resolution is expected.

From this observation, it can be inferred that a control system for a logging company would not apply to another logging company unless they have identical specifications.

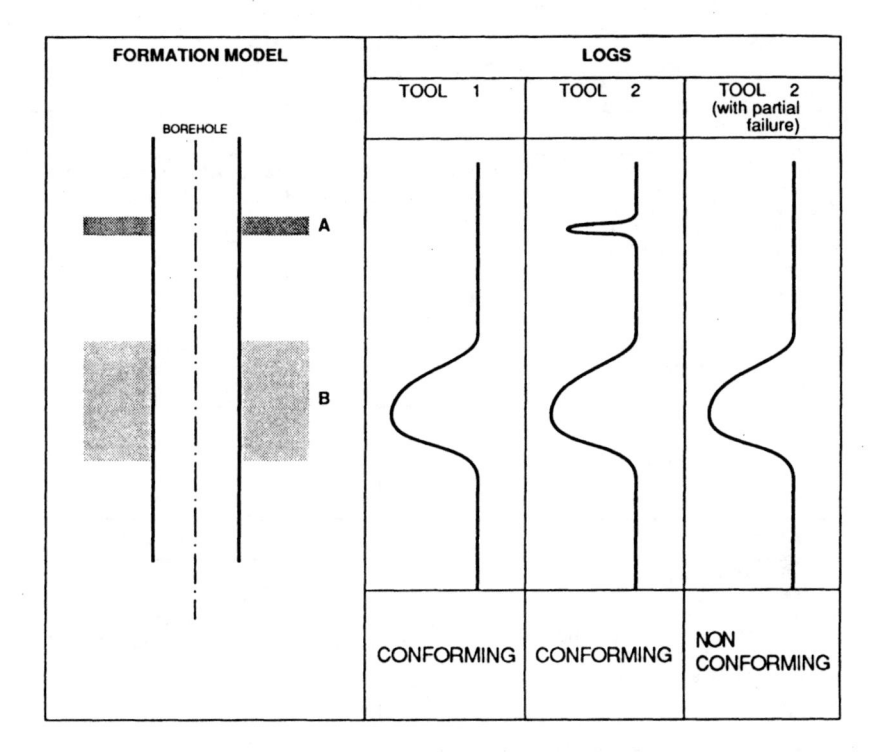

FIG. 21.2 : Rating of logs produced by tools with different specifications. Conformance is a matter of specifications, depending on the expected result and the obtained result.

21.3.3 Example of LQC form

The form shown in *Fig. 21.3* is currently used in some operating locations. Filling out the form is straightforward, **although it cannot be filled without a book of standards.** Every time a standard is not met, this corresponds to a **nonconformity**, which is reported on the form. To avoid bias, any deviation, minor or serious, is reported.

Figure 21.3 represents an example of a form filled after a dual laterolog/MicroSFL survey. The dual laterolog data collects a $B+$ rating, which means that the data was satisfactory except for occasional intervals with noise. The MicroSFL data receives a $C-$. The C^3 means that an interval displayed anomalous response that affected data quality. The $-$ means that not all prescribed operating procedures were followed.

---

[3] For more explanations on data quality ratings, see Section 25.3.

PRESENTATION				COMMENTS
1. HEADING COMPLETENESS/ACCURACY	X			Company Name misspelled
2. FILM/PRINT QUALITY				
3. CURVE/PIP/SCALES/INSERTS				
4. TD/FR/CSG IDENTIFICATION				
5. STANDARD PRESENTATION				
6. IMPORTANT CONSTANTS LISTED				
7. REMARKS				
8. WELL SKETCH/TOOL SKETCH				
9. OTHERS				

CALIBRATION	DLL	MSFL		COMMENTS
1. SHOP CALIBRATION VALIDITY				
2. BEFORE SURVEY ACCURACY				
3. AFTER SURVEY DRIFT				
4. OTHERS				

OPERATING TECHNIQUE				COMMENTS
1. DEPTH MATCHING				
2. LOGGING SPEED		X		3600 FPH +
3. CENTRALIZATION	X			No upper centralizer
4. RESPONSE IN KNOWN CONDITIONS		X		No caliper check in csg.
5. SETTING OF CONSTANTS				
6. PRESENCE OF STANDARD CURVES				
7. RELOGGING OF ANOMALIES		X		
8. TAPING QUALITY				
9. OTHERS				

DATA QUALITY				COMMENTS
1. OCCASIONALLY NOISY	3			Stuck at 9500'
2. ANOMALOUS RESPONSE		6		Did not relog anomaly
3. SERIOUS ANOMALOUS RESPONSE				
4. REPEATABILITY				
5. OTHERS				

Use the codes in the Data Quality section to explain anomalies.
1. ROUGH HOLE
2. BOREHOLE FLUID
3. TOOL STICKING
4. ANISOTROPY/FRACTURE
5. GAS
6. PAD CONTACT
7. TOOL LIMITATIONS EXCEEDED
8. TOOL FAILURE
9. POOR OPERATING TECHNIQUE
10. OTHERS

FIG. 21.3 : Log quality control form.

21.3.4 Log quality data base

The purpose of the log quality control is to provide a structure for an improvement scheme. Typically, a log quality control database would be able to sort performance levels by location, engineer, customer and service.

Example

The first quality control audit in an area revealed the following weaknesses:

(a) incompleteness of log heading information

(b) incompleteness of auxiliary data

(c) substandard calibrations

(d) large effects of borehole condition on the logs, especially affecting pad contact.

These areas of concern were tackled once they were identified and diagnosed with a log quality control system.

Figure 21.4 summarizes the monthly log quality control database for a district. This type of record permits improvements in training and organization.

21.4 And, if a log does not conform?

Having built a system that detects and lists nonconformities, what actions can be taken when they occur:

(a) Nonconformity does not mean the data is worthless. A log without calibration, repeat section, etc. may be quite valid though there is no proof for it.

(b) The real issue is to check that the data is adequate in spite of the missing quality evidences or present out-of-tolerance information. Substandard data may be acceptable. This is the case for a logging interval covering an intermediate zone on which no detailed quantitative evaluation is performed.

(c) After the nonconformity is recognized, the logging engineer and the client representative establish if the data would be acceptable as is, or if the cost of rerunning the tool, or another tool, would be prohibitive compared to the incremental information it would bring.

Long-term consequences need to be considered. Sometimes the wrong decision (not rerunning the log) is taken because of operator fatigue, whereas the log is judged essential several weeks to several years later.

(d) In many cases, nonconforming logs can be repaired. A nonconformity can be removed by playing back the log. For instance an incorrectly calibrated log can be recalibrated just after logging. The new calibration values can be used to obtain a valid log.

(e) Detection of a nonconformity is part of a learning process. Reporting and follow-up on a problem increases the odds that the same problem will be tackled appropriately in the future.

Training is of utmost importance in this context.

21. Data quality plan

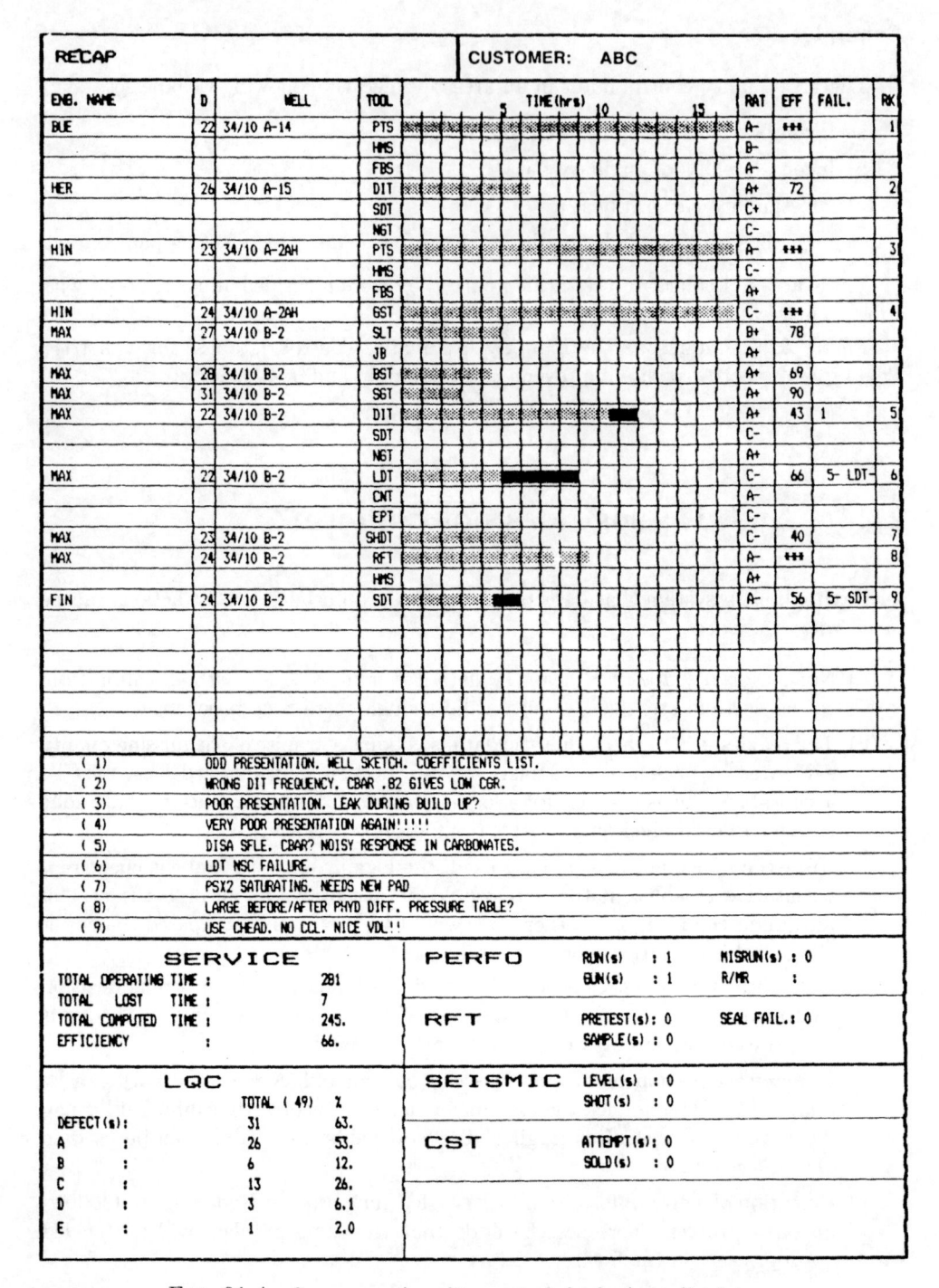

RECAP				CUSTOMER:	ABC			
ENG. NAME	D	WELL	TOOL	TIME(hrs)	RAT	EFF	FAIL.	RK
BUE	22	34/10 A-14	PTS	▓▓▓▓▓▓▓▓▓	A-	+++		1
			HMS		B-			
			FBS		A-			
HER	26	34/10 A-15	DIT	▓▓▓▓	A+	72		2
			SDT		C+			
			NGT		C-			
HIN	23	34/10 A-2AH	PTS	▓▓▓▓▓▓▓▓▓	A-	+++		3
			HMS		C-			
			FBS		A+			
HIN	24	34/10 A-2AH	GST	▓▓▓▓▓▓▓	C-	+++		4
MAX	27	34/10 B-2	SLT	▓▓	B+	78		
			JB		A+			
MAX	28	34/10 B-2	BST		A+	69		
MAX	31	34/10 B-2	SGT	▓	A+	90		
MAX	22	34/10 B-2	DIT	▓▓▓▓▓▓▓▓█	A+	43	1	5
			SDT		C-			
			NGT		A+			
MAX	22	34/10 B-2	LDT	▓▓█	C-	66	5- LDT-	6
			CNT		A-			
			EPT		C-			
MAX	23	34/10 B-2	SHDT	▓▓▓	C-	40		7
MAX	24	34/10 B-2	RFT	▓▓▓▓	A-	+++		8
			HMS		A+			
FIN	24	34/10 B-2	SDT	▓█	A-	56	5- SDT-	9

(1) ODD PRESENTATION. WELL SKETCH. COEFFICIENTS LIST.
(2) WRONG DIT FREQUENCY. CBAR .82 GIVES LOW CGR.
(3) POOR PRESENTATION. LEAK DURING BUILD UP?
(4) VERY POOR PRESENTATION AGAIN!!!!!
(5) DISA SFLE. CBAR? NOISY RESPONSE IN CARBONATES.
(6) LDT NSC FAILURE.
(7) PSX2 SATURATING. NEEDS NEW PAD
(8) LARGE BEFORE/AFTER PHYD DIFF. PRESSURE TABLE?
(9) USE CHEAD. NO CCL. NICE VDL!!

SERVICE		PERFO	RUN(s) : 1	MISRUN(s) : 0
TOTAL OPERATING TIME :	281		GUN(s) : 1	R/MR :
TOTAL LOST TIME :	7	RFT	PRETEST(s): 0	SEAL FAIL.: 0
TOTAL COMPUTED TIME :	245.		SAMPLE(s) : 0	
EFFICIENCY :	66.			

LQC			SEISMIC	LEVEL(s) : 0	
	TOTAL (49)	%		SHOT(s) : 0	
DEFECT(s):	31	63.	CST	ATTEMPT(s): 0	
A :	26	53.		SOLD(s) : 0	
B :	6	12.			
C :	13	26.			
D :	3	6.1			
E :	1	2.0			

FIG. 21.4 : Summary of quality control checks for a district.

References

1 Alixant, J. L., Storey, M., "First monitoring well in the Rabi field: design and data gathering," SPE 28399, SPE 69th annual technical conference and exhibition, New Orleans, 9-1994.

2 Alixant, J. L., Mann, E., "In-situ residual oil saturation to gas from time-lapse borehole gravity," SPE 30609, SPE 70th annual technical conference and exhibition, Dallas, 10-1995.

3 Anadrill, *Log quality reference manual*, Houston, 1998.

4 Brami, J., "Current calibrations and quality control practices for selected MWD tools," SPE 22540, *Trans.* SPE 66th annual technical conference and exhibition, Dallas, 10-1991.

5 Crosby, P. B., *Quality without tears*, McGraw-Hill, New York, 1984.

6 Duguid, S., "Quality in practice—a service company perspective," *The Log Analyst*, Vol. 35, No. 5, 9-1994.

7 Farnan, R. A., McHattie, C. M., "Log quality enhancement, a systematic assessment of logging company wellsite performance and log quality," paper L, *Trans.* SPWLA 25th annual logging symposium, 1984.

8 French oil and gas industry association, *Wireline logging tool catalog*, Éditions Technip, Paris, 1986.

9 Holt, O. R., "Log quality control," paper BB, *Trans.* SPWLA 16th annual logging symposium, 1975.

10 *Norsk standard, NS-ISO 8402*, Norsk Verkedsindustris Standardiseringssentral, 2-1989.

11 Schlumberger, *Log quality control system*, Tokyo, 1985.

12 Schlumberger, *Log quality control reference manual*, Paris, 1990.

13 Stebbing, L., *Quality assurance: the route to efficiency and competitiveness*, Ellis Horwood Limited, Chichester, 1986.

14 Theys, P., "Quality in the logging industry," *The Log Analyst*, Vol. 35, No. 5, 9-1994.

22

Completeness of information

During a typical field study, performed several years after logging, data collection often takes more than half the project duration and resources, leaving a relatively small share to data analysis and integration. This is because the required data is scattered in different files or has not been recorded at all. It may also happen that details which seem trivial a few days after data collection are lost forever. Gathering a data set that is as complete as possible is therefore the highest priority. The first step of quality control is to check that it has been done properly. Cases of missing information are easier to repair if the problem is recognized early.

When polled about logging data, oil and gas companies emit the contradictory complaints that they get too much **and** too little of it. It is correct that the perfect data set is in the same time complete **and** concise. Each bit of information must be present, and present only once.

In this chapter the issue of where the data is recorded or archived is not addressed, as the whole Chapter 23 is devoted to the flow of data to different supports and databases. The different sections review the data items that need to be collected.

22.1 Units

Before listing the data objects that need to be collected, it is necessary to define the unit system that will be used.

22.1.1 Unit systems

The units exchanged by the users have to be clearly understood [4]. There are many examples of financial disasters and life-threatening events caused by a mixup of units [1]. A company lost several hundreds of thousands of dollars by mixing kiloWatt hours and therms. A plane almost crashed because of a mixup between gallons and liters. This is all the more confusing as unit systems may share the same abbreviation for different items. The customary American unit system uses M for thousand. For Système international,[1] M means one million. A US gallon is $3.7854 \; 10^{-3} m^3$. A British or Canadian imperial gallon is $4.546 \; 10^{-3} m^3$. A 20-pu porosity in limestone units is worth 24 pu in sandstone units.

It is worth reminding that a US billion is 10^9 while a British billion is 10^{12}.

22.1.2 Abbreviating units

Rigorous rules control abbreviations of units [3]. One of them is that abbreviations are case sensitive. F is the abbreviation for farad, while ft is the one for foot. If at all possible, the physical dimension of the quantity so defined must be apparent. Pound expressed as a mass is lbm, while pound as a weight is lbf.

22.1.3 Conversion from one system to another

Conversion from one unit system to another can generate inaccuracies. Some companies have specific conversion factors that are different from the actual factor. For example, one foot is exactly 0.3048 m; i.e., the conversion factor could be written 0.3048000 with all subsequent numbers being zeros. When converting meters to feet, the inverse conversion factor is used: 3.280839895, with subsequent numbers not equal to zero. An organization may use only 3.28. A length of 8000 m corresponds to 26,246 ft 8.63 in., while the "reduced" conversion factor yields 26,240 ft, a length 6 ft 8.63 in. shorter.

22.2 References

Many geometrical and petrophysical parameters require linking to a reference. These parameters do not have much meaning unless the reference is spelled out and understood.

---

[1] Abbreviated as SI, it was created in 1948 and superseded CGS, MKS and MKSA.

22.2.1 Concentration

Chemical and physical concentrations are quantities that require special attention. They are often expressed in percentages (%), parts per thousands (ppk) or parts per million (ppm). It is critical to define accurately what is the quantity that is dealt with: weight, volume, ionic concentration. Is a barite content expressed in terms of weight or in terms of volume? Is the mud salinity expressed in equivalent NaCl or Na^+? Mud engineers may deliver the information in one format, but the environmental correction software may require the input in a different format. Significant errors result from a confusion between the two.

22.2.2 Depth

A depth reference is critical in the exploration and production of hydrocarbons. Potential targets are defined from seismic sections after a time to depth conversion has been interpreted. Then different structures may be used during the drilling and production process. An exploration well may be drilled from a floating vessel or semisubmersible. Downhole measurements may be referenced to the rotary kelly bushing, which incidentally varies with the heave.

Then development wells may be drilled from a fixed platform and now referenced to the drill floor. Finally the drilling rig would be dismounted and replaced by a production tree. Workover or production logging measurements may be referenced to a different point. In some cases, the field would subside and the platform may slowly sink into the sea.

An affirmative monitoring and reporting of the depth reference is necessary. Depth may come in a variety of flavors:

(a) depth from seismic processing after velocity modeling
(b) wireline depth corrected for some environmental effects such as mechanical and thermal stretch
(c) driller's depth, which is, by convention, the sum of the lengths of the drillpipes measured at surface on individual stands
(d) depth of drillpipe-conveyed wireline tools
(e) depth of coil-tubing conveyed tools
(f) depth measured with slickline tools
(g) depth measured with an inertial platform.

These depths cannot be identical. It is therefore critical to identify the type of depth being considered. Mistakes occur when a type of depth is tied to another type of depth. It is better to report a difference of depth between two recordings than arbitrarily edit the depth of one measurement to match the other one.

On a log recorded from a floating vessel, record whether a wave motion compensator is used. The amount of vertical displacement needs to reported.

In a situation where the overall geometry or relative position of the logging tool and a device at surface, for instance a seismic source, is important, it has to be ensured that the location of every item is precisely known. A remark such as **air gun placed in the mud pit** is useless.

22.2.3 Location

Coordinates of a point in a horizontal well are often expressed as displacements from the wellhead. When several wells come into play, as for an anticollision application, it is necessary to determine the absolute coordinates of a point in the wellbore.

Longitude and latitude are believed to be invariant. In fact depending on the **datum** selected, the same figures for longitude and latitude could mean a different physical location.

For instance, in the North Sea, three different references could be selected [2]: ED-50, OSGB-36, WGS-84. Differences larger than 100 m could be observed. *Table 22.1* describes a format that lifts no ambiguity to the location of a wellhead.

Knowing exactly which grid and grid type (e.g., UTM) are used is a prerequisite for an accurate directional survey [5]. The industry has many examples of gross errors resulting from miscommunication in this area. Until recently, the zero-azimuth reference was seldom clear on a directional survey: was it grid North, geographical North or magnetic North?

TABLE 22.1
EXAMPLE OF LOCATION COORDINATES.
Note: the actual numbers are not coherent.

Latitude	59°45 min 20.09 s N
Longitude	02°43 min 04.41 s E
UTM Northing	6744371.7 m N
UTM Easting	539052.5 m E
Reference	ED-50 European datum 1950
Projection	UTM, zone 31
	Central meridian 03°East
Meridian grid convergence	0.54°

22.2.4 Porosity

Porosity is computed for conditions of matrix and mud characteristics that must be defined (Section 7.1). The same piece of rock with a density of 2.35g/cm^3 has a porosity of 18.2 pu in a sandstone, fresh mud filtrate compatible scale, but a porosity of 22.4 pu in limestone, salty mud filtrate compatible scale.

22.2.5 Pressure

Zero pressure may correspond to vacuum for **absolute** pressure or to atmospheric pressure for **gauge** pressure; a suffix g—e.g., psig—is added to denote gauge pressure. Gauge pressure is 100 kPa (14.7 psi) larger than absolute pressure.

22.3 Key information

22.3.1 Client and product name

The client name must be correct. Logging companies can easily find out why this is so by breaking this rule.

Difficulties arise from the multiplicity of names used for the same product. The product may be identified to the tool that acquired the measurement. Local variations may appear. For training and quality control it is useful that product name and product presentation be as stable as possible.

Conversely, the specificity of the acquisition conditions are to be clearly indicated. It is critical to know if the data is real-time data or memory data. It is also crucial to indicate whether a log is referenced to measured depth or true vertical depth. Indications in the depth track are useful.

22.3.2 Field and well name, run number

An item as simple as the well name can generate a large amount of confusion. The well name is the key attribute of a field database behind which all logging data is filed. In a well where multiple logging runs are performed, data may be directed to different areas of a database if the well name for the several runs is (even slightly) different. Discrepancies in well names may arise from

(a) spelling mistakes
(b) numbering mistakes
(c) abbreviations

- W. instead of West
- DW#2 instead of Dry Well #2
- Explo instead of Exploration

(d) side tracks
(e) multilaterals.

22.3.3 Date

American and European notations are conflicting. 8-10-99 means August 10th in the US and October 8th elsewhere. It is recommended to use an alphabetical abbreviation of the month (e.g., Aug-10-99) to avoid ambiguity.

22.4 Other general information

22.4.1 Tracking software

(a) A coherent set of software programs needs to be used for a given logging job. All calibration and logging phases should be covered with compatible software versions. For instance, if a tool is calibrated with the June 1998 version, then logged with the January 1999 version, with an algorithm change from September 1998, a systematic shift may appear on the data.

(b) It is imperative that all changes in the constants during logging are reported along with the depth (or time) where they were performed. These changes can have a disastrous effect on the value of the data and only an in-depth scrutiny of the print or field tape can detect them. As an example, all calibrations parameters should be frozen and the software should make their modifications **on-the-fly** during logging impossible or obvious.

(c) In the event of a computer crash, the status and values of the logging constants should be noted before restarting the operations. Such system crashes should be reported on the film.

(d) Level of processing. The same name is used for a parameter that can be used at different levels of processing.

The multiplicity of outputs presently allows the user to obtain the most suitable information, but it may also become a source of confusion.

The software handling thermal neutron porosity acquisition gives the option of three different thermal neutron curves—NPHI, TNPH and NPOR—with three different characteristics:

- **NPHI** is identical to the output delivered by 1970s panels. It is computed from instantaneous off-depth counting rates. The rates are not dead time corrected. The so-called mod-8 porosity transform is used. NPHI is not changed if a different nuclear source is used.
- **TNPH** is computed from dead time-corrected and depth- and resolution-matched counting rates. Mod-8 porosity transform does not apply. TNPH may be shown after application of environmental corrections—temperature, pressure, salinity, standoff, mudcake and mud weight.
- **NPOR** is the result of enhanced resolution processing.

22.4.2 Other information

(a) Complete and correct casing information (depth, weight, size). This information is needed to verify caliper accuracy.
(b) Complete and correct bit information (depth, size).
(c) Listing of the logging services run during the same trip to the well.

22.5 Remarks

Only logging related items are usually included in the **remarks** section. Operating engineers should not be afraid to add drilling circumstances or mud characteristics. Drilling conditions (for instance, BHA changes) may directly impact logging data by the bias of hole condition. Examples of important information to be reported in the remarks section are listed:

(a) Hole data (deviation, kickoff point, under-reaming, drilling conditions that might have affected hole condition).
(b) Anomalous readings; explanation of the extra repeat sections required to validate data.
(c) Any deviation from the operational procedures dictated by the oil/gas or the logging company. If one procedural step had to be cancelled (for instance the logging passes could not be repeated because of high sticking risks), the reasons must be clearly stated.
(d) Report whether the film is a field original or a playback. In the second case, report the reasons for playing back data.[2]

22.6 Logging parameters

22.6.1 Mud data

Mud parameters and circulation timing are essential in the understanding of invasion, which affects most logs.

Many measurements are sensitive to changes in mud characteristics (type, density, viscosity, pH, salinity, temperature, presence of barite, presence of potassium). Every

---
[2]On a *Schlumberger* film, a playback film is recognizable by larger gaps between time marks.

time the mud is changed in weight or composition, a new set of mud parameters must be reported. Mud-related resistivities are measured with a mud tester along with the temperature of the measurement.

As some tools cannot perform a satisfactory measurement in oil-based mud or salt-saturated mud, these conditions are to be reported.

Some details on the mud circulation timing need to be collected: Time when circulation was stopped and time the sensors were at bottom. Extra information on mud circulation is useful to analyze static and dynamic filtrations, which control invasion.

22.6.2 Environmental data

As the accuracy of a measurement strongly depends on the control of the environmental corrections, whether or not environmental corrections have been applied must be clearly stated. The inputs to the corrections are important. Eventual changes with depth and time are to be reported.

A good way to find the necessary parameters is to review the inputs to charts that are required in the current logging company chartbook.

Examples

- How is the dual laterolog sonde positioned in the hole (centered or eccentered)?
- What are the barite and potassium contents in the mud?

22.6.3 Signal processing characteristics

Signal processing may be beneficial to the raw measurements by reducing the amount of noise. Conversely, excessive filtering tends to obliterate the fine features of the formation. For these reasons, the characteristics of filtering need to be made known to the data user.

22.6.4 Temperature

It is important to use three bottomhole thermometers on each run. Differences between the three readings may be anticipated and even strictly identical readings could be suspicious. Maximum temperature should differ from one survey to another as the time span between the end of the mud circulation and the moment the thermometer reaches total depth differs between logging runs. Temperature information is critical to obtain true formation temperature through techniques using Horner plots. Geologists use the geothermal gradient in the reconstruction of a basin's burial history.

Most logging companies combine the listing of the tool equipment numbers with a tool sketch. Distance to the zero of the tool string are also indicated.

22.7 Tool description

Tool sets should not be identified by color code, but all the set components should be identified in detail (*Fig. 22.1*). Equipment numbers should agree with the corresponding information on calibration tails. The last alphabetical character of the tool name helps identify the type of equipment (*Table 22.2*).

Auxiliary equipment to be listed

(a) stabilizers size and location on the string
(b) centralizers
(c) excentralizers
(d) standoffs.

TABLE 22.2
EQUIPMENT ABBREVIATIONS

Letter	Type	Example
S	Sonde	IRS, SLS
C	Cartridge	IRC, SLC, CNC
M	Module	DLM, NLM
H	Housing	PDH, HDH
T	Tool	SHDT, DLT

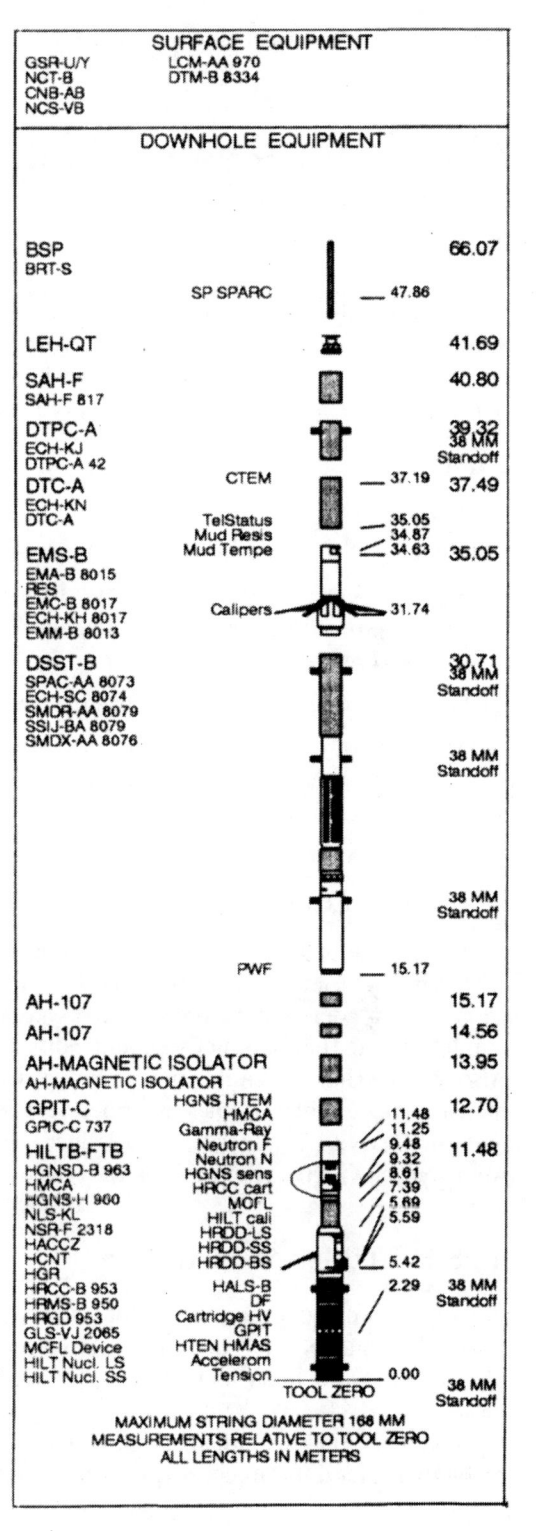

FIG. 22.1 : Example of tool sketch.

22.8 Indexed data

22.8.1 Guidelines

The bulk of the information is called dynamic information and is indexed versus depth or time. Data objects in this category are called channels or arrays. For conciseness, only the canonical channels should be archived. This is not done at the expense of completeness as every bit of information can be restored from these channels.

Example

A resistivity device is calibrated with a gain and an offset. It is then averaged with a three-level (0.25, 0.5, 0.25) filter. The data objects that need to be preserved are

(a) the raw uncalibrated resistivity data channel, R_{raw}
(b) the two calibration parameters, gain G and offset O
(c) the three filtering weights, $a_1 = 0.25$, $a_2 = 0.5$ and $a_3 = 0.25$.

From this simple data set, all desired presentations can be created:

(a) a calibrated unfiltered resistivity, $R_{cal} = G \times R_{raw} + O$
(b) a final—filtered—resistivity:

$$R_{final_d} = 0.25 \times R_{cal_{d-1}} + 0.5 R_{cal_d} + 0.25 \times R_{cal_{d+1}}$$

(c) a conductivity, $C_{final} = \frac{1000}{R_{final_d}}$.

It is not necessary to clutter the database with a calibrated channel and a conductivity channel that can be recomputed by an application. Keeping only the canonical channels is a good way to preserve the consistency of a database. In the previous example, assume that conductivity and calibrated channels are present in the database. Also assume that the calibration gain is changed as a master calibration after logging indicates the original coefficient is incorrect. If gain G is modified, then the calibrated and conductivity channels are no longer consistent with the data set because they were computed with the previous calibration coefficients.

22.8.2 Time-indexed data

As reviewed in Chapter 10, the rawest form of data is indexed to time. A few irreversible and poorly tracked processes convert this data into depth-indexed data. To have full traceability along the data flow, it may be desired to file this time-indexed data.

The advantage of indexing information versus time is that there cannot be any ambiguity on a data object acquired at a given time, while this statement is not true for the same object acquired at a given depth. For instance, resistivity at depth 8465 ft can be acquired on several MWD runs, then with a wireline run on its main pass, and again, on the repeat pass (*Fig. 22.2*).

Time-lapse applications, such as monitoring invasion with time, are better served with time indexing.

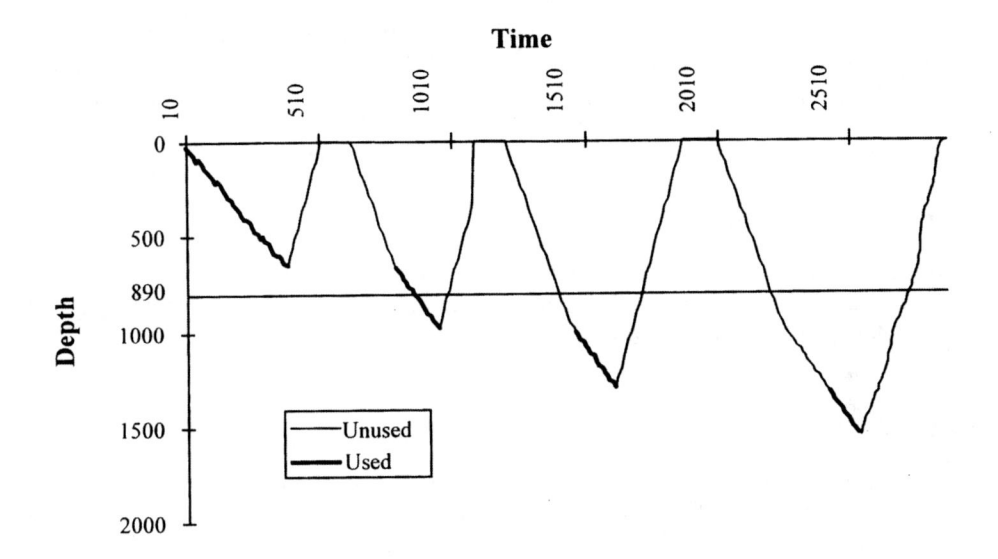

FIG. 22.2 : Time-lapse recording.
MWD tools are facing the same formation twice for each run. The shallow intervals are surveyed even many more times.

22.8.3 Caveat

22.8.3.1 Absent values

Some organizations use a special number for missing, absent, or unavailable values. (e.g., -999.25). This value has to be recognized by further processing as an absent value, not -999.25. Nine numbers around 200 and an absent value are listed. The average of these 10 values should be 200. If the software does not recognize the absent value, the computed average is 80.

22.8.3.2 Wrong values

A survey of the values collected by observatories for the Earth's magnetic field has highlighted that a database could have bytes but be void of information. Values for the Earth's field of 96,000 nT have been found (while this value should always be around 50,000 nT).

22.8.4 Auxiliary indexed information

22.8.4.1 Rate of penetration

Rate of penetration is analogous to cable speed in wireline logging. A fast drilling rate means fewer data points over a given interval. In severe cases, saw-tooth patterns may be observed. Repeatability of nuclear measurements is affected negatively by high rates.

22.8.4.2 Time after bit

In MWD, most sensors are not exactly at the bit. It takes a finite time for the drilled formation to be surveyed by these sensors. This time is required to understand the invasion and breaking-out process.

22.8.4.3 Sliding indicator

MWD tools are subjected to two drilling modes. In rotary mode, the drillstring and the bit are rotated from the surface. The sensors scan the formation in all directions around the axis of the sub. In sliding mode, only the bit rotates and the drillstring slides along after the bit. The sensors do not rotate. Some measurements may be invalidated by the lack of rotation (e.g., dip computation from resistivity images). The sliding indicator—or the continuous recording of drillstring rotation at surface, rpm—is useful for highlighting those intervals where the drillstring was sliding.

22.8.4.4 Tension and sticking indicator

As seen in Section 19.3, monitoring of tension is critical to identify zones where the logging tools are not moving. It is preferred to have surface tension and downhole tension curves. A sticking indicator displays this information in a flag. The sticking indicator may be used to alter the color or coding of the main curves.

22.8.4.5 Quality curves and flags

Chapter 18 insists on the requirement of monitoring and tracking quality curves (QC) and flags. QC curves are shown with limits where the main curves are deemed to be valid. These limits are fixed after thorough field testing and failure analysis. This information is useful to the data user.

Additional curves, containing diagnostic data, may shed some light on the validity of the main information, but they are generally useable only by tool designers or maintenance specialists. This information may not be required from the log supplier.

22.9 Additional quality features

22.9.1 Repeat sections and additional passes

As seen in Section 4.5, repeatability is an in-situ check of the precision of the measurement.

In addition, it is imperative that any anomaly is relogged except if it represents a major risk of sticking. A few minutes taken to rerun the log over a dubious zone could save considerable time later when a multiwell study is undertaken.

Often, only an optical record of repeat and additional passes is kept. This prevents this data from being fully used when needed. These passes must be recorded digitally.

22.9.2 Calibration parameters

Instead of cluttering the data set with calibrated channels in addition to raw data, it is more efficient to preserve calibration parameters digitally.

22.9.3 Directional surveys

As evoked in Chapter 20, directional surveys also belong to the data set.

22.9.4 Quality control tail

This insert can be completed by the data collector or the data user and kept attached in order to avoid duplication of control later. It can be made as a detachable component so that it can be separated from the log and addressed to the logging company organization.

22.9.5 Time marks

Time marks are used to verify that a correct recording speed has been used during logging. Generally, a specific tick is shown every minute (*Fig. 22.3*). The distance in meters [feet] between two time markers corresponds to the speed in m/min (ft/min).

The rate of penetration in MWD can be checked with two graphical presentations, the ROP curve, a critical element of any MWD log, and the time marks. A time mark, or tick mark, is present every time a data sample has been collected (*Fig. 22.4*).

338   22. Completeness of information

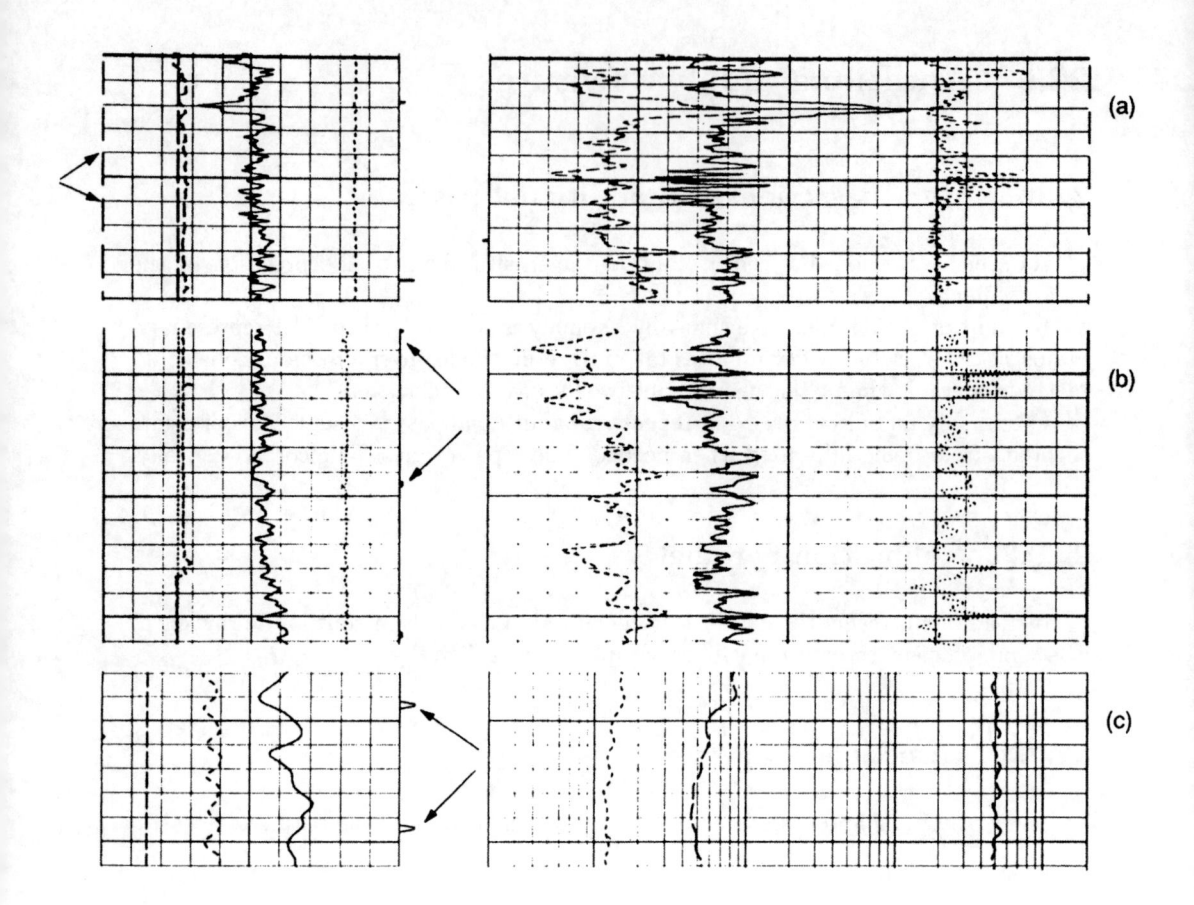

FIG. 22.3 : Examples of time markers.
a. *Schlumberger.* b. *Halliburton Logging Services.* c. *Western Atlas.*

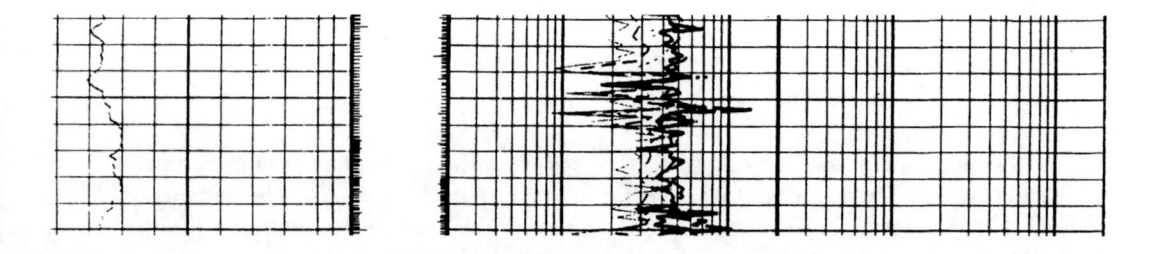

FIG. 22.4 : Examples of tick marks and ROP.

22.9.6 Depth editing

For the time being, the amount of cranking (Chapter 19) performed to correct depth is not captured in a logging record.

22.9.7 Dialogue between the logging engineer and the surface system

In some cases, the changes of important parameters during logging are not recorded on tape. They used to be available only from the scroll, which gathers the dialogue between the logging engineer and the surface acquisition computer.

22.10 Housekeeping recommendations

22.10.1 Be consistent

As a general rule, one input should be keyed in once and only once.

Examples

1. The hole is cased or open; it cannot be cased for one sensor and open for the other.

2. The scales should be limestone compatible for density-derived **and** neutron porosities.

3. The density ρ_b and its correction $\Delta\rho$ should have the same scale.

22.10.2 Don't throw anything away

Data collection requires that all pieces of information are kept to help understand the circumstances of the logging job.

Examples

1. Logging passes run because of the previous observation of anomalies are to be kept in optical and digital form. It has often occurred that the most important information was available on the repeat passes, not on the main pass.

2. Overlaps from one run to another are to be kept to ensure the traceability of the depth information.

3. One dangerous process is to compose separate runs as if they were a single one. In that case, overlapping intervals are cutoff and data between different runs is even interpolated (even though, in the two runs, environmental conditions are drastically different).

22.10.3 Avoid cosmetic surgery

The purpose of logging is to get data loaded with information, not data that looks good and smooth. The oil industry has condoned manipulation of data, while data should be left as clean and pristine as possible.

These common cosmetic effects are in common use:

1. inappropriate filtering of data

2. interpolating to avoid the visibility of missing data

3. normalizing to values expected by the data user

4. omitting the presentation of quality flags that indicate nonconformance

5. removing of anomalies linked to the formation because they are thought to originate from tool malfunction.

22.10.4 Maintain experts' presence at the wellsite

Even with optimal reporting and improved data transfer technology, it is impossible to completely document the facts taking place at the wellsite. The vibrations of the rig when the bit has such and such behavior can only be recorded and transmitted with difficulty. Still, they are felt by the driller—and by the logging company engineer—and used to make decisions. The conditions at the wellsite are so special that they can hardly be guided from a distance.

22.11 Consequences of incomplete information

The following questions need to be answered:

1. What is the real meaning of the curve displayed?

2. Which constants are the most critical during acquisition?

3. Which constants are critical for proper environmental corrections?

4. Where is it possible to find the calibration on the tape?

5. How can multiple sampling rates be identified?

6. Has the data been filtered?

7. How much data is lost as a result of irrecoverable parity errors?

8. In the case of data transmission, is what we sent being received?

The lack of information may lead

1. mistaking one curve for another

2. performing environmental corrections twice

3. getting systematic errors as the result of incorrect constants selection.

Example

Two parameters are related to the choice of lithology in thermal neutron logging. MATR relates to the lithology corresponding to the scaling of the input data. For instance, a choice of LIME gives limestone-compatible porosity units. POUT relates to the scaling of the output data. *Table 22.3* indicates the value of the output porosity for different choices of MATR and POUT starting from NPHI equal to 20 pu.

TABLE 22.3
EXAMPLE OF INCORRECT
CONSTANT SELECTION

POUT →	SAND	LIME
MATR ↓		
SAND...	20.0	15.8
LIME ...	24.4	20.0

A poor understanding of the impact of the constants could lead to catastrophic results in terms of formation evaluation.

References

1 Duller, P., personal communication, 1997.

2 Jamieson, A., personal communication, 1998.

3 Moureau, M., *Guide pratique pour le système international d'unités (SI)*, Éditions Technip, Paris, 1980.

4 Society of Petroleum Engineers, *The SI system of units and SPE metric standard*, Richardson, 1984.

5 Wright, J. W., "Directional drilling azimuth reference systems," IADC/SPE 17212, *Trans* IADC/SPE drilling conference, Dallas, 1988.

23

Data management

> If you don't check it, it might be garbage.
>
> *Louis Vermeulen*
>
> I have made this [letter] longer than usual, only because I have not had the time to make it shorter.
>
> *Blaise Pascal, 1657*

23.1 Introduction

After following all the recommendations of the previous chapter, a complete and optimal data set is available on the surface system. It is composed of strings of bits that cannot be used directly. Logging information is delivered

1. on screens, as real-time displays

2. as a digital data storage record (tape, floppy, CD-ROM)

3. as a digital file transmitted directly from the wellsite surface system to another computer

4. as a print or film

Because important decisions can be taken from screen displays (e.g., steering the well, stop drilling to set casing or to initiate coring), it is useful to keep screen digital dumps for future reference and traceability. This enables a learning process.

The digital physical record (tape, floppy, CD) is a way to move large data blocks from one computer to another; only a subset of the complete data set can be transmitted because of bandwidth limitations. The print is still favored for the display of logging data. In the past, prints contained information that was not available in a digital form. In the 1960s, all information was only on print, then in the 1970s basic channels were encoded to tape although variable density waveforms and master calibrations were not yet imported.

Today, all raw data (waveforms, arrays, etc.) has the potential to be ported to a digital support. Digital records are more and more popular for the following reasons:

1. Films and prints occupy important volumes.

2. They require trained archivists to file and retrieve data.

3. They incur inconvenient and costly procedures when moved.

4. Data on magnetic/optical supports occupy small volumes.

5. They can be transferred and easily retrieved, quickly loaded on an interactive graphics workstation and efficiently upgraded with high power megapixel/megaflop computers.

This preeminence is not yet obvious in log quality control. Log quality control procedures are mainly, if not uniquely, directed toward print and film. In the future, software may be developed that checks logging data solely from tape, in a way similar to personal computer diagnostics programs.

In this digital age, it seems that the management of logging data should be a well-controlled and slick process. Reality is different. Problems arise from the confusion on whether data should be kept or not, **which** data should be kept, which format should be used. The following issues concerning data are addressed in this chapter:

1. transfers

2. storage media

3. models

4. formats.

23.2 Data transfers

Data can be transferred in two different ways, manually (handwriting or keying an entry on a computer) or digitally (bit by bit). The first mechanism (*Fig. 23.1a*) is error prone but allows some form of control and filtering. The second is error free, (*Fig. 23.1b*), but garbage information is faithfully transferred without limitation.

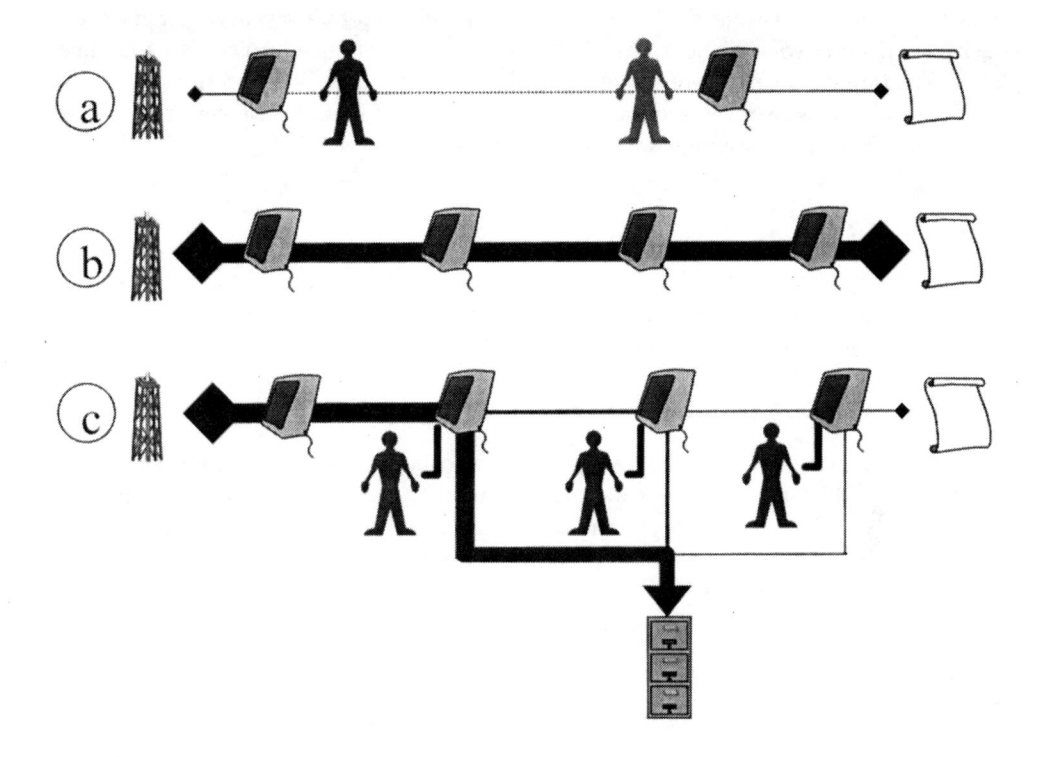

FIG. 23.1 : Three data transfer schemes.
- **a.** Data flow is interrupted by manual transfers.
- **b.** All data is digitally transferred.
- **c.** The optimal process is described. All data is transferred digitally. In addition, it is controlled at every stage of the process. Data that is not immediately needed is archived.

23.2.1 Reality

The first case described by *Fig. 23.1* represents reality or, at least, the most common scheme of data flow. Data is transmitted digitally to surface, then some data (typically the directional survey data) is keyed in manually into a different computer. Other data may be filtered and a limited data set is transferred on a floppy disk. Following the requirements set by the oil and gas data user, the field engineer selects the data objects that are ported to the diskette. Finally, some other channels may be selected to go to a DAT tape or CD-ROM. The rest of the data corresponding to an acquisition run is stored in an **internal** digital record or thrown away. Internal records may be erased a short time after the acquisition job.

The overall process is subject to human errors and is not controllable. In addition, the persons selecting the objects to be kept or thrown away may not have the background knowledge or be aware of the implications of their choice. For instance, for a directional survey, there could be as many as 25 manual transcriptions of the information from sensors to a data archive. This type of manual transfer is also used to key in calibration parameters and errors often result.

23.2.2 Digital transfer of data

The second process corresponds to the digital transfer of all data from the wellsite to the decision-maker.

Example of digital transfer: calibration of sensors

One shop system is used to calibrate a sensor. The software derives the set of calibration coefficients. These coefficients are then transferred digitally into the tool memory. The calibration information is then transferred back to a wellsite surface system. In the whole process, there is no human intervention in the data flow.

When used for every bit of data, this process reaches limits because of the shear volume of information. The decision-maker is overwhelmed and confused.

23.2.3 Sorting and archiving data

An improved process is depicted in *Fig. 23.1c*. All data transfers are digital. No data is thrown away. It is sorted in directly usable sets and sets that can be archived. At different steps in the data flow, there is human intervention—no editing, but data control and interpretation—so that information reaches the decision-maker in a digested and controlled way. The original data is not modified but is given added value at every step.

23.3 Data storage media

23.3.1 Historical summary

The first magnetic recordings of logging data were performed by the *Humble Oil Company* on dipmeter information. In the early 1960s, *Schlumberger* introduced the Truck Quantizer for processing dipmeter field tapes with *IBM* computers. A few years later, the application of tape recorders was extended to other logging data. The tape storage was of limited flexibility. The introduction of wellsite computers permitted storage of all information related to a logging job.

23.3.2 Current data storage media

Data storage media technology is changing rapidly. Currently available media are listed in *Table 23.1*.

TABLE 23.1
DIGITAL DATA STORAGE MEDIA

Medium	$\dfrac{\text{Size}}{\text{(megabyte)}}$	Storage
3.5-in. floppy [5]	1.4	Equivalent to one set of prints
600-ft 1600-bpi tape...	10	Current standard format for logs
3600-ft 6250-bpi tape..	160	Sonic waveform, images, borehole seismic
WORM[1] magnetic disk	600	Entire exploration well log data
WORM optical disk...	2000	Complete data base (old logs)
DAT[2] tapes...........	1200	Entire exploration well log data (old type)
DVD	9600	Complete data base of modern logs

In selecting the appropriate storage medium, a number of questions needs to be addressed:

(a) What is the durability of the format?[3] How vulnerable is the support to X-rays, magnetic fields, dust, heat, humidity and handling damage?
(b) Is there any backup if the primary support is destroyed or damaged?
(c) What is the storage cost?
(d) What are the error checking and recovery procedures?
(e) Is there a program or even a piece of hardware (e.g., tape reader) to recover the data?
(f) Is all data available on a single data format?

23.3.3 Archiving of data

Archiving needs to be done by a professional organization. Dust, temperature and humidity are conditions that need to be controlled. Keeping two independent archives is also condoned.

---

[1] Write once, read many.
[2] Digital audio tapes.
[3] Most logging companies are committed to store data on a digital medium during a limited time, from several months to a couple of years. It is recommended that oil/gas companies store data in a more durable way.

23.3.4 Tape quality checks

The tapes and floppy disks being the most used media, it is useful to review a few precautions about their handling.

(a) A tape reel or floppy label has to be completed.
(b) The tape or floppy is protected in a shielded case when it is sent by mail or courier. On the package, a sticker showing prominently that the parcel contains magnetic data is added.
(c) One of the best ways to check the validity of a tape or floppy is to read it. Generally, this operation produces a verification listing that is a useful document even if it is not too friendly at the first glance.
(d) Using scratch tape or floppy to record new data sets is not recommended. Previously recorded data has a good chance of being decoded and loaded along with the new data, which may cause confusion.

23.4 Data models

Data cannot be stored on a medium at random. It must be organized so that it can be efficiently stored and then retrieved. The structure of a data base is defined by a data model. Building a data model takes years of multidisciplinary teamwork. It requires full understanding between the groups designing logging tools and the ones using the data. Classification of data cannot be performed without an in-depth knowledge about the data flow and the precise purpose of the information. Examples of data models are:

(a) SDM, the Schlumberger data model
(b) GDM, the Geoshare data model[4]
(c) Epicentre, the POSC[5] data model.

Data models are built in a way that information is classified in different categories so that it is easy and intuitive to fill the data base in an orderly manner. Unclassified data goes in a "Others" folder. Indeed, an optimal balance is to be struck so that not too many nor too few items go in this category.

---

[4]The changes of GDM are managed by a specific group, the Geoshare Users Group (GUG), whose members meet regularly to consider modifications.
[5]POSC is the Petrotechnical Open Software Corporation, whose mission is to facilitate integrated business processes and computing technology for the exploration and production segment of the international petroleum industry.

Example

The Geoshare data model has no place for calibration, correction or filtering parameters. The paradigm of Geoshare is that the data stored in Geoshare is controlled and qualified as "validated." Calibration, correction and filtering parameters may be stored as "others." An application designed to recalibrate a sensor or improve corrections may require some manual search of the "others" folder as the required parameters are not identified.

23.4.1 Data objects

Data objects could be parameters, digital channels, optical curves. They could be scalar, vectors or arrays. Each data object is generally described by a short name (mnemonics) and by a long name (description).

Data usability has been restricted by poor management of these objects. Typical problems are

- the lack of guideline in the naming of objects
- obscure mnemonics
- multiple descriptions for a single data object
- the use of the same description by distinct data objects.

23.5 Data formats

The most significant problem in tape retrieval is negotiating the maze of data formats. Sometimes, a data user attempts to decode data with a software incompatible with the recording software (most of the time, recording software has many more features than the reading software). It is therefore essential that compatibility of recording and reading software be checked long in advance.

Available tape formats are:

(a) ASCII
(b) LAS
(c) *Schlumberger* LIS
(d) *American Petroleum Institute* DLIS
(e) TIF.

23.5.1 ASCII

ASCII is the **American Standard Code for Information Interchange**. It is a way of representing printing and control characters as binary values. An ASCII file usually means that the file makes some sense when the bytes it contains are interpreted as ASCII codes. If an ASCII file of log data is requested, it is necessary to define precisely what is required: how many columns of data, what characters are used to separate the columns, how many digits in each value, etc. This can be difficult to pin down in detail. Failure to define this structure in advance guarantees the file will be unreadable.

23.5.2 LAS

The **Log ASCII standard** (LAS) format [4] has been created and defined by the *Canadian Well Logging Society*, CWLS. It is a form of ASCII file with a better defined structure.

Originally introduced in 1989, its intent was to standardize the organization of digital log curve information for floppy disks. Version 2.0, introduced in 1995, has standardized well information, clarified the location of units and improved documentation.

23.5.3 LIS

The **Log Information Standard** (LIS) was designed by *Schlumberger* [6] as a mechanism for storing log data in digital form. It has been revised on a couple of occasions, but retains the same characteristics. It defines various "record types," which in the early days corresponded to physical records stored on magnetic tape (one movement of a magnetic tape recorder, for example). LIS does not say what medium is used to store the records, hence LIS may exist as a continuous stream of bytes in a disk file. In that case, some mechanism must be added to the LIS data to define record boundaries (e.g., TIF). These added bytes do not form part of LIS.

LIS distinguishes three types of information associated with well logging:

(a) **Data frame**. It is a collection of sensor readings entered in conjunction with an index value. This index may be of two types:

- For tapes recorded in the field, the index is recorded only once at the beginning of a sequence of data frames.
- For tapes recorded with a computing center software, the first data channel of each frame is dedicated to the index.

An example of the quantity of data contained in a data frame is shown in *Fig. 23.2*. The record, related to a density/thermal neutron porosity/spectral gamma ray combination, includes 90 channels.

```
-------------------------------------------------------------
VERIFICATION OF ONE DATA FRAME --- TYPICAL CSU EDIT TAPE
-------------------------------------------------------------


Data record type   0 DEPT 1187700  .1IN 9897.5 FT 3016.758 M
-------------------------------------------------------------
BS    12.25     TOD 256619439   TIME 8824.      ETIM 8.829     CS   203.8      DIFF -0.6
TENS  2554.     MARK 0.         RCAL 0.         CALI 12.25     IHV  1.228      ICV  0.4697
SHVD  1310.     LHVD 1270.      RLLL 39.75      RLUL 161.4     RLLU 130.4      RLUU 2.484
RSLL  89.44     RSUL 145.3      RSLU 135.4      RSUU 18.63     LL   134.5      LU   261.3
LS    265.      LITH 29.88      SS1  207.5      SS2  273.5     PARI 0.         LSHV 1270.
SSHV  1310.     FFSS -0.004883  FFLS -0.03223   RLL  134.5     RLU  261.3      RLS  265.
RLIT  29.88     RSS1 207.5      RSS2 273.5      SLDT 17        PEF  -72.81     S1RH 2.367
LSRH  2.43      LURH 2.447      QRLS 0.6147     QRSS 0.751     QLS  -0.04834   QSS  0.09424
RHOB  2.453     DRHO 0.0166     DPHI 0.1499     IRHO 1.593     RNRA 3.936      FCNL 477.
NCNL  1878.     NRAT 3.66       NPHI 0.3643     SCNL 0         RHGA 0.         V1M3 0.
V2M3  0.        V3M3 0.         MDM2 0.         DPL  0.        NPL  0.         PHIA 0.
U     0.        UMA  0.         CNPH 0.         RHGX 2.986     RW5N 6.211      RW4N 6.211
RW3N  17.39     RW2N 55.88      RW1N 149.       RSGR 62.25     MRES 0.09863    MTEM 213.5
DTEM  0.        W5NG 6.211      W4NG 6.211      W3NG 17.39     W2NG 55.88      W1NG 149.
THOR  0.        URAN 0.         POTA 0.         SGR  0.        CGR  0.
HTEN
          242.6           248.3         251.         251.          248.3           245.4           251.           248.3
          251.            248.3         242.6        248.3
-------------------------------------------------------------
```

FIG. 23.2 : Data frame format.
The data originates from the field.
Courtesy of Schlumberger

(b) **Transient information**. It consists of the dialogue between the system and the logging engineer, in addition to comments and messages.

(c) **Static information**. It consists of information about the logical structure of the reel or disk file. They are used to describe how data frames are formatted.

23.5.4 DLIS and RP66

Digital Log Interchange Standard (DLIS) was originally developed by *Schlumberger* as an expansion of LIS. It has been accepted by the *American Petroleum Institute (API)* as an industry standard under the name RP66 [2, 5 and 7]. It is supported by the MAXIS* acquisition system and has not been implemented on the CSU* system. Converters exist on the MAXIS system to convert between LIS and DLIS, though with some limitations.

The rationale behind DLIS is that when more complex logging tools were introduced, people handling log data were slowed by the limitations of the existing tape formats. One of the major problems is the wide variety of data types (among others, waveforms and arrays) and record length (from a few bits to several thousand bits). The existing formats have difficulty handling variable sampling rates, which can eventually be recorded during the same logging run.

The DLIS format was developed to address these limitations [5]. It has the following features:

(a) ability to contain both standard and auxiliary logging data
(b) possibility to merge, splice and flip log data
(c) presentation of channels regardless of sampling rate and dynamic range
(d) allowance for complex forms of data including arrays
(e) record of indefinite length
(f) textual data capabilities
(g) encryption capabilities.

23.5.5 Encapsulation

23.5.5.1 TIF

Tape Image Format (TIF) is a method taking any physical record that is normally written to a magnetic tape and writing it to a continuous stream of bytes in a disk file. The physical records may or may not contain LIS information. Basically, TIF adds 12 bytes to each physical record (a back pointer, forward pointer and end-of-file counter, each a 4-byte integer). TIF description is used when writing LIS data to floppy disks for use in computers using PC-based interpretation software. The expression "a TIF file" may sometimes be used to mean "an LIS file with TIF encapsulation." TIF may also be used to encapsulate SEG-Y files.

23.5.5.2 Other

There are many other ways to create an archive domain (set of files): TAR (more appropriate for Unix operating system, NTbackup (NT), VMSbackup (VMS). In all cases, it is necessary to define precisely the format of the files to be bound together.

23.5.6 Graphics

The **Picture Description Standard** (PDS) enables the encoding of images into a sequential steam. This standard has gained popularity with the availability of an application, PDSVue, that is built for viewing PDS files.

The **Computer Graphics Metafile** provides a format for capturing and retrieving 2D pictures. This picture information can be exchanged between different software systems and graphical devices. *POSC* has refined the requirements for Exploration and Production applications into the CGM Petroleum Industry Profile, CGM*PIP.

23.6 Attempts of standardization

Considering the confusion on the data model, the data medium, the data format and possibly the preferred archiving /encapsulation process, a number of organizations have attempted to set standards on all these aspects of data management:

(a) CDA covers log data acquired in the United Kingdom.
(b) MMS applies to the USA.
(c) CANOGGIS [3] governs standards for Canada.
(d) Rules have been set by the *Norwegian Petroleum Directorate* for Norway (DR-2 HQLD-2) [8].

Examples of requirements are as listed:

(a) All data should be on 8-mm exabyte cassette in Unix tar format (floppy diskettes are no longer accepted).
(b) Data format is LIS-TIF, except for wave train data, which requires DLIS.
(c) Optical scans should be in compressed TIFF.[6]
(d) VSP data shall be in SEG-Y format.
(e) Discrete well path data should be related to RKB[7] and reported in ASCII format. Continuous wellpath should be shown by increments of 6 in. converted to meters, on a UTM projection with grid North as reference.

23.7 Prints and films

Even though logging data can be analyzed on a screen, it seems that the "log format," a piece of film or paper 23 cm (9.25 in.) wide, will stay popular. It is important to be aware that digital data and print/film data are almost always different. A complete digital data set contains several orders of magnitude more information than a print.

(a) The print does not contain any raw data that is helpful to recompute usable information (recalibration, change in filtering, new tool response).
(b) Print information is decimated because the printer cannot restitute the definition of the raw data. For instance, the original data set could be sampled every 5 cm (2 in.). The data sets obtained with the microresistivity and electromagnetic propagation tools belong to this category. *Figure 23.3* shows data originating from a tape and from a field print.
(c) The print does not contain auxiliary information such as the accelerometer data that is used for speed correction.

---

[6]Different from TIF, described in Section 23.5.5.1, TIFF is the Tag Image File Format.
[7]RKB is the rotary kelly bushing.

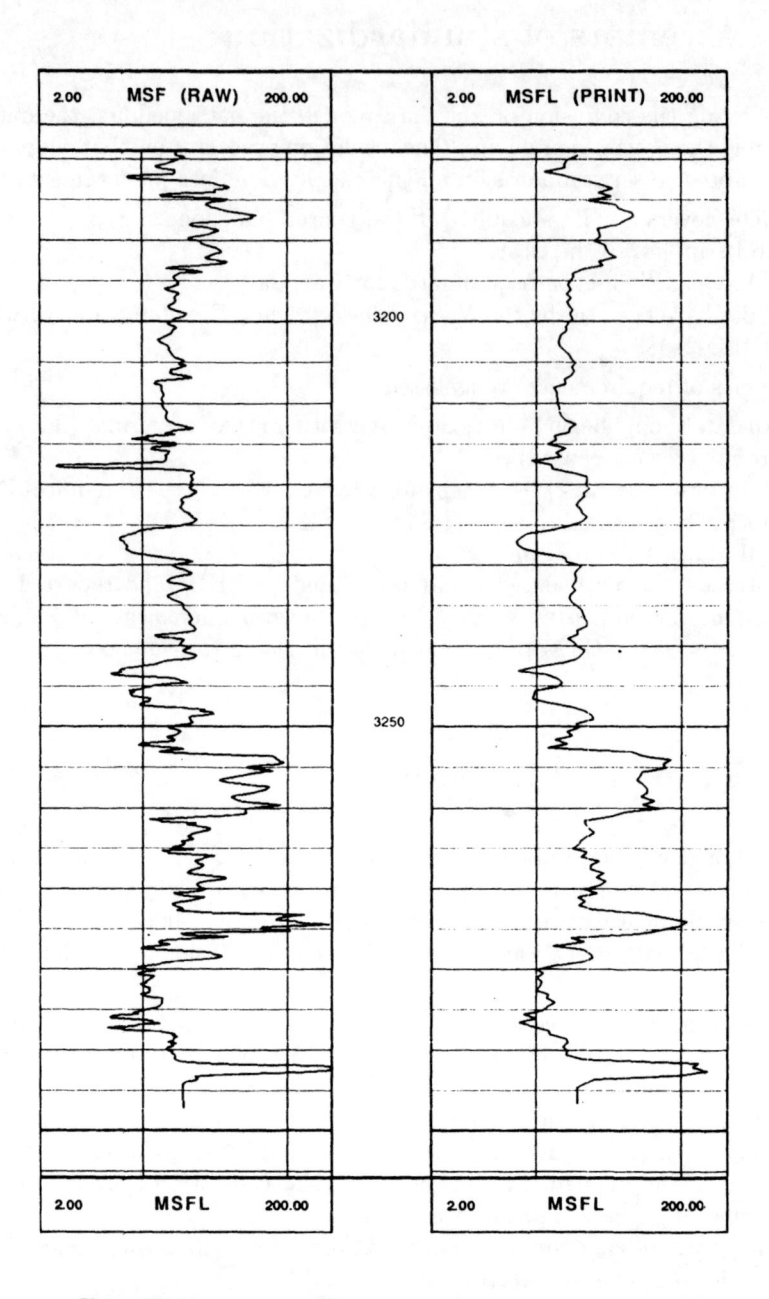

FIG. 23.3 : Raw (taped) and film data.
The raw data has no smoothing and contrasting with the print data, is not decimated.

(d) Array tools may contain dozens of transmitter-receiver couples (28 for the AIT tool). The optical curves are the result of a complex combination of these arrays. The number of final curves is limited (5 for the AIT tool).
(e) Formation images cannot be practically delivered in analog format. The usual scales are 1/5 and 1/50, which give a very large volume of paper printout. The choice of scale and format scale depends on the use of the data, which may be difficult to know in advance. This type of data is better handled in a workstation environment.
(f) Formation tester pretests or sample buildups are not shown on print/film on a small enough scale for detailed permeability analysis.
(g) High-resolution data, for instance sampled every 3.05 (1.2 in.), cannot be properly discerned on a standard 1/200 (or 1/240) depth scale.

23.7.1 Standard presentation

In spite of these limitations, the print/film is an essential document for quality control. It is recommended that the print/film is presented in a consistent and standard manner to allow fast and easy recognition of missing elements. *API* has developed such a standard, RP31 [1]. It facilitates quality control because the auditor always finds the log components in the same sequence. This sequence is selected jointly by the oil and service companies so that their needs are covered.

While flexibility is advocated because it allows the satisfaction of specific needs, some degree of standardization is encouraged. Prints/films are exchanged between partners or between local branches and headquarters. A special presentation request made locally may cause some embarrassment to the subsequent users.

The recommended components, starting from the bottom, are

(a) quality control form
(b) tool quality records
(c) shop calibration(s)
(d) before survey verification(s)/operational check(s)
(e) after survey verification(s)
(f) sections repeated because of suspicious readings
(g) casing check
(h) repeat section
(i) additional passes
(j) down log (optional)
(k) main log
(l) tool configuration sketch
(m) parameter/constants/environmental corrections summary
(n) well sketch
(o) heading.

An example of presentation is given in *Fig. 23.4*.

356   23. Data management

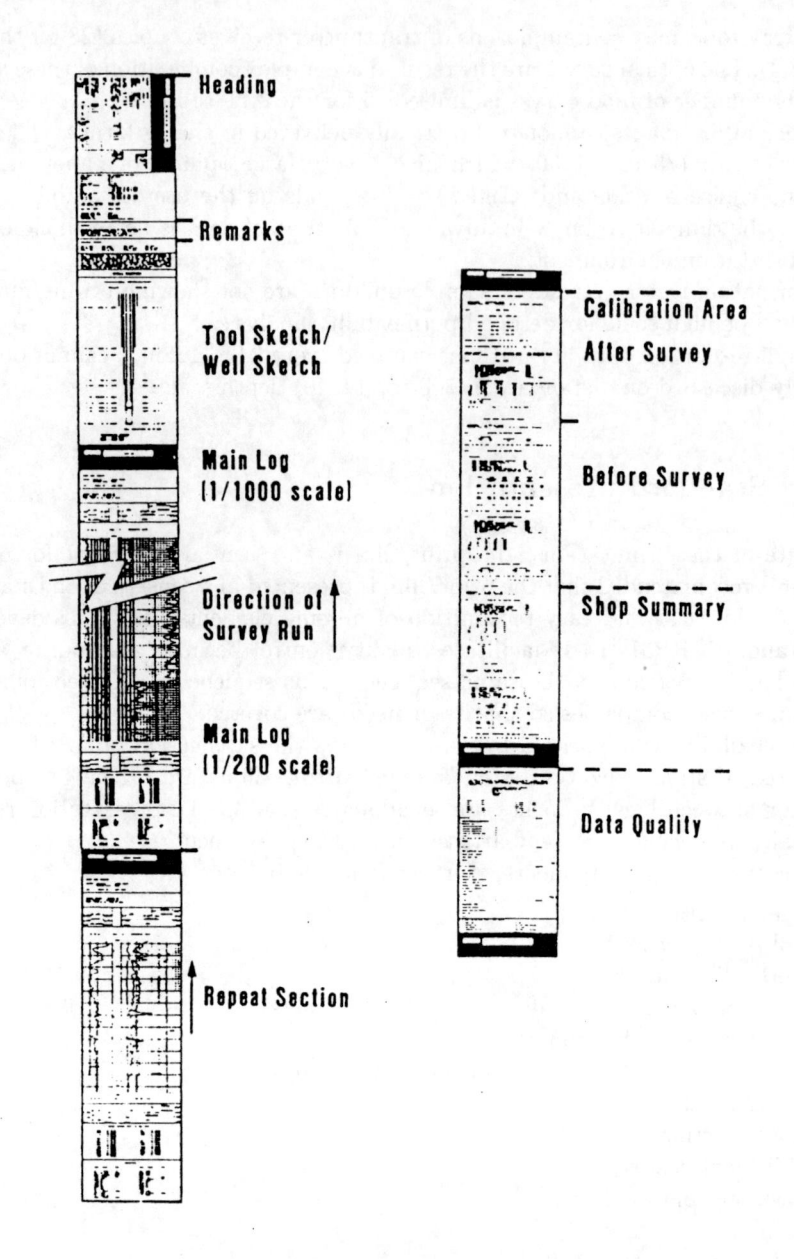

FIG. 23.4 : Standard log presentation.

Courtesy of Schlumberger

References

1 American Petroleum Institute, *Recommended practice RP 31*, 8-1995.

2 American Petroleum Institute, *Recommended practice RP 66, version 2*, 8-1997.

3 Austin, B., "CANOGGIS—Canadian oil and gas geo information systems," pp. 56-57, *The Log Analyst*, Vol. 35, No. 5, 9-1994.

4 CWLS floppy disk committee, "LAS, a floppy disk standard," pp. 395-396, *The Log Analyst*, Vol. 30, No. 5, 9-1989.

5 Froman, N. L., "DLIS, the API digital log interchange standard," pp. 390-394, *The Log Analyst*, Vol. 30, No. 5, 9-1989.

6 Schlumberger, *LIS-A customer manual*, , Austin, 1984.

7 Marett, G., "The digital interchange standard," 1990.

8 (The) Norwegian Petroleum Directorate, *Provisions relating to digital transmission of geological and reservoir technical data in connection with the final report - Drilling regulations, section 12*, ISBN 82-7257-476-4, 11-1995.

24

Log quality checks

Errare humanum est. Perseverare diabolicum.

The operating procedures that impact or help validate the accuracy and precision of data are
- (a) monitoring the logging speed or rate of penetration
- (b) monitoring the tool rotation
- (c) configuring the tool string or BHA
- (d) performing repeat passes.

24.1 Logging speed/rate of penetration

As seen in Chapter 10, logging speed (or in MWD, ROP) directly impacts measurement precision. In *Fig. 24.1*, two wells in the same field are logged with a density-neutron porosity combination string. Well A is surveyed at 14 ft/min as a spectral gamma ray log is run in combination and well B is scanned at 30 ft/min. Well A log displays
- (a) better bed definition
- (b) more precise readings because the curves make deeper incursions in low and high values
- (c) smaller depth-matching discrepancies.

Recommended logging speeds are listed in *Table 24.1*. When a combination of tools is run, it is obvious that the speed to be used is the one of the slowest service.

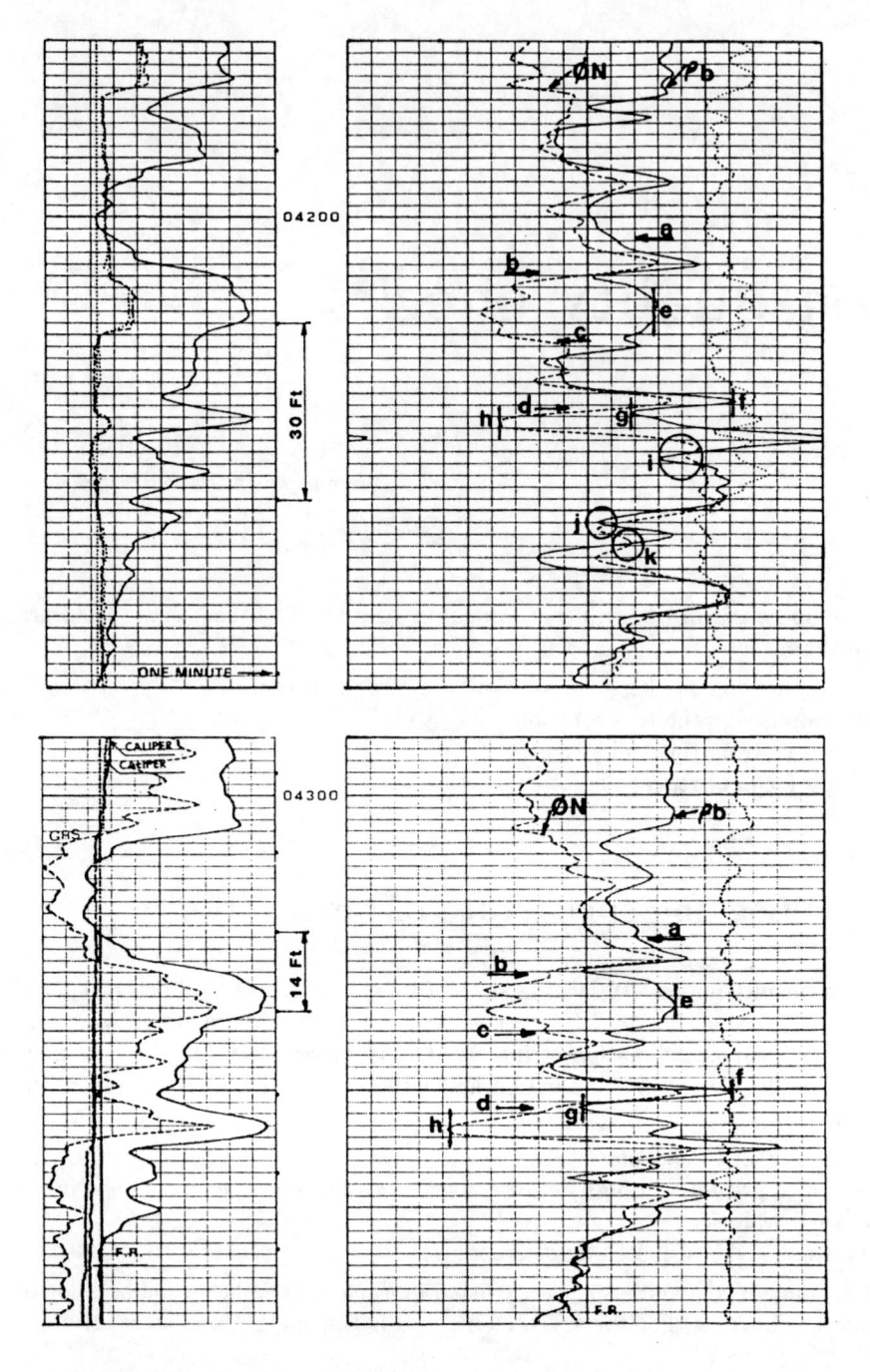

FIG. 24.1 : Comparison of logging speeds.
Top log is recorded in well A, bottom log in well B. Arrows *a* to *k* identify the same geological features.

TABLE 24.1
RECOMMENDED LOGGING SPEEDS

Survey	Logging speed			
	(ft/h)	(ft/min)	(m/h)	(m/min)
Spectral natural gamma ray....	900	15	270	4.5
Density........................	1800	30	540	9
Neutron porosity...............	1800	30	540	9
R_{xo}	2000	33	600	10
Induction......................	5000	83	1500	25
Sonic..........................	3600	60	1100	18
Imaging tool...................	3000	50	920	15

24.2 Tool rotation

24.2.1 Wireline logging

In wireline logging, the tool angular speed or rate of tool rotation is linked to logging speed. Excessive angular speed is detrimental as it causes processing difficulties for a dipmeter or a scanning tool (borehole televiewer, Formation MicroScanner) whereas insufficient rotation prevents the tool from efficiently scanning the formation. The maximum rotation rate for a multipad tool is one turn per 100 ft (30 m).

24.2.2 MWD

While drilling, sensors **need to rotate**. Otherwise, only part of the borehole is scanned, and inaccurate measurements and borehole images are collected. This is the case when the BHA slides. Density sensors, for instance, are facing one portion of the hole only (it is likely that this will be mud) and provide density readings that are too low. The dips derived from a button electrode are erroneous. The sliding indicator helps detect these intervals (Section 22.8.4.3).

24.3 Tool configuration

24.3.1 Selection of the correct auxiliary equipment

The correct choice of ancillary devices is critical in the acquisition of accurate data. The performance of a logging tool loaded with high technology features may be degraded by the selection of inadequate or substandard centralizing equipment.

For instance, some tools may have different backup arms options. For the density tool, a long arm is necessary to log holes larger than 30 cm (12 in.). But the use of the same arm in a smaller size hole causes inadequate pad contact (*Fig. 24.2*).

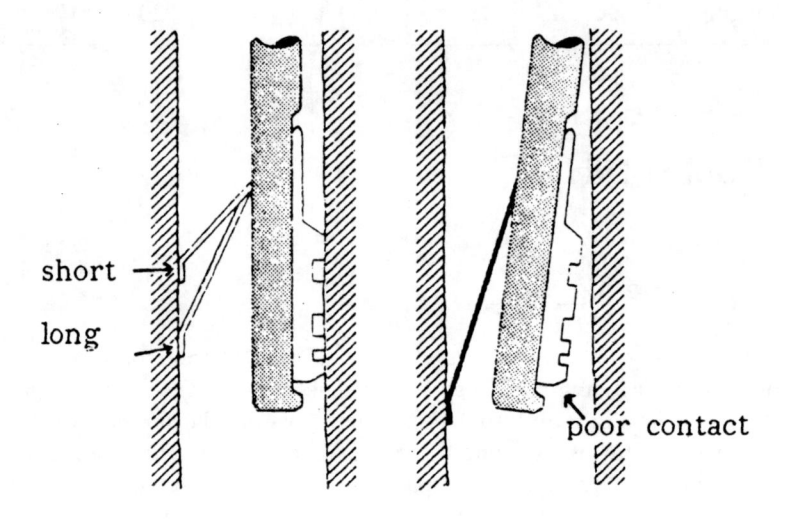

FIG. 24.2 : Effect of incorrect backup arm selection.
Courtesy of Total

24.3.2 Combination of tools

The combination of tools in the same string could affect individual measurements. Tool combination and centralization have to be adapted to the environment (poor hole condition, deviation, swelling shales, etc.). *Figure 24.3* is an extreme example where the quality of the dual laterolog is considerably affected by tool movement because it is run on a string including pad tools despite poor hole conditions.

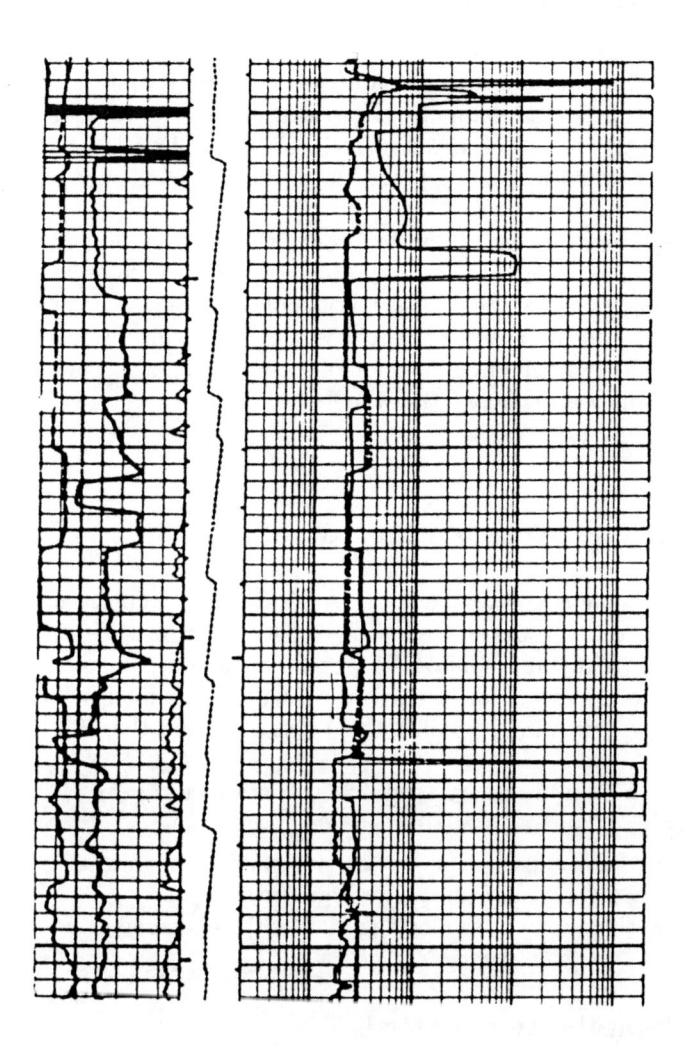

FIG. 24.3 : Effect of tool configuration.
The resistivity curves in tracks 2 and 3 are anomalous. The tension curve in the depth track indicates irregular tool movement. In fact, the tool string stops and slips continuously. Consequently, the log is a sequence of stationary measurements. Running the laterolog tool by itself would have salvaged the data.

Courtesy of Schlumberger

24.4 Importance, advantage and limitations of repeatability checks

Repeatability is a necessary condition of the validity of a measurement, but it is not a complete assurance as there are some cases of good repeatability of erroneous measurements [1]. For instance, if the master calibration has been done incorrectly, a systematic error would be superimposed to the totality of the log, including repeat and main sections.

Repeatability checks are important, but they must be given the correct perspective. The logging tool may not give the same value over the same interval. This could be the result of tool rotation between runs, erratic tool displacement on one of the passes, etc. In some cases, the two passes are equivalent even though the curves do not read the same because of different environmental effects (for instance, when standoff varies between runs). It may also happen that the formation varies quickly if the hole caves or if invasion is rapid.

24.4.1 Current checks of repeatability

A survey of log quality control methods reveals that repeatability is usually verified by superimposing the film of the first run over the image of the second run as it is displayed on the logging unit monitor. This method cannot be more than qualitative. Four repeat sections and one main log run over the same interval are displayed in *Fig. 24.4*. In these examples, it is difficult to determine whether the logs are within standards.

For this reason, it is necessary to design a more quantitative check. A rigorous repeatability check requires
  (a) the setting of a standard, a well-defined number representing the threshold between acceptable and substandard quality
  (b) a quick method relating the logging data with this number.

24.4.2 Repeatability standards

Repeatability depends on
  (a) the physics of the measurement (induction logs are expected to repeat better than nuclear logs)
  (b) the environmental corrections: heavier mud and bigger hole size usually degrade repeatability

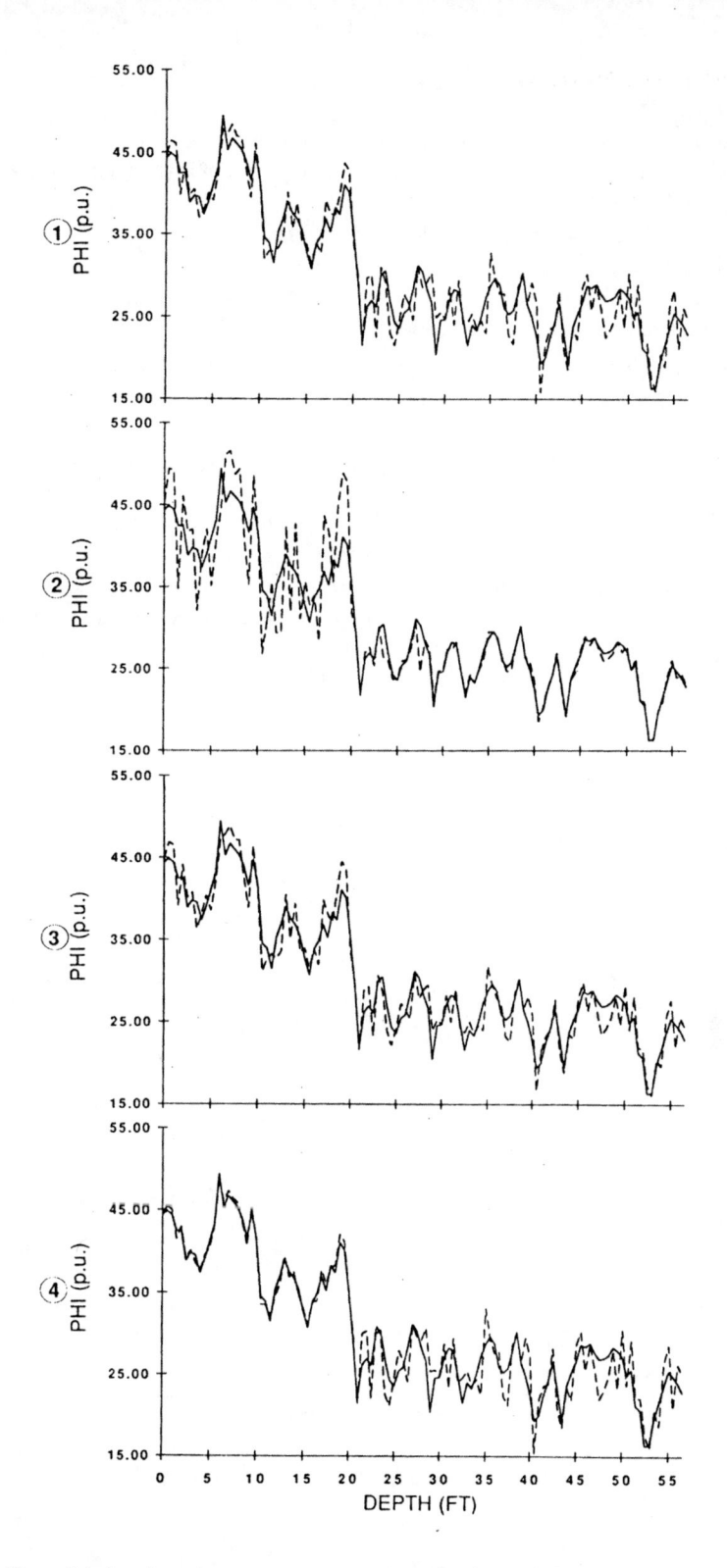

FIG. 24.4 : Are these curves repeating satisfactorily?
Figures 1 to 4 represent a neutron porosity main log (coded in solid) along with a repeat pass (each time as a dashed curve). Do all these runs conform? The answer is given in Section 24.4.5.

(c) the tool trajectory in the hole: unless the formation and environment are perfectly isotropic, different measurements will be made if the tool follows different paths
(d) the range of the measurement: density repeatability is expected to be better at low density than at high density
(e) the degree of filtering and the logging speed: smoothed data recorded at lower logging speeds is expected to repeat better.

A single value (for instance, 0.01 g/cm^3 is quoted for density) cannot account for all these effects. A repeatability standard needs to be associated with a number of conditions.

Such a standard could be: *Repeatability of the density measurement is 0.012 g/cm^3 at 2.2 g/cm^3 in a 20-cm (8-in.) round hole with a mud density of 9 lbm/gal with a logging speed of 1800 ft/h and no filtering. This value has been ascertained by laboratory measurements and confirmed on a random sample of 800 logs run worldwide.*

The repeatability specification should be a function varying according to the range of the measurement. For most types of detection, a single number is not suitable for the whole range of measurement. For instance, the neutron porosity repeatability changes significantly above 45 pu (*Table 24.2*).

TABLE 24.2
REPEATABILITY STANDARD (IN pu) FOR NEUTRON POROSITY

Range	Expected repeatability
0 - 20	1.0
20 - 30	2.0
above 45	6.0

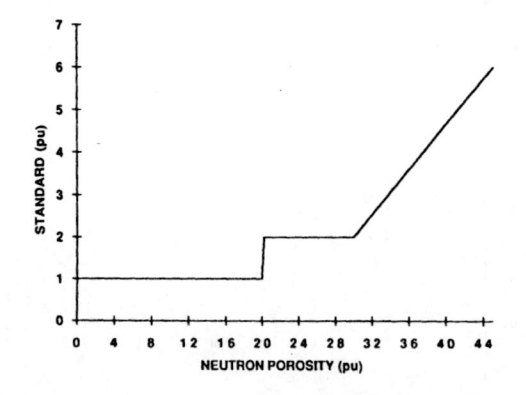

FIG. 24.5 : Repeatability standards for *Schlumberger* thermal neutron porosity log.

Table 24.2 can be represented graphically by a plot (*Fig. 24.5*) where the repeatability standard is shown as a step function of porosity, $s(\phi)$ (*Table 24.3*).

TABLE 24.3

$s(\phi)$

Standard	Function	Range
$s(\phi)$	1	$0 < \phi < 20$
$s(\phi)$	2	$20 < \phi < 30$
$s(\phi)$	$(4\phi/15)-6$	$30 < \phi$

24.4.3 Normalized repeatability check

Once a repeatability standard has been fixed, its correct usage must be defined. Taking a main log value (measurement X_i at depth i) and a repeat log value (measurement X'_i), both assumed to belong to the range where the repeatability standard, R, is valid, how should these two values relate to the standard? Does this mean:

(a) "$(X_i - X'_i)$ should never be greater than R (go/no-go approach)"

(b) or "Taking a large number of values X_i and X'_i values, the population of $(X_i - X'_i)$ should follow a Gaussian distribution with a standard deviation R."

The selection of the second definition implies that

(a) 68.3% of the $(X_i - X'_i)$ values are less than R

(b) 95.5% of the $(X_i - X'_i)$ values are less than $2R$.

This second definition allows a log to be classified as good even if there are (a few) large discrepancies between the main and repeat log values (*Fig. 24.6*).

Combining the variable standard and the statistical approaches, the repeatability check would be to plot function y, defined as

$$y = \frac{\phi_{\text{main log}} - \phi_{\text{repeat log}}}{s(\phi_{\text{main log}})}.$$

It is clear that the standard deviation of this function, plotted over the interval common to the main and repeat logs, should be less than 1.0 if the three populations of porosities (0 to 20 pu, 20 to 30 pu, and above 30 pu) are within standards.

In practice, some compensation may happen if one group of data is slightly better than standard and another group slightly worse, the overall result being a satisfactory log.

The neutron porosity log has been taken as an example. It is obvious that similar methods can be implemented for all logging tools whose specifications vary with the range of the measurement.

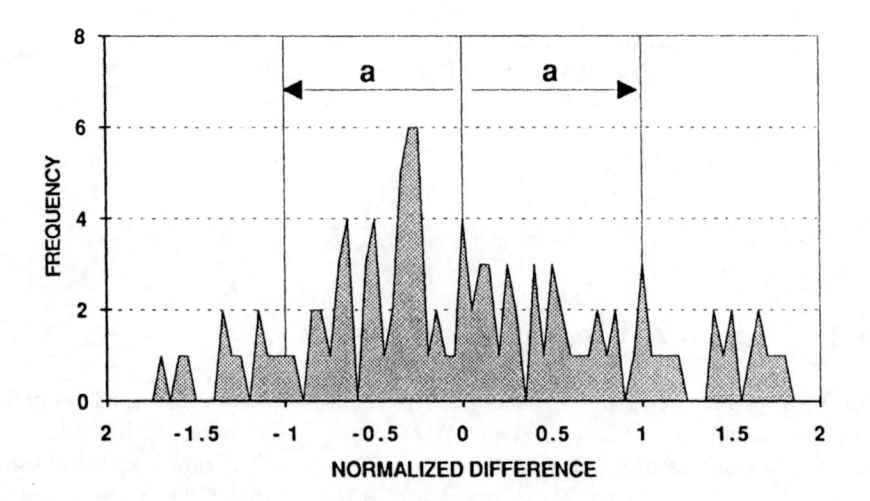

FIG. 24.6 : Go/no-go check versus statistical check.
With a go/no-go check, the log fails the repeatability check because there is a number of values outside interval $[-a,a]$. With a statistical check the log passes (standard deviation is 0.95).

24.4.4 Practical hints for the repeatability check

Precise depth matching

The curves of the repeat and main sections must be carefully depth matched before the function y is calculated and derived. *Figure 24.7* shows raw curves before (*a*) and after depth matching (*b*).

Selection of the repeat section interval

In a well, logging achieves the following objectives by order of priority:
(a) Acquire data in the zone of interest (the reservoir).
(b) Acquire data over the whole openhole section.
(c) Confirm the overall validity of data by running a repeat section where it is safe.
(d) Confirm data over the zone of interest.

Depending on the reservoir position, the main log and the repeat section may be recorded in different orders. The best compromise obtains a relevant repeat section in the shortest time.

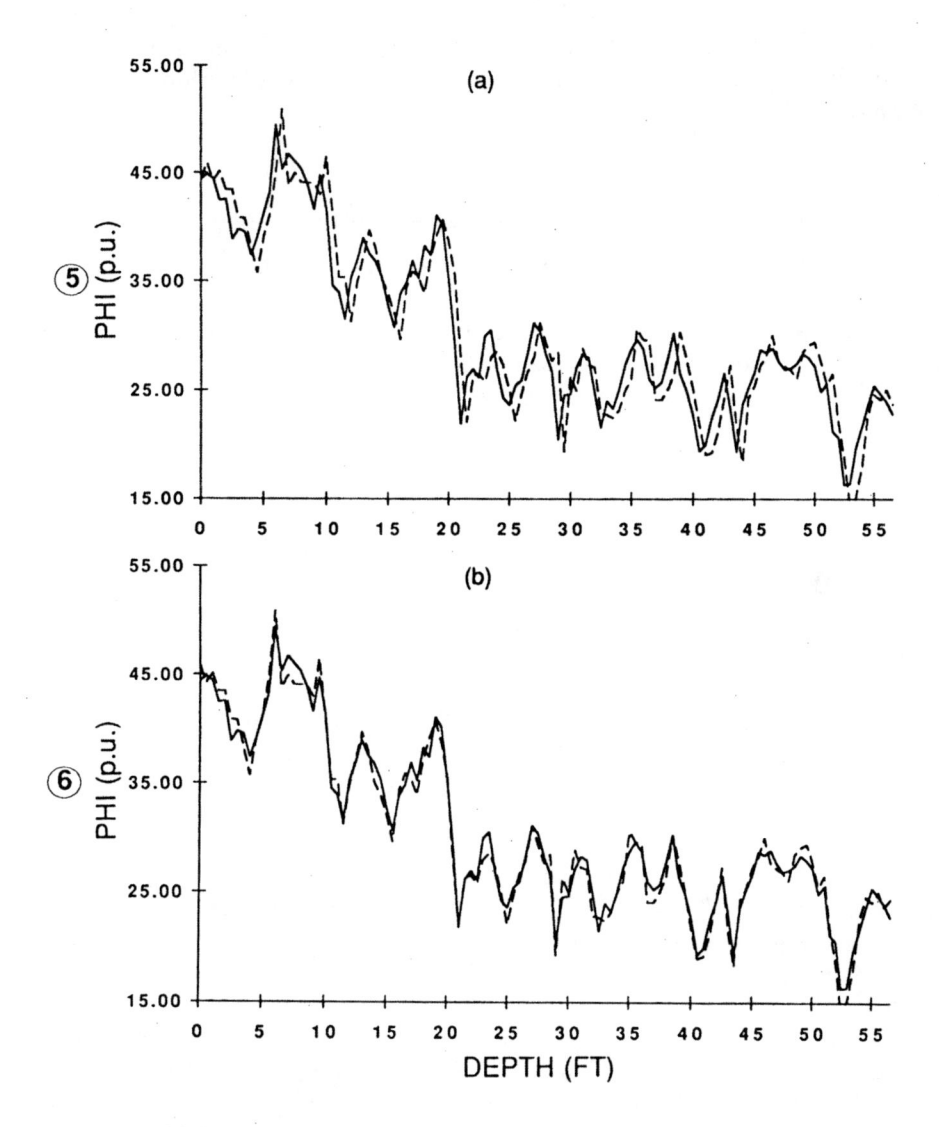

FIG. 24.7 : Importance of depth matching in the repeatability check.
a. Field log.
b. Personal computer based depth matching.

Other requirements

The repeat section needs to be representative. Therefore, it is recommended that it be at least 65 m (200 ft) long. This excludes any pick-up interval (the tool not moving, but the surface computer recording dead curves). At least one depth figure should be included on the repeat section film.

When the information under scrutiny is subjected to signal processing with a long sliding window (e.g., spectral gamma ray yields with active filtering), the length of the repeat section must be increased.

24.4.5 Numerical application of the normalized repeatability method

Using the variable standard defined in *Table 24.3* and performing a statistical analysis of the neutron porosity data sets shown in *Figs 24.4* and *24.7*, it is possible to obtain at the quantitative results gathered in *Table 24.4*.

TABLE 24.4
NORMALIZED REPEATABILITY ANALYSIS
FOR FIGS 24.4 AND 24.7
Subscript 1 is for the main log, subscript 2 for the repeat log.

	Units	1	2	3	4	5	6
$\phi_2 - \phi_1$ Average	(pu)	−0.028	−0.051	−0.051	−0.035	−0.046	0.130
Standard deviation	(pu)	2.165	1.914	3.106	2.181	2.724	1.136

	1	2	3	4	5	6
$(\phi_2 - \phi_1)/s(\phi_1)$ Average	−0.004	−0.029	0.019	0.001	−0.020	−0.096
Standard deviation	1.140	0.854	0.825	1.056	1.285	0.554

Results summary

The logs with the greatest degree of repeatability are the ones with the lowest standard deviation on $(\phi_2 - \phi_1)/s(\phi_1)$, in decreasing order: 6, 3, 2, 4, 1, and 5. Logs 6, 3 and 2 conform while logs 4, 1 and 5 do not. Log 5 is not conforming because of the lack of adequate depth matching.

24.4.6 Repeatability standards for nuclear logging tools

Repeatability specifications are collected or derived from *Schlumberger* literature and are used to illustrate the concept of quantitative repeatability analysis. It is recommended to consult logging company personnel to obtain specific and updated information.

24.4.6.1 Density log

The expected repeatability continuously varies with the range of measurement. The function shown in *Fig. 10.1*, can be approximated by

$$s(\rho_b) = 0.013\rho_b - 0.006.$$

Therefore, y is:

$$y = \frac{\rho_b - \rho_b'}{s(\rho_b)}.$$

24.4.6.2 Gamma ray log

The gamma ray repeatability standard is

$$s(GR) = \frac{G - G'}{0.07 \times G}$$

where G is the gamma ray reading of the main log and G' that of the repeat log.

References

[1] Theys, P., "A serious look at repeat sections," *Trans.* SPWLA 35th annual symposium, Tulsa, 1994.

25

Data quality evaluation

Once completeness of information has been controlled and log quality checks performed as specified in the previous chapter, it is necessary to have a global look at data.

25.1 First look

The sensors scan the same volumes of underground formations; therefore, the changes observed with different measurements should correlate as the sensors face the boundaries between these volumes.

This is true also on a field scale. A reasonable amount of correlation should appear between wells traversing the same geological formation. Comparison with neighbor wells is still one of the most powerful quality control techniques (*Fig. 25.1*).

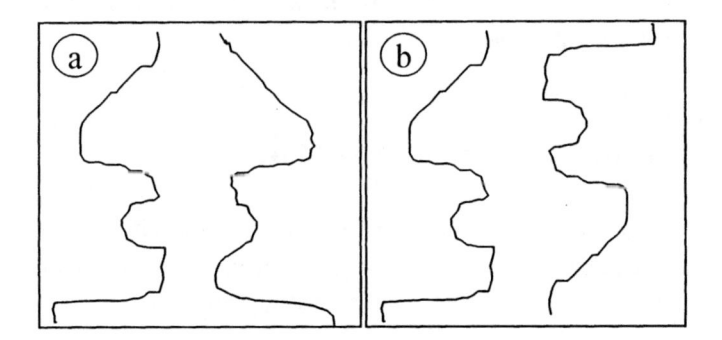

FIG. 25.1 : Correlation between curves.
 a. Curves correlate. This is the minimum requirement for a correct log.
 b. Curves do not correlate. This log looks suspicious.

25.1.1 Quick interpretation to discriminate zones of interest

The action resulting from the discovery of an anomaly on a log will differ depending on whether the interval is "a zone of interest." It is therefore important to recognize such zones. The method selected does not have to be highly sophisticated and could be limited to the following:

(a) R_{wa}

When resistivity and porosity measurements are available, it is possible to combine them to give a crude indicator of water-bearing, hydrocarbon-bearing, clean and shaly zones. R_{wa}, the apparent water resistivity is defined as

$$R_{wa} = \frac{R_t}{F}.$$

In most cases, F equals $1/\phi^2$, ϕ being extracted from a sonic log or the combination of the density-neutron. In simple cases, R_{wa} is equal to the formation water resistivity in clean water-bearing zones, R_w, and to a distinct value called bound water resistivity, R_{wb}, in shales. In a hydrocarbon-bearing zone, R_{wa} is several times larger than R_w. For instance, R_{wa} is equal to 4 R_w for $S_w = 50$ su.

(b) **Gamma ray or spontaneous potential**

The gamma ray curve and the spontaneous potential curves are directly used to recognize clean and shaly zones in a qualitative way. In simple cases, they are sufficiently effective in distinguishing zones of interest.

(c) **Separation of the density-thermal neutron porosities**

Large separation between the density and neutron curves ($\phi_D < \phi_N$) is often indicative of a shaly zone. In a known area, this separation helps identify zones of little reservoir potential.

25.2 Looking at a larger picture

The objective of logging, to acquire valid formation data, is not achieved when

(a) **Logging procedures are not properly followed**. This would be the case with inferior calibration, erroneous operating parameter selection or inappropriate logging speed.
(b) **Tool design is not adapted to the environment**. This may be observed if the tool is too small for the borehole or if auxiliary equipment (for instance the caliper arm) is unable to ensure proper application of the sensors against the formation.
(c) **The physics of the measurement has reached its limits**.

Examples

- Extremely deep invasion—more than 2.5 m (100 in.). In that case, it is difficult to accurately assess the value of the virgin zone resistivity.
- Measurement of the photoelectric effect constant if the hole is filled with barite mud.

If limits are reached in a certain field of physics, it is necessary to investigate alternative concepts. Before the advent of conductivity or sonic-based sensors in dipmeter logging, it was difficult, if not impossible, to detect resistive changes in oil-base mud to derive a dip.

Participation of the operating and logging companies in acquiring relevant data in the three cases mentioned above is gradually changing. In the first category, the responsibility to correct and improve data rests definitely with the logging company. In the third case, the operating company must display some flexibility since no physical principle can be used to obtain the information. The acquisition of the data may require a change in the drilling operation, if it affects the mud program or the drilling parameters. In the second case, efforts are expected from both parties and good communication should provide a path to the acquisition of relevant logs. In all cases, the logging company can help in providing historical data. Log quality control is then no longer a punitive and negative lever to keep the logging contractor on his toes, but a constructive and positive way to diagnose past problems, wherever they originate, and thereby improve future acquisition.

25.3 Data quality control system

While traditional methods of log quality control only check the proper execution of the logging procedures, the data quality control system analyzes the logging data directly. It is recommended that both approaches be applied simultaneously. *Tables 25.1* and *25.2* list the associated severity ratings and nonconformity categories.

TABLE 25.1
SEVERITY RATINGS IN DATA QUALITY EVALUATION

A	No problems
B	Minor problem, but data is usable after correction
C	Data is lost, but no serious effect on formation evaluation
D	Data is lost with serious effect on evaluation
E	Extreme case of D

TABLE 25.2
CAUSES OF DATA QUALITY
DETERIORATION

1	Rough hole
2	Borehole fluid limitations
3	Tool sticking
4	Anisotropy/fractures
5	Gas
6	Inadequate pad contact
7	Tool limitation exceeded
8	Poor operating technique
9	Others

Item 8 is completely under the control of the logging company. Items 4 and 5 are related to the formation being surveyed. They cannot be modified. Items 1, 2, 3, 6 and 7 are under the combined control of the oil/gas and logging companies.

25.4 Noise

25.4.1 Spontaneous potential

The spontaneous potential curve can be affected by the following sources of noise:

(a) **Magnetism**: It appears as regular cyclic disturbances on the spontaneous potential curve. It disappears when the cable stops and is generated by one or several parts of the unit or truck transmission that are magnetized (*Fig. 25.2*). To get rid of it, the logging cable and its drum must be demagnetized. Sometimes it may also be necessary to demagnetize the chain linking the logging unit engine and the drum separately.

(b) **Bimetallism**: In some cases, a DC potential is superimposed on the spontaneous potential curve. It is generated when two distinct metals are in contact with in an electrolytic solution, which most muds constitute, and form a battery.

The noise is reduced when the metallic parts originally exposed to the mud are insulated.

(c) **Telluric currents**: Naturally occurring Earth's currents are induced by solar activity and usually subside at night, except for the northern lights (aurora borealis). They are observed on the spontaneous potential curve as a slow random drift.

The cure is to run a **differential SP** with a downhole return.

(d) **Random electrical disturbances**: These are caused by any electrical anomaly around the ground return, such as welding, generators, cathodic protection circuits on offshore rigs, etc.

This illustrates the necessity to place the fish away from the rig and to stop all electrical welding by rig personnel during logging.

(e) **Cable noise**: It is caused by the making and breaking of the bimetallic cell constituted by the cable armor and the casing.

The cure is to put the fish at some distance from the rig and to prevent the cable from rubbing against any metal (rotary table, etc.).

25.4.2 Sonic

Sonic readings may be affected by sharp skips that yield too long or too short transit times:

(a) **noise resulting in too long sonic time**
  - triggering on noise on a near receiver
  - cycle skipping on one or two far receivers

(b) **noise resulting in too short sonic time**
  - triggering on noise on far receiver
  - cycle skipping on near receiver.

Figure 25.3 shows an example of cycle skipping.

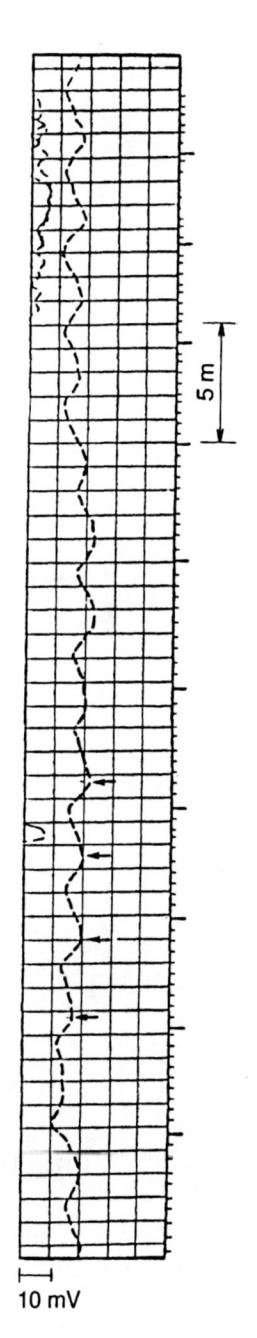

FIG. 25.2 : Noise on spontaneous potential. The noise is due to a magnetized drum.

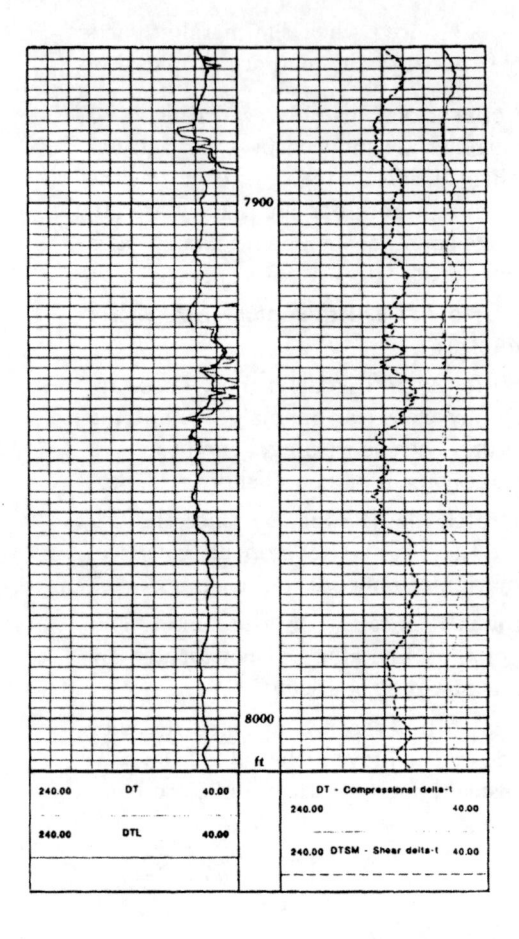

FIG. 25.3 : Sonic cycle skipping.
On the left, the two sonic curves, DT and DTL, recorded with a standard long-spacing tool exhibit skips in both directions. On the right, the sonic curve recorded with a digital device indicates actual formation transit times unaffected by skipping. Δt_{shear}, $DTSH$, is shown for reference.

25.4.3 Dipmeter

Noise affects the dipmeter measurement by producing simultaneous anomalies on all microresistivity curves. These anomalies generally appear as sharp spikes which, when correlated, result in a high-confidence computed dip. As the spike appears at the same measured depth for all correlatable curves, the resulting dip corresponds to a plane perpendicular to the borehole axis. As a consequence, the dip and the hole orientation display similar patterns (same dip magnitude) but their respective azimuths would be opposite (with a difference of 180°). Hence, the name of these artificial dips: **mirror images**.

These noise-generated dips should be carefully discarded prior to structural or stratigraphic interpretation. *Figure 25.4* shows a field-recorded dipmeter log affected by noise with the corresponding computing center processing.

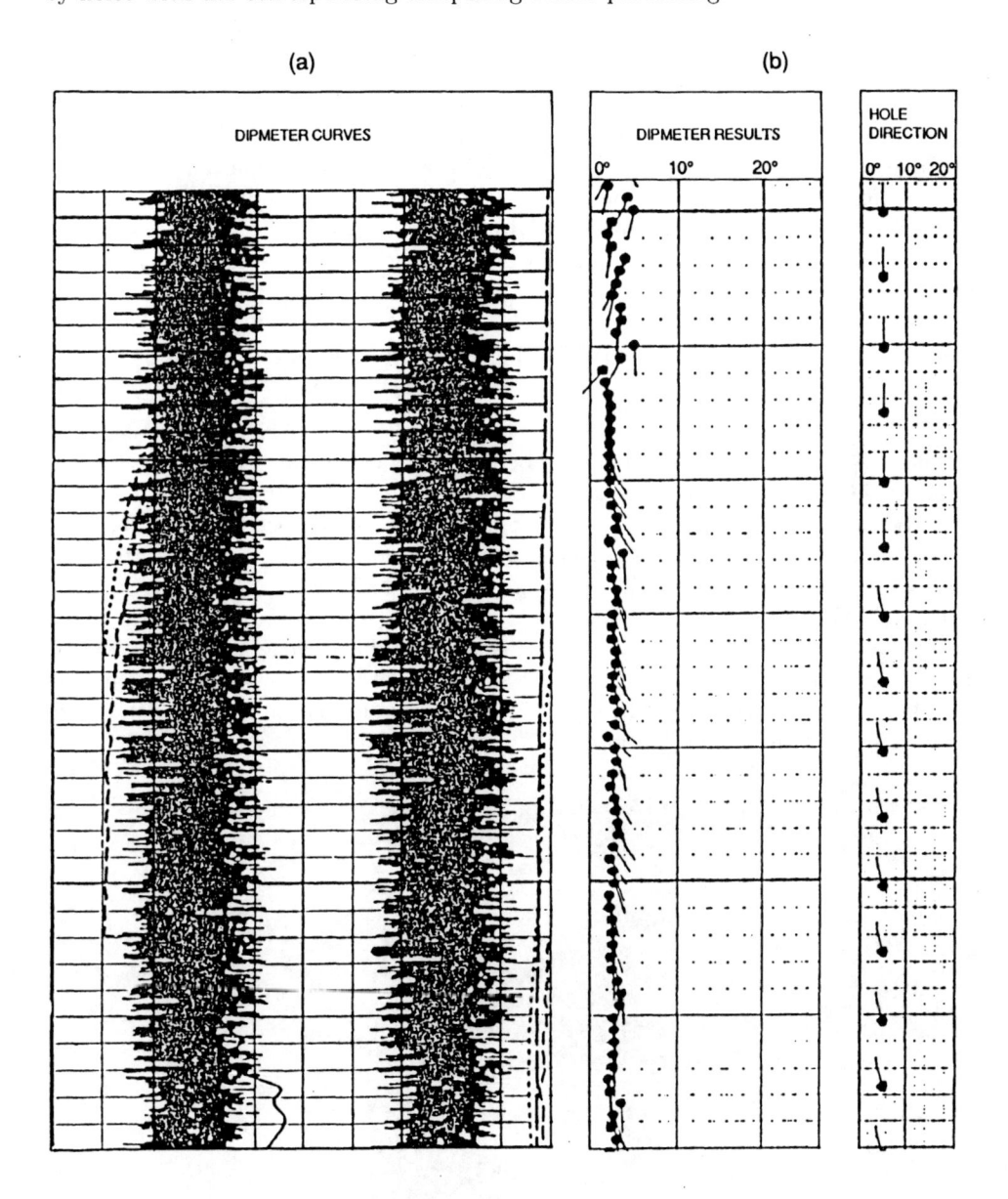

FIG. 25.4 : Noise on the dipmeter log.
   a. Raw data.
   b. Resulting processing. The dip and the hole direction are mirror images.

25.5 Anomalous response

25.5.1 Hole conditions

While most logging tools are designed to log smooth and round holes, such a favorable environment, as seen in Chapter 15, is far from being the rule. The occurrence of log readings affected by adverse hole conditions would therefore be frequent. An accurate report of past logging in such environments allows the development of some remedies, short of being able to convince the operating company to drill in a different way.

This section examines some logs severely affected by hole conditions. To quantify these effects, logs surveyed with auxiliary hardware are shown in comparison.

Example 1

The logs in *Fig. 25.6* represent two density surveys run in a 31-cm (12.25-in.) diameter hole. Run 1 was made with the tools connected in a standard configuration. From the caliper measurement, it is evident that the tool string settled along the long axis of the borehole, which resulted in an artificially low and noisy density reading. The tool pad slid over the mud-filled rugosity of the borehole.

Run 2 was performed with an auxiliary centralizer (*Fig. 25.5*). It is generally run with a swivel head adapter so that torque is released during logging. In this second run, the tool string was oriented along the short axis of the borehole along a surface much less affected by rugosity. The density values reflect formation characteristics and can be used in formation evaluation.

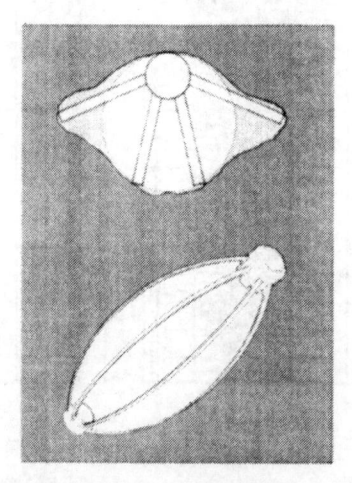

FIG. 25.5 : Short axis logging eccentralizer.
Courtesy of Schlumberger [VI]

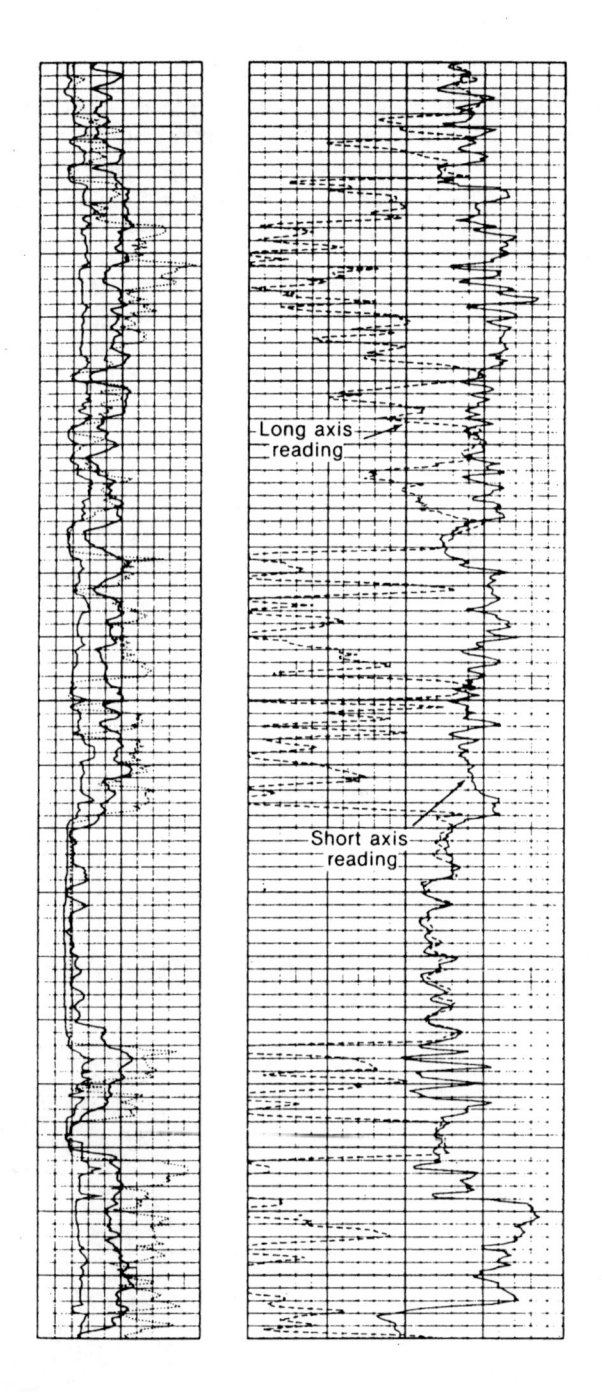

FIG. 25.6 : Density log run in difficult hole conditions.
Courtesy of Schlumberger [VI]

Example 2

Another approach [6] to logging a pad tool in poor hole conditions is to adopt a tandem tool string; in other words, combining two density sondes in the same tool string. Tandem logging (*Fig. 25.7*) is reported to improve data quality from 20-30% of usable measurements to 90%.

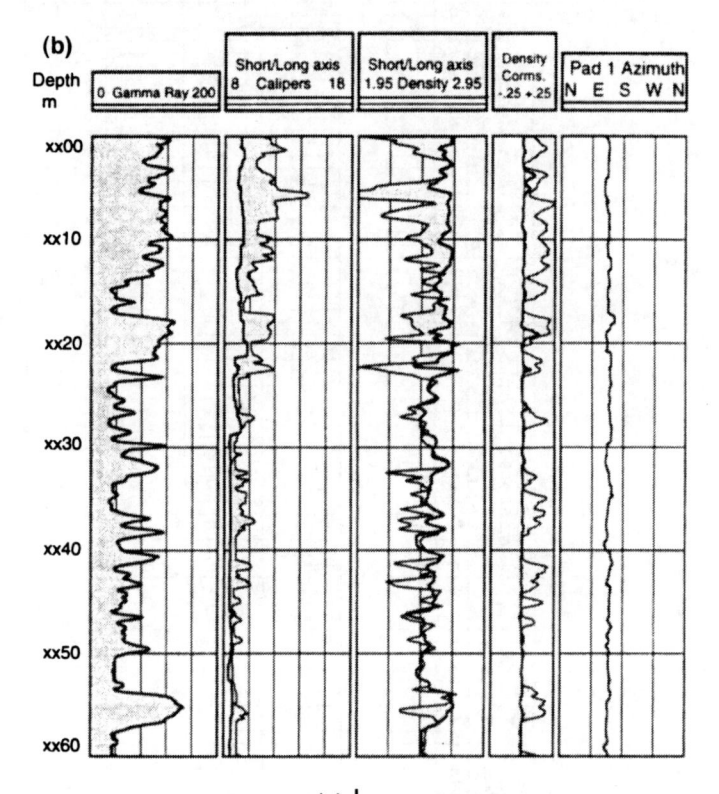

FIG. 25.7 : Tandem density logging.
a. Tool string.
b. Example of logs acquired with this tool string. The long axis pad remains in the same direction, aligned to the southeast.
c. Ultrasonic image of the borehole. Note the similarity with *Fig. 15.2*.

Courtesy of SPE [6]

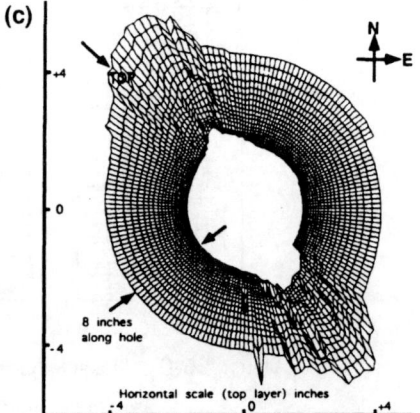

Example 3

The information shown in *Fig. 25.8* originates from dipmeter tools run with two different centralizing configurations. With the standard centralizer, the pads run along the long and short axes. The sensors on two pads are affected by rugosity (*Fig. 25.8a*).

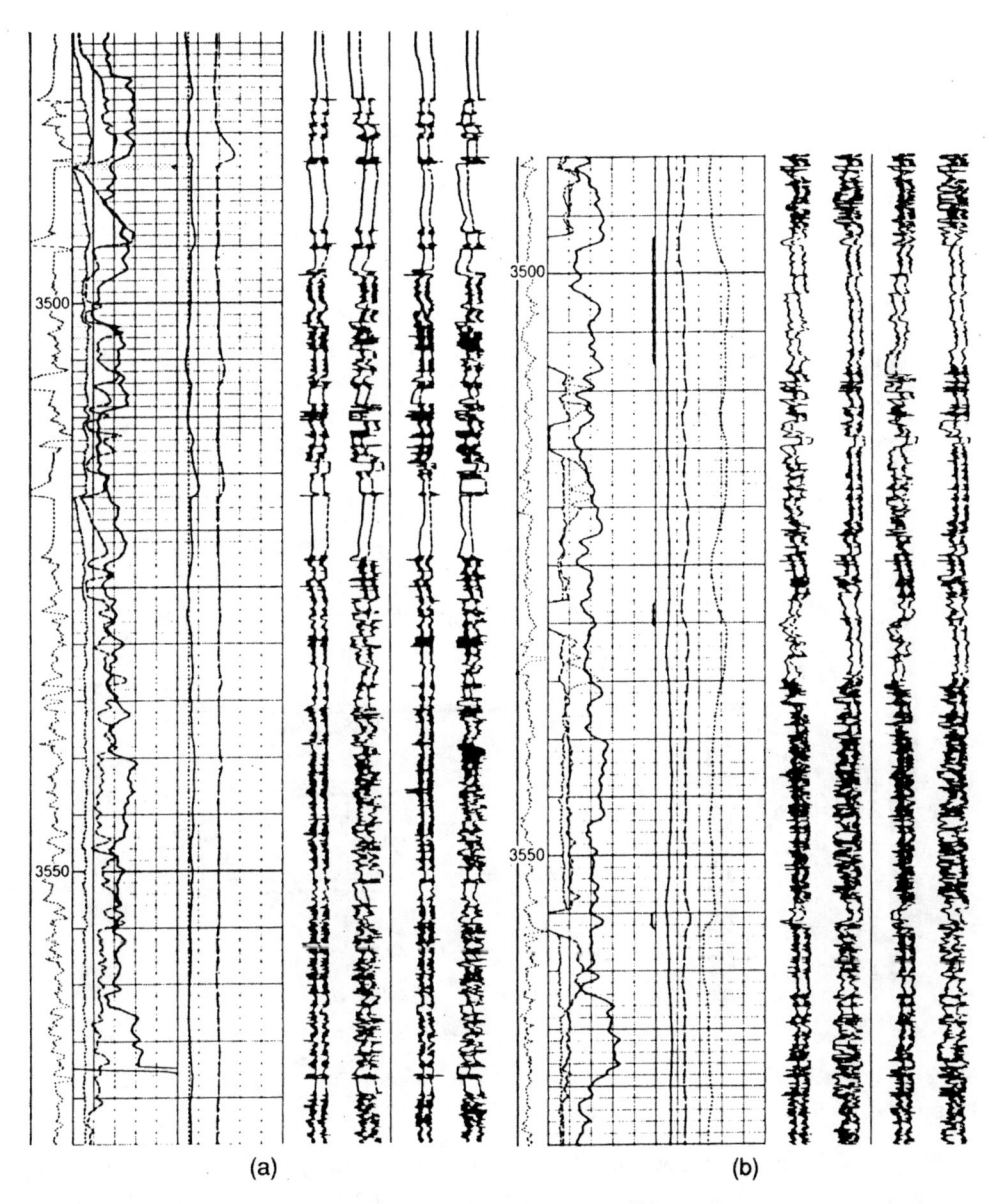

FIG. 25.8 : Dipmeter affected by borehole conditions.
Courtesy of Schlumberger [VI]

The second run (*Fig. 25.8b*) is performed with a special 45° centralizer (*Fig. 25.9*). The caliper values are now lower than on the first run. A marked improvement on the density and quality ratings of the derived dips results from this configuration.

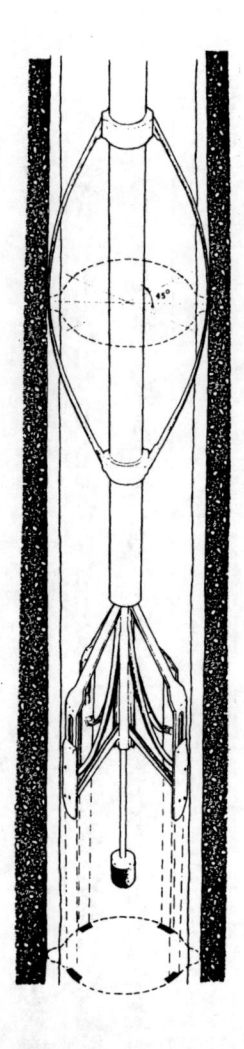

FIG. 25.9 : Special 45° centralizer. The large blades of the centralizer ride the long axis of the borehole, so that the dipmeter (or Formation MicroScanner pads) run over portions of the borehole that are less rugose.
Courtesy of Schlumberger

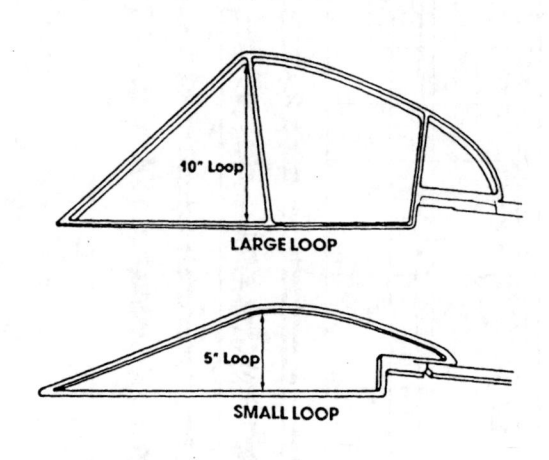

FIG. 25.10 : Big-hole logging kit.
Courtesy of Schlumberger [5]

Example 4

Big-hole logging: A few years ago, the cost of reaming a pilot hole was considered prohibitive. Logging companies have developed special hardware [4 and 5] to log successfully holes with bit sizes to 1 m (25 in.). A diagram of the equipment developed for the density-neutron combination is shown in *Fig. 25.10*.

25.6 Tool limitations

25.6.1 Geometrical limits

The most common situation where the logging tool is unable to collect formation data is when the configuration of the sensors does not match the geometry of the borehole. This occurs when borehole size is much larger than bit size or when the logging engineer is using equipment not adapted to the hole size. *Figure 25.11* gives an example of a density log run in a shallow well where the hole size is larger than bit size.

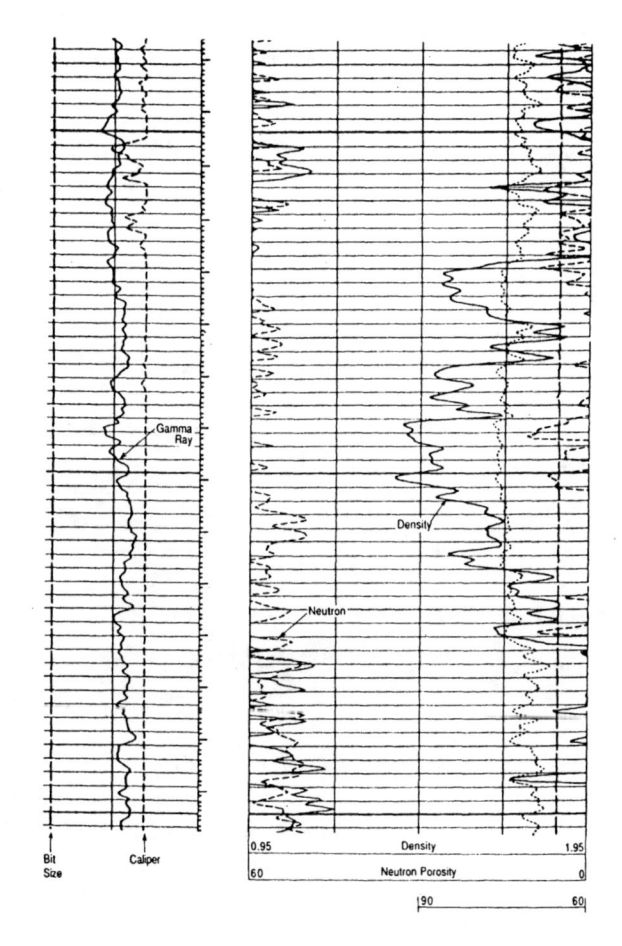

FIG. 25.11 : Example of caliper saturating.
The pad supporting the sensor is not applied against the formation and the information given by the density curve relates to the mud, not to the formation.

Figure 25.12 shows a sonic log where the transit time curve corresponds to the mud—610 μs/m (200 μs/ft).

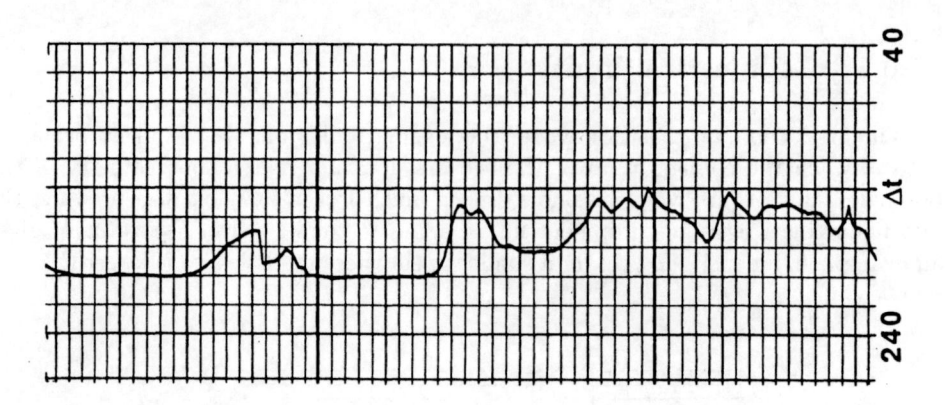

FIG. 25.12 : Sonic curve flattened while the tool is reading drilling mud. This observation can also be made on individual travel times, when all receivers are not simultaneously affected by the mud.

25.6.2 Pressure and temperature

Most logging tools are rated 20,000 psi in pressure and 350°F (175°C) in temperature. Specially designed tools are able to withstand temperatures to 500°F (260°C) and pressures to $170 \cdot 10^6$ Pa (25,000 psi). Use of a tool above its rating absolves the logging company of responsibility for tool failure.

25.7 Anomalous but correct readings: a formation troubleshooting

25.7.1 Delaware and anti-Delaware effects

The Delaware effect makes the deep laterolog measurement read too high when the tool, logged up, goes from a low-resistivity formation to a high-resistivity formation.

The anti-Delaware effect is the opposite, corresponding to a reduction of resistivity as the tool goes from low to high resistivity.

These effects, mentioned here for historical reasons, are no longer observed, since the electrode configuration of the laterolog tool has been changed to minimize these effects.

25.7.2 Squeeze and antisqueeze

Squeeze and antisqueeze are observed on the deep laterolog [3]. The squeeze geometry corresponds to the case of a bed of low resistivity R_b sandwiched between two beds with higher resistivity R_s. The log reading is too high. It is recommended to use the shallow laterolog measurement instead for R_t evaluation.

The antisqueeze geometry relates to a high-resistivity bed located between two low resistivity beds. The current beams associated with these two configurations are represented in *Fig. 25.13*.

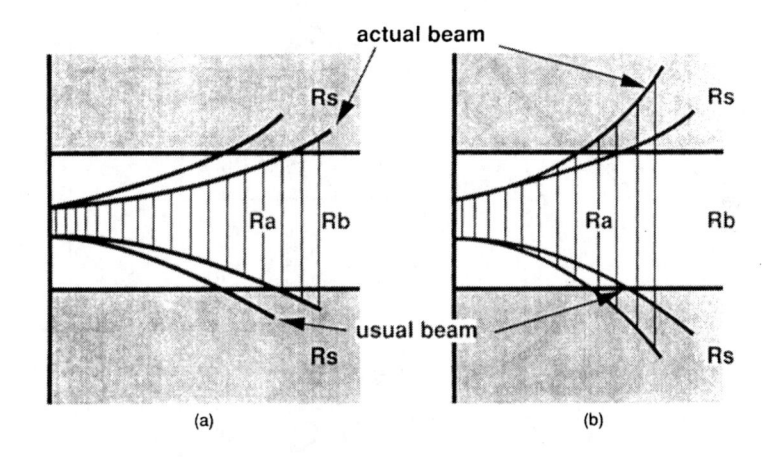

FIG. 25.13 : Change of shape of the current beams.
a. Squeeze ($R_s > R_b$), overfocusing.
b. Antisqueeze ($R_s < R_b$), underfocusing.

25.7.3 Groningen effect

This effect was first observed in a field near *Groningen*, a city in the Netherlands. The deep laterolog shows an increase of resistivity as soon as the electrode located on the bridle[1] enters a resistive overlaying bed. This effect is observed under these conditions

(a) in uncased wells, if the resistive bed is very thick (*Fig. 25.14*).
(b) below a casing, where it is generally of larger amplitude. It may cause a shift of 10 Ω/m on the resistivity (*Fig. 25.15*).

---

[1] A piece of rubber with several electrodes located at the top of the toolstring.

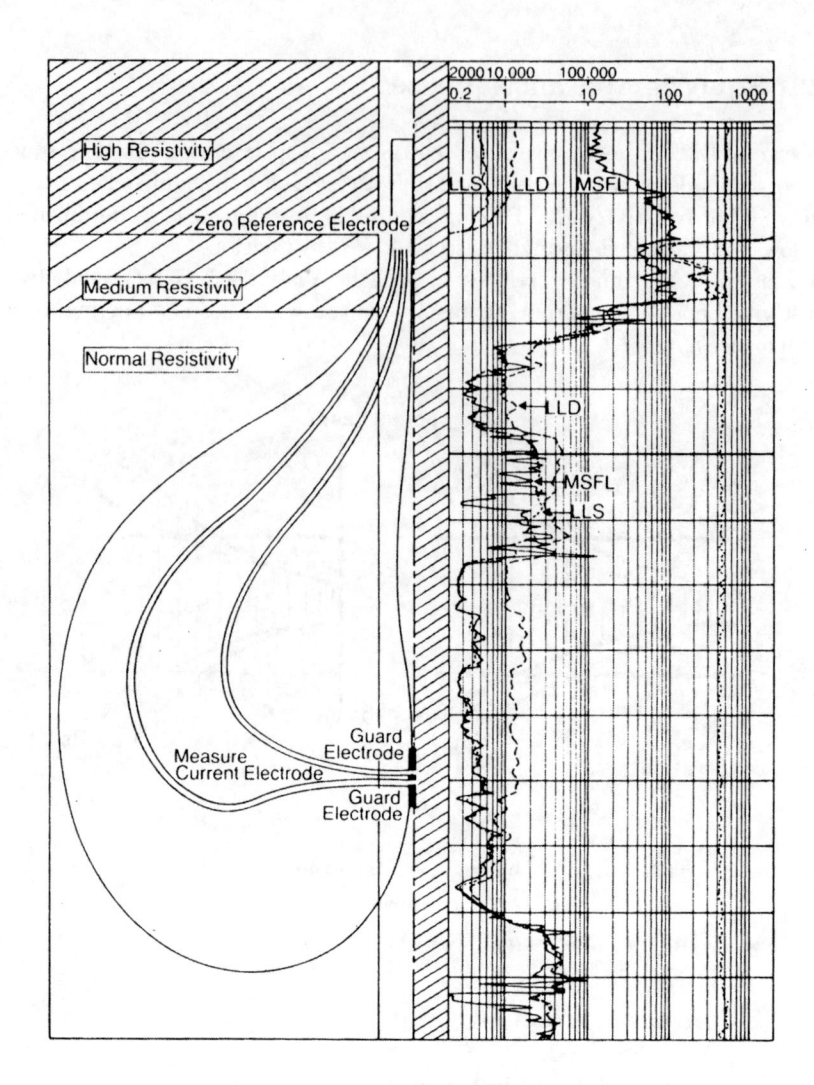

FIG. 25.14 : Groningen effect without casing.
Courtesy of Schlumberger

How is it possible to detect the Groningen effect?
(a) Since only the deep laterolog is affected, the separation between the shallow and deep readings is most of the time diagnostic.
(b) The anomaly starts at a well-defined distance from the casing shoe or the bottom of the high-resistivity zone. This distance corresponds to the spacing between the electrodes on the tool and the bridle. It is 100 ft on the example shown *Fig. 25.15*.

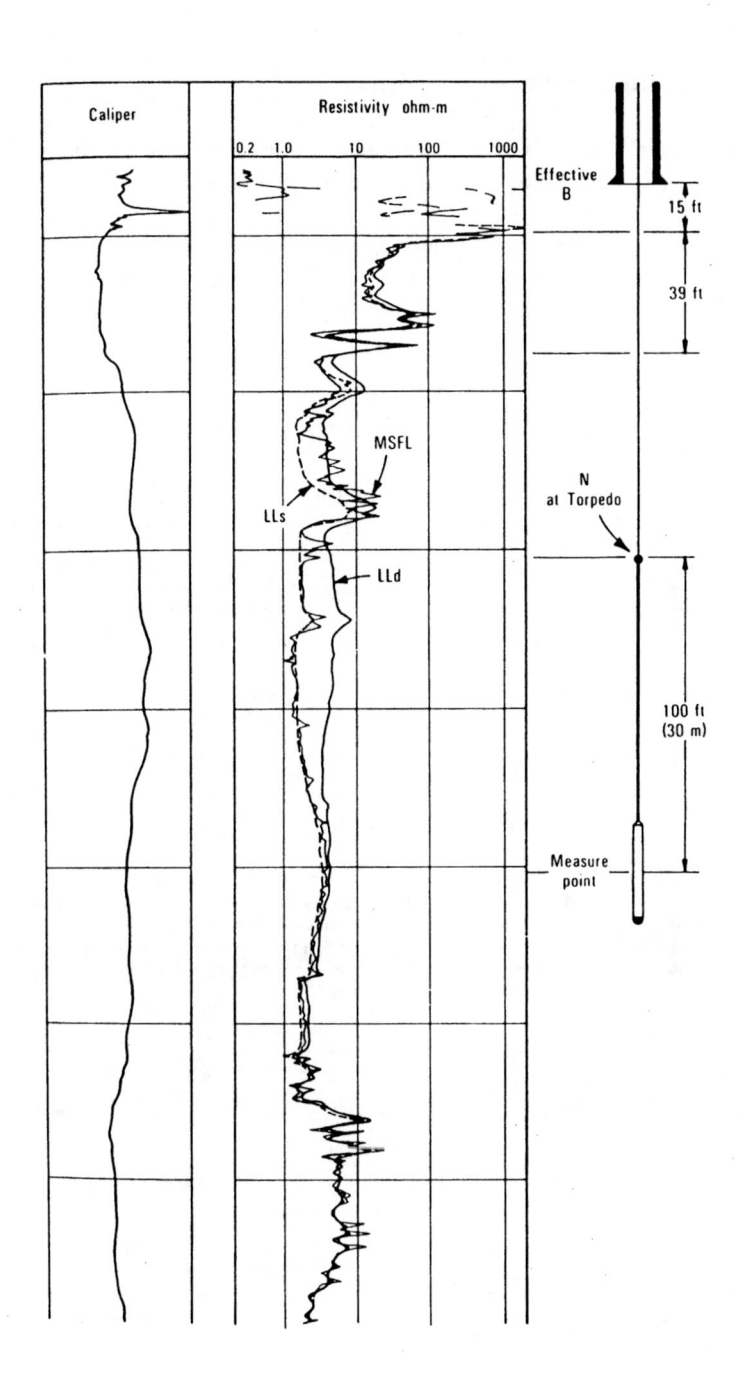

FIG. 25.15 : Groningen effect with casing.
Courtesy of Schlumberger

25.7.4 Conductive anomalies

Figure 25.16 shows the erratic response of the medium and deep induction tools in a North Sea well drilled with oil-base mud. These anomalies are repeatable and cannot be attributed to tool malfunction.

These anomalies have been modeled. They are related to conductive anomalies. *Figure 25.17* shows the modeled response of the induction to a succession of 15-cm (6-in.) thick conductive beds at 0.5 Ω/m while mud resistivity reaches 1000 Ω/m.

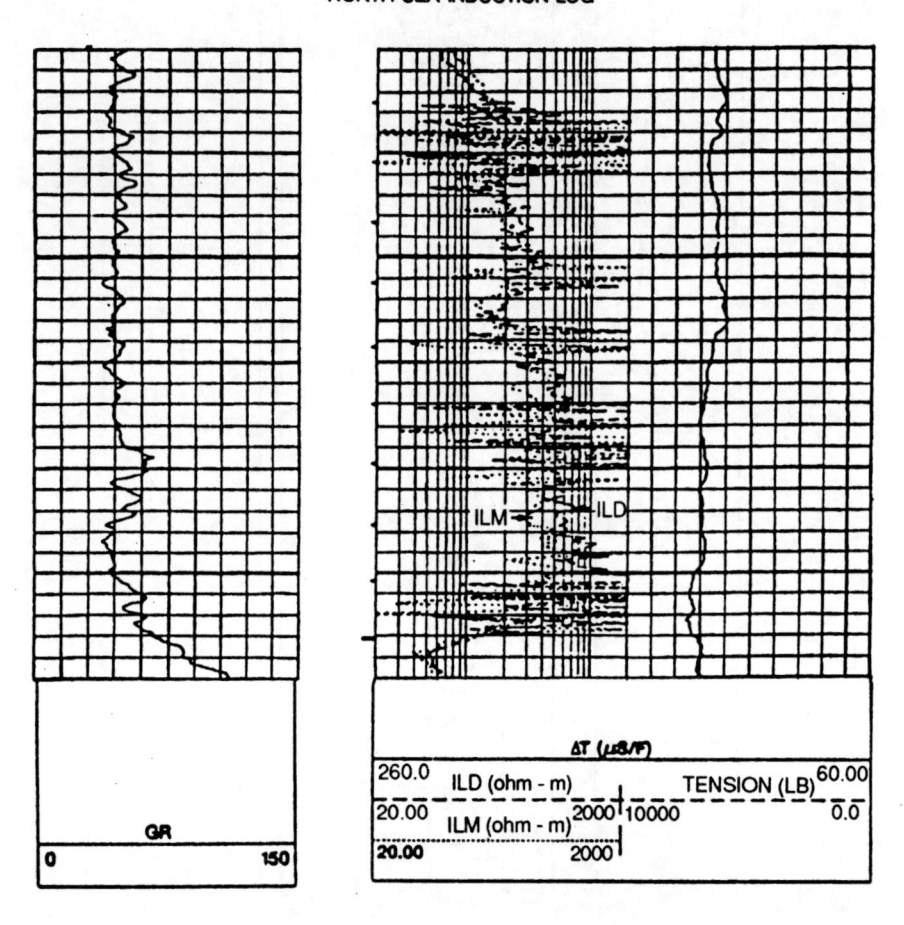

FIG. 25.16 : Induction anomalies in a North Sea well: log example.

Courtesy of SPWLA [1]

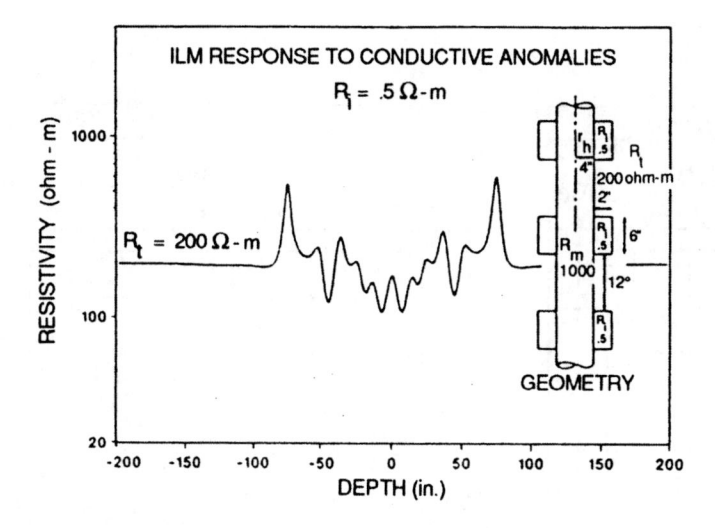

FIG. 25.17 : Induction anomalies in a North Sea well: formation model.

Courtesy of SPWLA [1]

25.7.5 Cycles

Figure 25.19 shows periodic sine wave oscillations in a high-resistivity zone. Over the same interval, the caliper shows that the borehole contains grooves about 0.25 in. (6 mm) deep and 2 ft (60 cm) long. This geometry has been simulated by mathematical modeling, which gives the response actually observed (*Fig. 25.18*).

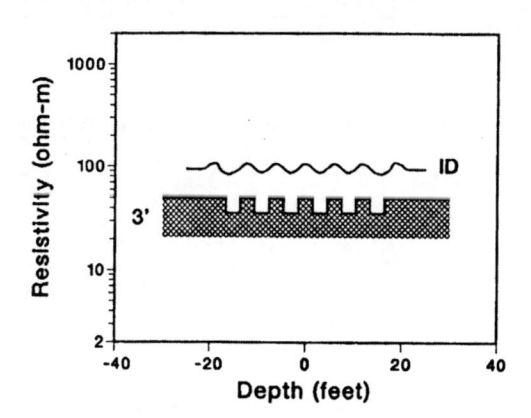

FIG. 25.18 : Cyclic modes on induction: borehole model.

Courtesy of SPWLA [1]

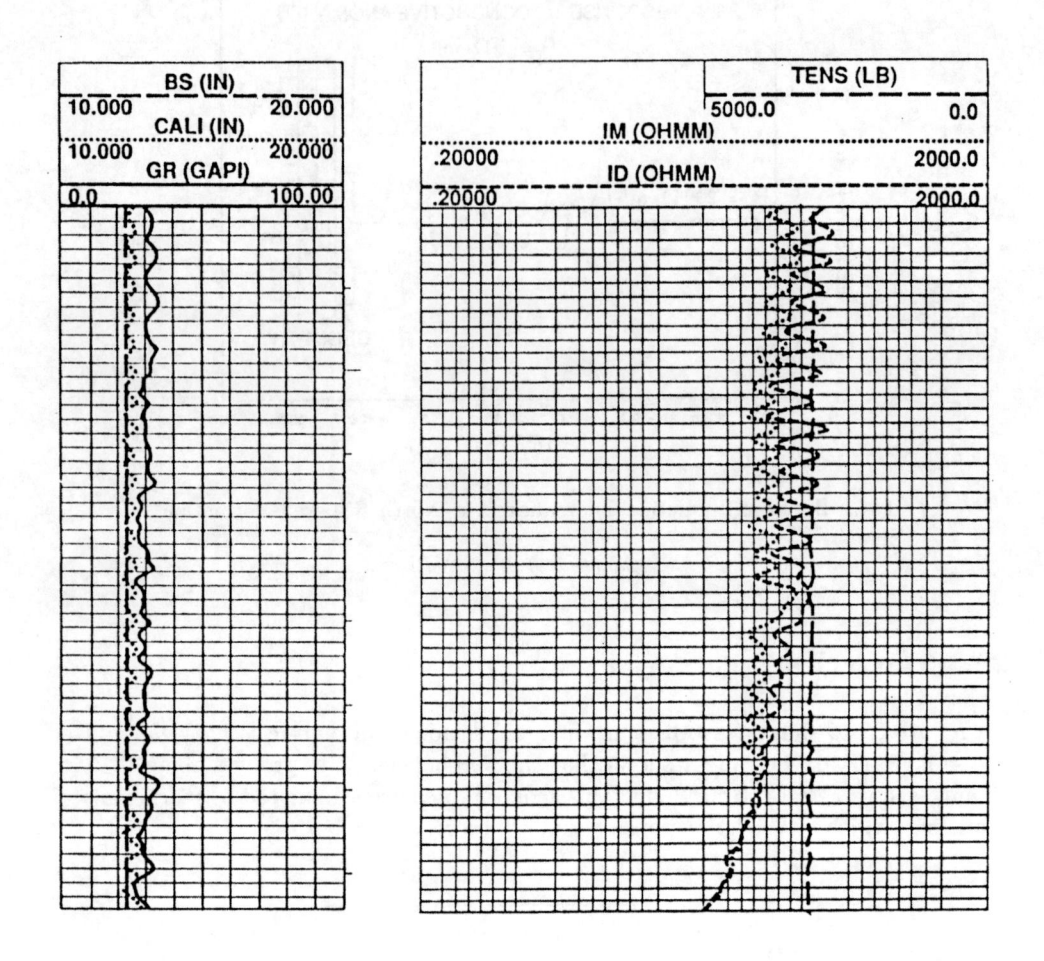

FIG. 25.19 : Cyclic modes on induction: log example.
Courtesy of SPWLA [1]

25.7.6 Lithology

Small amounts of specific minerals may cause anomalous readings on the logs. For instance, minerals with elements with a large thermal neutron capture cross section display high thermal neutron porosity readings, even though the culprit is present in terms of parts per million in the formation (*Table 25.3*). Gadolinium (Gd) is an extreme example. Its capture cross section is more than 1000 times the one of chlorine (Cl). These effects are not observed if the neutron-derived porosity is based on the observation of epithermal neutrons.

TABLE 25.3
CAPTURE CROSS SECTION
OF ABSORBING ELEMENTS

Element	Capture cross section (cu)
Gadolinium	49000.
Boron	759.
Lithium	71.
Chlorine	33.2
Hydrogen	0.3

Radioactive elements are also the cause of some extraordinary readings.

Logs run over the *Woodford* shale, in Oklahoma, exhibits gamma ray curve peaks at 500 API. The *Granite Wash*, an oil-producing clean horizon, displays readings of 200 API.

In cased hole, large scale deposition of celestite ($SrSO_4$) causes large gamma ray readings.

Similar readings in other geographical areas or conditions would certainly be attributed to tool failure.

25.7.7 Pyrite

Pyrite(Fe_2S) is the most common of the heavy minerals associated with sedimentary rocks.

Pyrite has little effect on the sonic and neutron logs, but, because of its density ($\rho_b = 4.985$ g/cm^3), it can be detected by the density log.

One percent (by volume) of pyrite shifts the density by 0.023 g/cm^3, which corresponds to 1.4 pu.

Because it appears in low concentration, pyrite distribution is generally of little importance to the petrophysicist except for the electrical conductivity anomalies it causes.

These effects are noticeable at concentrations greater than 7%.

They consist of sharp conductive anomalies of apparent thickness equal to or greater than the vertical resolution of the resistivity device. *Figure 25.20*, next page, shows an example of formations with important concentrations of pyrite.

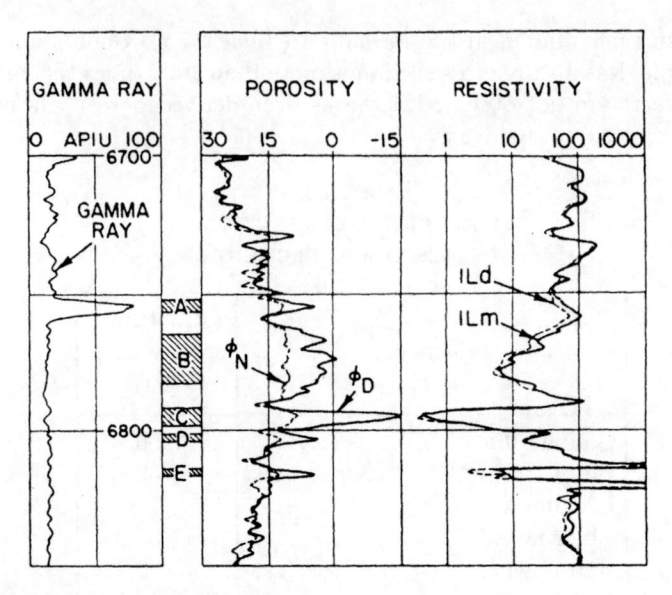

FIG. 25.20 : Formation with pyrite.
The zones are split in three categories: Zone A has a high clay content. Zones C, D and E have apparent pyrite contents of 16%, 4.5% and 6% as confirmed by the large drop in resistivity. Zone B has a variable content (4.5% to 6%).
Courtesy of SPWLA [2]

References

1. Anderson, B., Barber, T., "Using computer modeling to provide missing information for interpreting resistivity logs," paper H, *Trans.* SPWLA 29th annual logging symposium, 1988.

2. Clavier, C., Heim, A., Scala, C., "Effect of pyrite on resistivity and other logging measurements," paper HH, *Trans.* SPWLA 17th annual logging symposium, 1976.

3. Crary, S. F., Smith, D. J., "The use of electromagnetic modeling to validate environmental corrections for the dual laterolog," paper C, *Trans.* SPWLA 31st annual logging symposium, 1990.

4. Gianzero, S., Konen, C. E., Chemali, R. E., Strickland, R. W., "Sondes for big hole logging," SPE 15542, *Trans.* SPE 61st annual technical conference and exhibition, New Orleans, 1986.

5. Kienitz, C., Flaum, C., Olesen, J. R., Barber, T., "Accurate logging in large holes," paper III, *Trans.* SPWLA, 27th annual logging symposium, 1986.

6. Kimminau, S. J., Edwards, J., "Tandem density logging in rugose, tectonically stressed boreholes, Oman," SPE 28408, 69th annual technical conference and exhibition, New Orleans, 9-1994.

26
Images and nuclear magnetic resonance

In the nineties, borehole imaging and nuclear magnetic resonance, NMR, have emerged. Their popularity justifies a special chapter.

26.1 Borehole imaging

Borehole imaging technology inherited from dipmeter devices [6, 8, 11 and 15]. In the late 1970s, several software programs extracted stratigraphic information from dipmeter microresistivity curves. The dipmeter tool was also successfully run in pancake formations where there was not much dip to compute to define formation microfeatures and fractures. The first imaging tool had two arrays of 27 resistivity buttons, soon replaced by a four-pad, 64-button device providing a 40% coverage in an 21.6-cm (8.5-in.) borehole. The Formation MicroImager tool has four additional flaps, corresponding to a total of 192 buttons and 80% hole coverage.

26.1.1 Job preparation

26.1.1.1 Mud program

Electric imaging tools send current into and sense current in the formation. Therefore, mud resistivity should not exceed 50 Ω/m; neither should it be too conductive, as in that case, the current tends to flow in the borehole, reducing the sharpness of the images. Contrast in resistivity between the mud and the formation should be below 10,000. Electric images can be recorded in oil-base mud if the water content is at least 30%.

Acoustic image quality depends much on the density and homogeneity of the mud. The data user should check the specifications of the tool on these two parameters.

26.1.1.2 Borehole coverage and repeat passes

Considering the hole and pad sizes, additional runs after a tool rotation of 45° may be required for electric images.

In any case, in zones of interest, a repeat section is strongly recommended to validate the geological reality and representativeness of the features.

26.1.1.3 Tool configuration

In vertical holes (less than 10°), tools require centralizers. Oblique positioning of the tool or imperfect pad contact may cause blurred images.

26.1.2 Acquisition issue: tool movement

Wireline imaging tools are pad tools. Hence, they are subject to yo-yo and stick-slip effects. Speed correction, by measurement with accelerometers or by processing through the redundancy of the curves, is critical.

26.1.3 Processing guidelines

Postacquistion processing converts button current intensities or acoustic transit times and amplitudes into variable gray-scale or color images.

26.1.3.1 Speed correction

For electric images, depth shifts between rows of buttons are corrected with input from an accelerometer.

26.1.3.2 Equalization of electric buttons

There are some slight imbalances in button sensitivities and electrode application against the formation. They are handled by equalization that keeps different current measurements within a given mean and standard deviation.

26.1.3.3 Static and dynamic normalization

Static normalization establishes a discrete correspondence between current intensities and color for the entire recorded interval. It is well suited for large-scale resistivity, amplitude or transit-time variations. When small variations are recorded, special image enhancement programs can be used on selected intervals or sliding windows. It is important to observe that these techniques may attribute different colors to similar geological events surveyed on different runs.

26.1.4 General guidelines for image quality control

(a) Dead buttons on electric images are observed as straight lines over a long interval.
(b) Tool rotation should be lower than one turn per 12 m (40 ft).
(c) Intervals where pad tools stick cause data degradation after conventional speed correction. These intervals are easy to identify as they result in saw-tooth images. Interpretation is not affected except in excessive stick-and-slip conditions.
(d) In MWD, drillstring rotation needs to be checked. A sliding indicator or rpm curve is required.
(e) Mud smears and heavy oil may stick to pad tools and blind the sensors.
(f) Halos may reduce the contrast when a high-angle fracture is crossed by the borehole (*Fig. 26.1*).

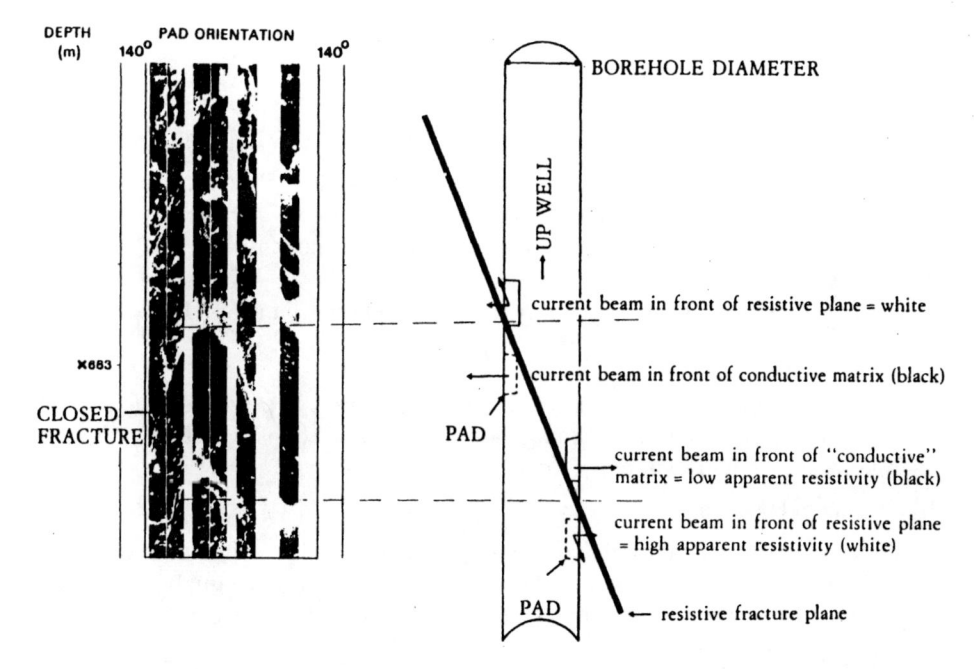

FIG. 26.1 : Explanation of the halo effect.
A strong resistivity contrast is observed between the two lips of the fracture. The current beam tends to concentrate in conductive beds. When a resistive bed is crossed, the current diffuses in the less resistive part of the formation.
Courtesy of Schlumberger [16]

26.2 NMR

26.2.1 Behavior of hydrogen proton spins

This section does not go into the details of NMR physics or interpretation; several books, [1 and 14], and articles are available for additional material, [4 and 17].

Oil industry NMR focuses on the behavior of hydrogen proton spins. In the absence of magnetic field, an equal number of spins are positive and negative. Under the influence of a magnetic field, spins shift from one state to the opposite one—this phase is called polarization. This creates an imbalance in the spin population, which results in a net magnetization. This magnetization can be measured. After removal of the magnetic field, the spins relax back to their original state.

Things become complex when several magnetic fields are imposed (e.g., a static field and an oscillating field). The antenna used to detect the spin population magnetization can also be used to pulse the formation spins. The most efficient pulse sequence is referred to as CPMG (after Carr, Purcell, Meiboom and Gill).

26.2.2 Relaxation times and mechanisms

The rates of change in spin population are characterized by T_1 and T_2.

T_1 is the spin-lattice relaxation time. It characterizes the ability of the spin system to change from a random state to a polarized state.

T_2 is the spin-spin relaxation time. It characterizes the way magnetic energy—stored in the spins—is dissipated.

The values of T_1 and T_2 depend on NMR relaxation mechanisms [10].

(a) Bulk relaxation:

only observable in fluids, bulk relaxation is important in presence of large pores, of high concentration of paramagnetic ions, of fine solids in suspension in the mud and for highly viscous fluids.

(b) Surface relaxation:

collision of a fluid molecule containing hydrogen with a solid surface provides an opportunity for spin-spin relaxation. Surfaces are not equally effective in relaxing hydrogen proton spins and are characterized by their relaxivity.

(c) Diffusion:

diffusion occurs if the static magnetic field varies and, hence, presents a gradient.

Relaxation mechanisms act in parallel. Therefore:

$$\frac{1}{T_1} = \frac{1}{T_{1B}} + \frac{1}{T_{1S}}$$

$$\frac{1}{T_2} = \frac{1}{T_{2B}} + \frac{1}{T_{2S}} + \frac{1}{T_{2D}}.$$

Subscripts B, S and D characterize bulk, surface and diffusion relaxations. The longitudinal relaxation has no diffusion component as it is strictly a dephasing process.

Typical relaxation mechanisms are gathered in *Table 26.1*. The large differences observed on the relaxation times of the formation components (solids, bound fluids, free fluids, viscous fluids, water, gas, oil) is the basis of NMR interpretation.

NMR measurements require a few corrections as compared to other logs [3]. They can often be directly used for interpretation. This is one of the reasons for the success of this logging method.

TABLE 26.1
RELAXATION MECHANISMS

	T_1	T_2	T_1/T_2
Mineral hydrogen	10-100 s	10 - 100 μs	$\sim 10^6$
Water (in clastics)	Surface dominated	Surface dominated	$T_1/T_2 \sim 1.5$
Water (in vugs)	Bulk	Bulk/diffusion	$T_1 \geq T_2$
Medium to heavy oil	Bulk	Bulk	$T_1 = T_2$
Light oil	Bulk	Bulk/diffusion	$T_1 \geq T_2$
Gas	Bulk	Diffusion	$T_1 \gg T_2$

Courtesy of SPWLA [10]

26.2.3 NMR acquisition sequence

26.2.3.1 Wait time

The wait time is the time interval during which spins are polarized. A rule of thumb indicates that wait time should be about $3 \times T_1$. In sandstones, a wait time of 1.3 s is frequently used; wait times of the order of 3 s are used in limestones; these figures also depend on the mud type.

Incomplete polarization results from insufficient waiting time. Incompletely polarized nuclei do not fully contribute to signal. Incomplete polarization effects can be corrected for and monitored on quality control curves. Incomplete polarization can also be used on purpose to increase logging speed or to intentionally disregard the contribution of some fluids in the formation [7].

26.2.3.2 Pulse echo spacing and number

Echo spacing (T_E) and number of echoes acquired (n) are critical. Short echo spacings minimize effects from diffusion and rock internal gradients. Moreover the short decay times that represent water residing in smallest pores and within clays often appear in such echo trains. MRIL[1] typically uses a T_E around 1 ms and 1200-echo stacks. CMR T_E is commonly 280 µs, but can be as low as 200 µs. n varies between 200 and 8000, with 1800 a typical value. High n values are usually acquired in stationary mode.

Tool ringing, linked to the quality of the electronics, is more problematic with short T_E.

26.2.3.3 Multiple shells

Some NMR logging devices are able to excite distinct volumes of formation—called shells [13]. This is equivalent to run several NMR tools in combination. The oscillating magnetic field of such tools operates at different frequencies.

With such tools it is possible to have one formation shell subjected to a given wait time and pulse echo spacing, another shell to a different wait time and echo spacing, etc.

Differential methods—differential spectrum, DSM, and shifted spectrum, SSM—that use different wait time and pulse echo spacing, respectively, can be implemented on a single logging pass. Multiwait and multi T_E methods increase the sensitivity of the collected information to small pore size, improve precision and enable more accurate polarization corrections.

The ability to excite different shells widens the operating options in terms of pulse sequences and reinforces the need for careful job planning. It also enables higher logging speed without sacrificing precision.

26.2.3.4 Measurement sequence and logging speed

Once the wait time and pulse echo spacing and number have been selected, it is possible to compute the duration of one measurement sequence:

$$\text{Measurement duration} = 2 \times [\text{wait time} + n \times T_E].$$

With multiple excitation, simultaneous sequences are lead in parallel. The measurement duration can still be computed.

The resulting logging speed results:

$$\text{Logging speed (ft/h)} = 300 \times \frac{\text{Depth sample rate (in.)}}{\text{Measurement time (s)}}.$$

---

[1] Two types of logging tools supply NMR measurements: the one from the *Schlumberger* Combinable Magnetic Resonance CMR* tool—with an antenna mounted on a pad—and the *Halliburton Numar* Magnetic Resonance Imager Log, MRIL—derived from a mandrel device.

26.2.4 Other NMR specifications

26.2.4.1 Magnetic field gradient

The gradient of the magnetic field created by the sonde is important as diffusion effects depend on its magnitude. A typical value is 17 gauss/cm. In addition, the magnetic field in the formation changes as a result of tool motion.

26.2.4.2 Vertical resolution

The measurement resolution is controlled by the vertical aperture of the radio frequency antenna. The MRIL-C features a 1.09-m (43-in.) aperture, the MRIL-D a 61-cm (24-in.) aperture, the CMR a 15.2-cm (6-in.) aperture.[2]

Vertical resolution is also linked to the sampling rate and to the stacking option. CMR proposes 3-, 6- and 9-in. samplings. Common stacking options are 3 and 5. While the stationary vertical resolution of the tool is 6 in., the actual resolution of the information is equal to the product of the sampling interval times the number of sampling intervals that are stacked. It could be as high as 45 in.

26.2.4.3 Precision

The quoted precisions for CMR are 1.0, 0.5 and 1.2 pu for porosity, free fluid volume and bound fluid volume [9]. In contrast with nuclear porosity measurements, the precision varies little with the parameter that is estimated.

26.2.5 Job planning

26.2.5.1 Prediction of formation relaxation times

NMR interpretation is considerably easier when prior knowledge of the properties of formation fluids is available. Reference [10] contain charts and tables that enable the determination of T_1 and T_2 for various fluids from their temperature, pressure, API gravity, gas-oil ratio and viscosity. This information can be summarized in a fluid properties form (*Table 26.2*).

---

[2]The CMR tool investigates a volume of about 1.3 to 3.2 cm (0.5 to 1.25 in.) into the formation in front of the pad. The MRIL investigates cylindrical shells around the borehole. The volume of one shell is around $0.75 \; 10^{-3}$ m^3.

TABLE 26.2
FLUID PROPERTIES FORM

TOE is expressed as $(1 - e^{-w/T_1})$, where w is the wait time.

Property	Value
Temperature	95°C (200°F)
Pressure	6100 psi
Gas hydrogen index	0.48
Gas T_1	4.5 s
Gas TOE	0.25 (waiting time: 1.3 s)
	0.59 (waiting time: 4 s)
Gas D	8×10^{-4} cm^2/s
Gas T_2	0.5 s ($T_E = 0.32$ ms)
	0.05 s ($T_E = 1$ ms)
Oil gravity	30°API
Gas-oil ratio	400 cf/B
Viscosity	1 cP
Oil T_1	1.1 s
Oil TOE	0.69 (waiting time: 1.3 s)
Oil T_2	0.9 s ($T_E = 1$ ms)

Courtesy of SPWLA [10]

26.2.5.2 Choice of the acquisition parameters

Most NMR services have job planners that facilitate the operating conditions of a specific survey. The following parameters or options must be selected [12]:

(a) wait time
(b) number of echoes
(c) T_E
(d) pulse sequence options[3]
(e) expected noise or precision
(f) logging speed
(g) resolution
(h) number of raw measurements to be stacked.

Preliminary interpretation requires the selection of a T_2 cutoff that allows the distinction between free and bound fluids. A value of 33 ms is commonly selected in sandstones; a value of 100 ms in carbonates.

---

[3]Measurements may be completed with different wait times and pulse echo spacings. These measurements can be performed simultaneously with multishell devices.

26.2.6 Quality control curves and parameters

Schlumberger NMR standard log presentation includes quality curves and flags.

26.2.6.1 Bad hole flag

NMR pad tools are sensitive to hole rugosity and washouts because NMR porosity readings can be affected by borehole mud signal. The derived NMR porosity may exhibit spikes associated to fast relaxation times.

The bad-hole flag highlights this condition.

26.2.6.2 Polarization correction curves

Two curves $CMRP_{min}$ and $CMRP_{max}$ corresponding to minimum and maximum T_1/T_2 ratios are computed and displayed with the CMR porosity. When the wait time is insufficient the three curves differ significantly.

26.2.6.3 High-voltage current and system gain

The current going to the transmitter is monitored. It may change with temperature, operating frequency and mud conductivity, but should not be erratic.

System gain is a parameter that is computed from the test loop signal measured during logging and during master calibration. System gain decreases in conductive muds and is affected by temperature.

26.2.6.4 ΔB_0

The magnetic field strength, B_0, is estimated by a Hall probe and temperature sensor. The difference between the two estimates is shown as ΔB_0. This value relates to the amount of metal debris on the magnets. If ΔB_0 exceeds 0.1 mT, the tool should be retuned—to the Larmor frequency of hydrogen at ambient temperature.

26.2.6.5 Signal, noise and standard deviation

Spin echoes are acquired in two channels using quadrature detection. From this can be derived (a) a channel that contains signal plus noise (b) a second channel that contains only noise. This second channel is used to estimate the noise and standard deviation of the NMR porosity.

Noise and standard deviation can be decreased by stacking, but to the detriment of vertical resolution.

26.2.6.6 Gamma

A regularization method is used to select a T_2 distribution that is consistent with the spin echo sequence. The method computes a parameter γ that depends on the signal-to-noise ratio and the shape of the underlying T_2 distribution.

Figure 26.2 depicts the QC display of the *Schlumberger* CMR tool.

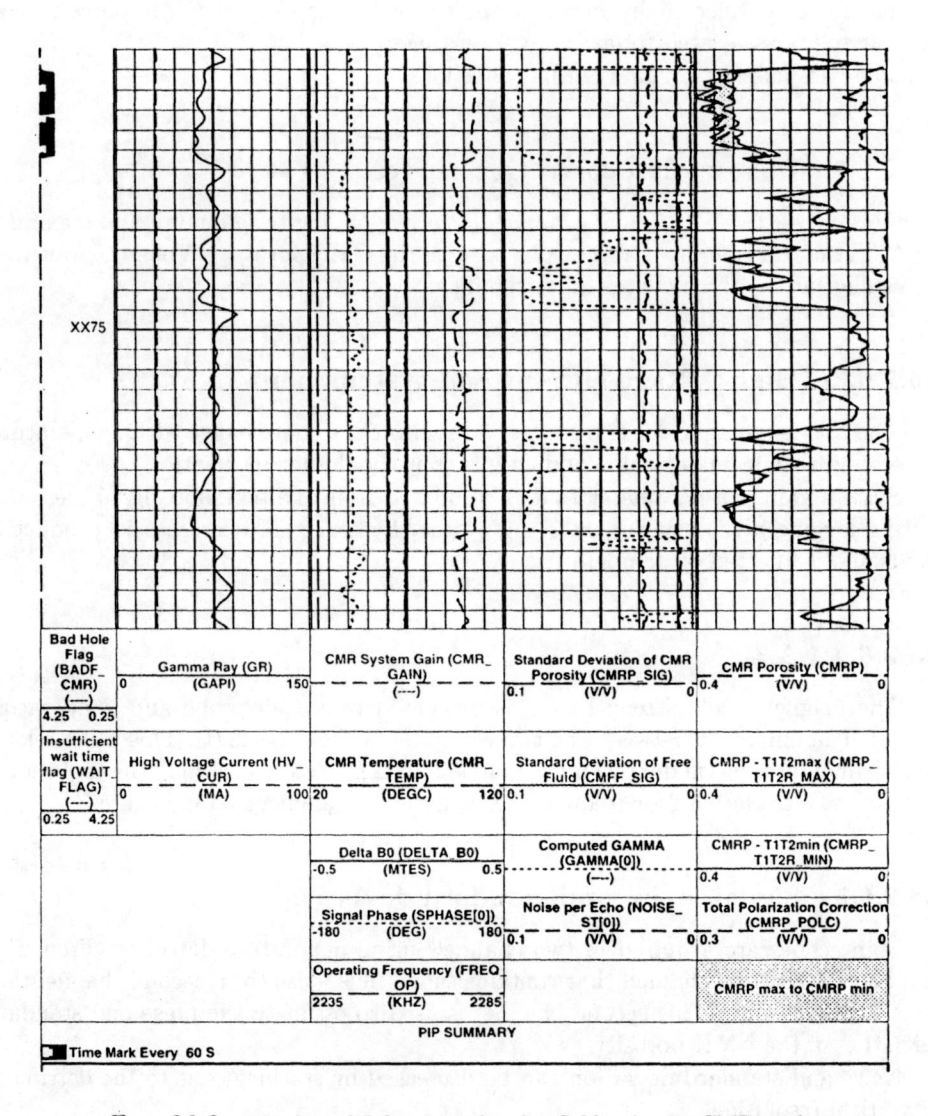

FIG. 26.2 : Display of QC curves for the *Schlumberger* CMR tool.

Courtesy of Schlumberger

References

1 Abragam, A., *Principles of nuclear magnetism*, Oxford University Press, Oxford, 1983.

2 Akkurt, R., Prammer, M. G., Moore, M. A., "Selection of optimal acquisition parameters for MRIL logs," *The Log Analyst*, Vol. 37, No. 6, pp. 43-52, 1996.

3 Allen, D., Crary, S., Freedman, B., Andreani, M., Klopf, W., Badry, R., Flaum, C., Kenyon, B., Kleinberg, R., Gossenberg, P., Horkowitz, J., Logan, D., Singer, J., White, J., "How to use borehole nuclear magnetic resonance," *Oilfield Review*, Vol. 9, No. 2, 7-997.

4 Brown, R. J. S., Gamson, B. W., "Nuclear magnetism logging," *Trans*. SPE 34th annual technical conference and exhibition, 10-1960, Dallas.

5 Chandler R. N., Drack, E. O., Miller, M. N., Prammer, M. G., "Improved log quality with a dual-frequency pulsed NMR tool," SPE 28365, SPE 69th annual technical conference and exhibition, New Orleans, 9-94.

6 Ekstrom, M. P., Dahan, C. A., Chen, M, Y., Lloyd, P. M., Rossi, D. J., "Formation imaging with microelectrical scanning arrays," *The Log Analyst*, Vol. 23, No. 3, 5-1987.

7 Flaum, C., Kleinberg, R. L., Bedford, J., "Bound water volume, permeability and residual oil saturation from incomplete magnetic resonance logging data," paper UU, *Trans*. 39th SPWLA annual logging symposium, Keystone,1998.

8 Foulk, L., "Letter from the special issue editor," *The Log Analyst*, Vol. 38, No. 6, 11-97.

9 Freedman, R., Morriss, C. E., "Processing of data from an NMR logging tool," SPE 30560, SPE 68th annual technical conference and exhibition, Dallas, 10-95.

10 Kleinberg, R. L., Vinegar, H. J., "NMR properties of reservoir fluids," *The Log Analyst*, Vol. 37, No. 6, pp. 20-32, 1996.

11 Lacazette, "The STAR (simultaneous acoustic and resistivity) imager," London Petrophysical Society, 1996.

12 Morriss, C. E., Deutch, P., Freedman, R., McKeon D., Kleinberg, R. L., "Operating guide for the combinable magnetic resonance tool," *The Log Analyst*, Vol. 37, No. 6, pp. 43-60, 1996.

13 Prammer, M.G., Bouton, J., Chandler, R. N., Drack, E.D., Miller, M. N., "A new multiband generation of NMR logging tools," SPE 49011, SPE 73rd annual technical conference and exhibition, New Orleans, 9-98.

14 Schlicter, C. P., *Principles of magnetic resonance*, Springer-Verlag, Berlin, 1980.

15 Seiler, D., King, G., Eubanks, D., "Field test results of a six-arm microresistivity borehole imaging tool, paper W, *Trans*. 35th SPWLA annual logging symposium, 1994.

16 Serra, O., *Formation microScanner image interpretation*, Schlumberger Educational Services, Houston, 1989.

17 Timur, A., "Pulsed nuclear magnetic resonance studies of porosity, movable fluid and permeability of sandstones," SPE 2045, *Journal of Petroleum Technology*, Vol. 24, No. 6, 6-1969.

27
Comparison of logged data with other information

Some conditions encountered in the well are known in advance. This known information can be used to check the validity of the newly acquired data. Then, the data set just acquired can be compared with logging data collected from wells drilled in the same field and with core and production data originating from the same well.

27.1 Response in known conditions

Any measurement made and controlled downhole is useful as it provides an in-situ validity check at conditions (pressure and temperature) close to those found when the tool was logging.

27.1.1 Casings

The casing allows the check of the sonic transit time, Δt, which should read close to 187 μs/m (57 μs/ft). It is sometimes difficult to record this value if the casing is well cemented.

Manufacturing of the casing is closely controlled so that it meets precise specifications: the internal diameter does not vary more than 0.75%.[1] This makes the casing

---

[1] API specifications for casings include tolerances on outside diameter (0.75%) and weight (−3.5%, +6.5%).

the place to check accuracy of the caliper(s) and proper functioning of the hole volume integration. *Table 27.1* lists the inside diameter of the most common casings. The corresponding hole volumes are given in the same table.

TABLE 27.1
CASINGS CHARACTERISTICS

Outside diameter	Weight couplings	Inside diameter	Integrated hole volume		Outside diameter	Weight couplings	Inside diameter	Integrated hole volume	
			200 ft $\frac{}{(\text{ft}^3)}$	60 m $\frac{}{(\text{m}^3)}$				200 ft $\frac{}{(\text{ft}^3)}$	60 m $\frac{}{(\text{m}^3)}$
(in.)	(lbm/ft)	(in.)			(in.)	(lbm/ft)	(in.)		
4.5	9.5	4.090	18.3	0.51	8.625	24.0	8.097	71.5	1.99
4.5	11.6	4.000	17.5	0.49	8.625	28.0	8.017	70.1	1.95
4.5	13.5	3.920	16.8	0.47	8.625	32.0	7.921	68.5	1.91
5	11.5	4.560	22.7	0.63	8.625	36.0	7.825	66.8	1.86
5	13.0	4.494	22.0	0.61	8.625	40.0	7.725	65.1	1.81
5	15.0	4.408	21.2	0.59	8.625	44.0	7.625	63.4	1.77
5	18.0	4.276	19.9	0.56	8.625	49.0	7.511	61.5	1.71
5.5	13.0	5.044	27.8	0.77	9.625	29.3	9.063	89.6	2.49
5.5	14.0	5.012	27.4	0.76	9.625	32.3	9.001	88.4	2.46
5.5	15.5	4.950	26.7	0.74	9.625	36.0	8.921	86.8	2.42
5.5	17.0	4.892	26.1	0.73	9.625	40.0	8.835	85.2	2.37
5.5	20.0	4.778	24.9	0.69	9.625	43.5	8.755	83.6	2.33
5.5	23.0	4.670	23.8	0.66	9.625	47.0	8.681	82.2	2.29
6	15.0	5.524	33.3	0.93	9.625	53.5	8.535	79.5	2.21
6	18.0	5.424	32.1	0.89	10.75	32.8	10.192	113.3	3.15
6	20.0	5.352	31.3	0.87	10.75	40.5	10.050	110.2	3.07
6	23.0	5.240	30.0	0.83	10.75	45.5	9.950	108.0	3.01
6.625	17.0	6.135	41.1	1.14	10.75	51.0	9.850	105.9	2.95
6.625	20.0	6.049	39.9	1.11	10.75	55.5	9.760	103.9	2.89
6.625	24.0	5.921	38.2	1.06	11.75	38.0	11.150	135.6	3.78
6.625	28.0	5.791	36.6	1.02	11.75	42.0	11.084	134.0	3.73
6.625	32.0	5.675	35.1	0.98	11.75	47.0	11.000	132.0	3.67
7	17.0	6.538	46.6	1.30	11.75	54.0	10.880	129.1	3.59
7	20.0	6.456	45.5	1.27	11.75	60.0	10.772	126.6	3.52
7	23.0	6.366	44.2	1.23	13.375	48.0	12.715	176.4	4.91
7	26.0	6.276	43.0	1.20	13.375	54.5	12.615	173.6	4.83
7	29.0	6.184	41.7	1.16	13.375	61.0	12.515	170.9	4.76
7	32.0	6.094	40.5	1.13	13.375	68.0	12.415	168.2	4.68
7	35.0	6.004	39.3	1.09	13.375	72.0	12.347	166.3	4.63
7	38.0	5.920	38.2	1.06	16.0	55.0	15.375	257.9	7.18
7.625	20.0	7.125	55.4	1.54	16.0	65.0	15.250	253.7	7.06
7.625	24.0	7.025	53.8	1.50	16.0	75.0	15.125	249.6	6.95
7.625	26.4	6.969	53.0	1.47	16.0	84.0	15.010	245.8	6.84
7.625	29.7	6.875	51.6	1.44	20.0	94.0	19.124	399.0	11.11
7.625	33.7	6.765	49.9	1.39					
7.625	39.0	6.625	47.9	1.33					

Figure 27.1 represents caliper and sonic curves recorded as the logging string entered the cased section. Sonic curve reads 55.6 μs/ft (182 μs/m), a value within tolerance–2 μs/ft (6.5 μs/m)—of the nominal value. The caliper reads 6.31 in. The expected value of the internal casing diameter is 6.28 in. considering the outside diameter of 7 in. and the casing weight of 26 lbm/ft. The difference between the caliper reading and the inside casing diameter is 0.04 in., less than the published tolerance of 0.20 in. The caliper casing check is positive.

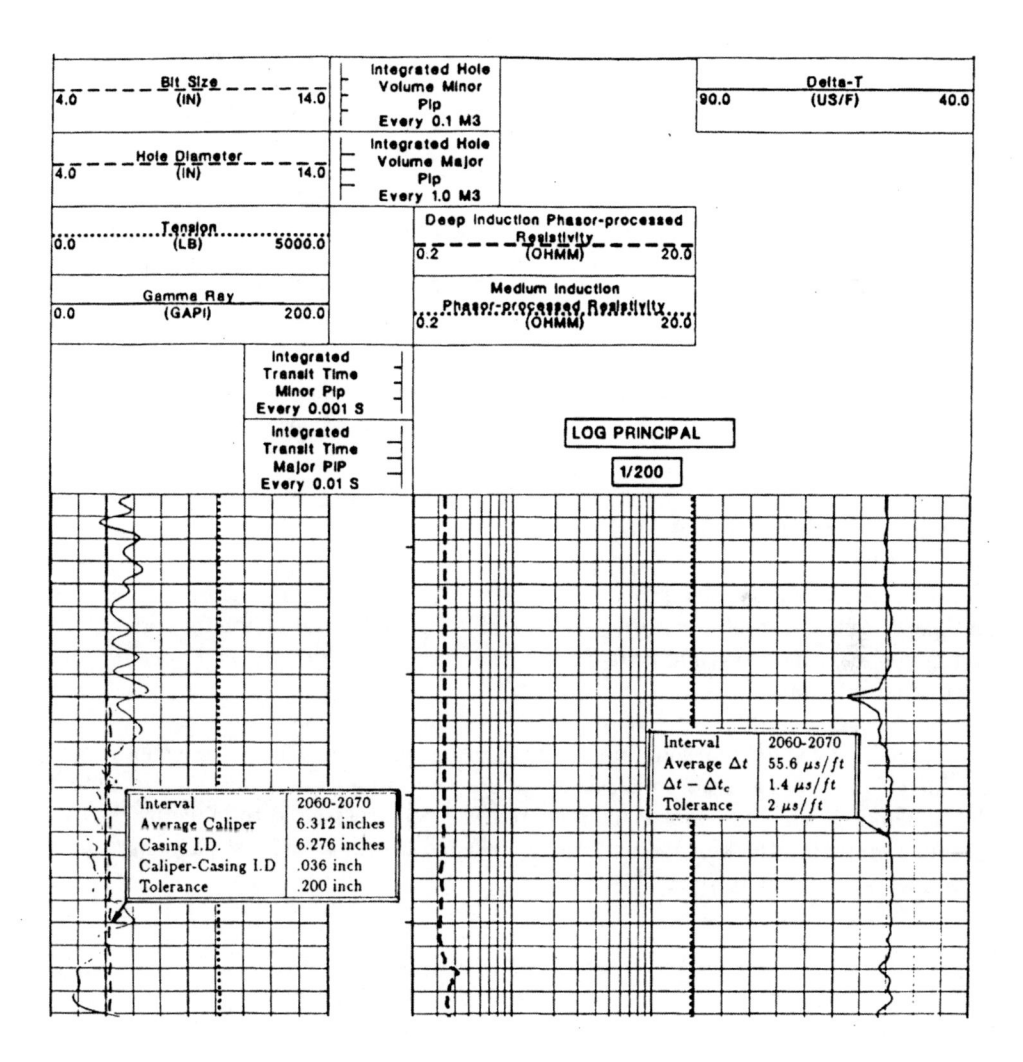

FIG. 27.1 : Sonic and caliper casing check.

Courtesy of Schlumberger

The casing response is also used in cement bond logging. The transit time from the transmitter to the receiver via the casing should belong to a narrow range of values. Similarly, the amplitude of the acoustic signal in an unbounded casing has values related to the casing size and weight (*Table 27.2*).

TABLE 27.2
EXPECTED RESPONSE IN FREE PIPES
This table is valid for the *Schlumberger* QA/NA type sonic sondes. Other tools show different values.

Casing		Amplitude	Transit time
Size (in.)	Weight (lbm/ft)	(mV)	(μs/ft)
5	13	77±8	240
5 1/2	17	71±7	247
7	24	61±6	271
8 5/8	38	55±6	295
9 5/8	40	52±5	313
10 3/4	48	50±5	331

Finally, the response of some electrical surveys can be checked in the casing. The electrical log and the laterolog resistivity curves should read zero in the casing.

27.1.2 Geological formations

Some rock formations lend themselves to in-situ control of logging tools. These formations have good lateral continuity and stable petrophysical characteristics. Moreover, good hole conditions and fieldwide environmental conditions should prevail across the marker interval.[2]

A variation of mud system may reduce the validity of the check. Shale and coal beds make stable geological markers, but they are often linked to inferior hole conditions.

Table 27.3 shows a list of minerals with good potential as reference for this type of check [2 and 8]. The list is not exhaustive but has been restricted to minerals that are commonly found in thick beds with good lateral extent. *Figure 27.2* shows a histogram performed in a low-porosity carbonate.

---

[2]Checking the response of a logging tool in downhole conditions is not a strict conformance test, since it is difficult to describe the mineralogical composition of the rock with certainty. Discrepancies between actual and expected readings raise the possibility of an incorrect tool behavior, which needs to be confirmed by other means.

TABLE 27.3
CHARACTERISTICS FOR THE MOST STABLE FORMATIONS

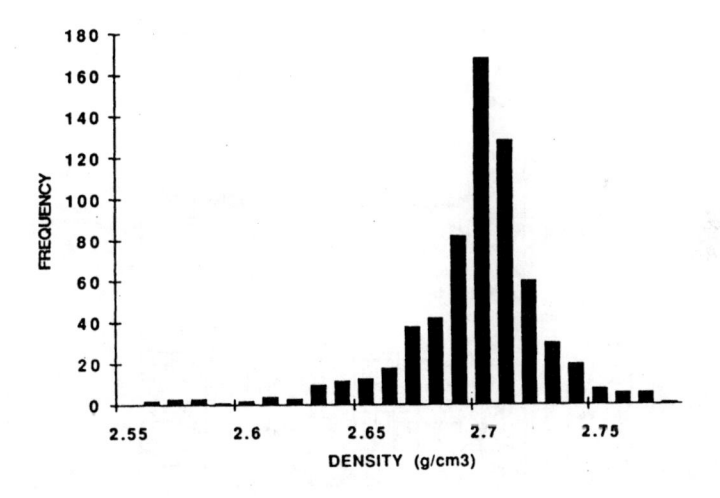

Mineral	$\frac{\rho_b}{(\frac{g}{cm^3})}$	$\frac{P_e}{(\frac{barns}{electron})}$	$\frac{\Sigma}{(cu)}$	$\frac{\phi_N}{(pu)}$	$\frac{\Delta_t}{(\frac{\mu s}{ft})}$
Halite.....	2.03	4.65	748.5	−1.8	67.0
Anhydrite .	2.98	5.05	12.5	−0.7	54.0
Gypsum ...	2.35	3.99	18.5	57.6	52.5
Tachhydrite	1.64	3.84	400.0	> 60	92.0
Bischofite..	1.57	2.59	328.7	> 60	100.0
Quartzite..	2.65	1.81	4.5	−2.1	50.5

FIG. 27.2 : Histogram of apparent density in a marker.
The apparent matrix density, $(\rho_{ma})_a$, derived directly from the bulk density and the thermal neutron porosity, is plotted on the x-axis versus frequency. Out of 662 levels, the average apparent matrix density is 2.698 g/cm³ and the standard deviation is 0.028 g/cm³, in line with the geologist's prognosis.

27.1.3 Muds and formation fluids

Some properties of mud are constant enough to be used to verify the stability of logging sensors. This is the case of a pressure gauge whose responses before and after a pressure test or a fluid sampling operation are controlled by reading the hydrostatic pressure. An example is shown in *Fig. 27.3*. The difference of reading of the hydrostatic column before and after is supposed to be less than a given tolerance (generally a fraction of a psi).

When such a check is performed, some care must be exercised so that the fluid level in the well is not changed.

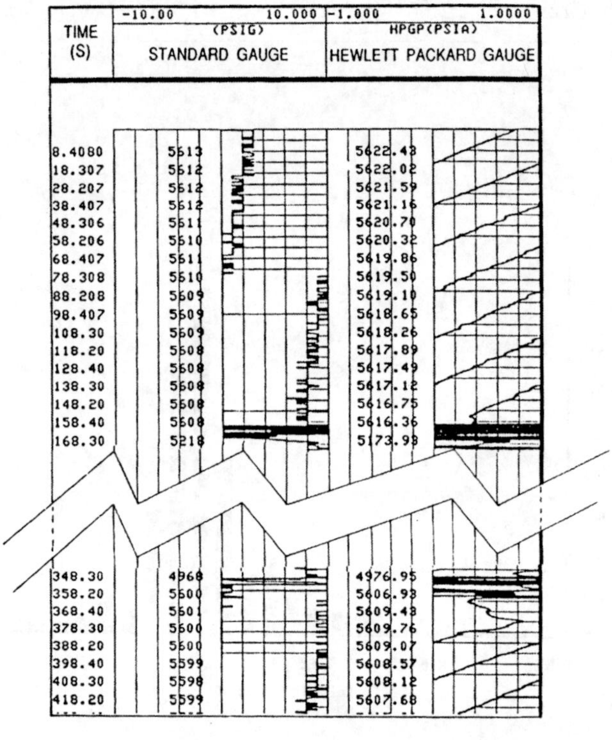

FIG. 27.3 : Pressure gauge stability.
The differences of hydrostatic pressure readings before and after pretest are large, 14 psi on the strain gauge (SGP) and 14.5 psi for the *Hewlett-Packard* gauge (HPGP), and beyond stability standards. The similarity of the differences observed on the two gauges tends to indicate the fluid level has changed in the well.

27.2 Comparison of data acquired on the same well

Data from different sources is continuously compared, repeat pass to main pass, MWD runs to wireline logs, real-time data to recorded mode data, core information to logs. Comparison can be fruitful, but it needs to be performed in a structured and informed way. Apples need to be compared to apples.

27.2.1 Caveat

Before data sets from different origins are compared [7 and 9], the following questions must be answered:

(a) Are units and references identical? For instance, when two gauges are run on the same pressure tester, are the pressures they measure both expressed in psi **absolute**?
(b) Are these data sets related to the same volume of formation?
(c) Are logging speeds, rates of penetration, measurement durations equivalent?
(d) Are compatible signal-processing schemes applied?
(e) What about the sampling rate?
(f) Were the well environments similar at the time of acquisition?

27.3 Multiwell analysis

A coherence check can be performed on logs run in different wells in the same field. The method is a check independent of calibrations and tool monitoring. Sonic, neutron and density logs are well adapted to these checks since they reflect porosity and lithology variations that are often of limited magnitude within the same geological unit. In contrast, the resistivity logs vary substantially as the nature of the fluid in the rock changes.

The coherence method concentrates on simple formations that are stable throughout the field. Excellent candidates are anhydrite, salt, pure calcite and dolomite, but any other stable formation can be used.

27.3.1 Choice of markers

The logs are analyzed to determine if several intervals are suitable. Statistical analysis is used. All values of a selected logging parameter are plotted on a histogram. Means and standard deviations are computed for the intervals of interest. The log values corresponding to the same interval vary within a limited range unless a major geological change takes place between the wells. Such geological changes are unlikely to happen simultaneously on all markers, because of their different age.

When a systematic and consistent deviation from the field average is observed on a given well, the log is reported as "anomalous."[3]

Table 27.4 represents a collection of average readings over five markers, A to E, in nine different wells of the same field, for neutron porosity, density and sonic logs. *Table 27.5* lists the differences between the values read at the wells and the average for the field.

The analysis confirms that the density and sonic logs are in line. However, it underlines discrepancies with the neutron porosity log. On wells 1 and 5, readings below average suggest that the caliper correction was not switched on. This is particularly noticeable in zones A and E where the hole diameter is significantly different from the bit size. Neutron porosity on well 2 is consistently larger, by about 4 pu (*Fig. 27.4*). This finding was followed by a closer look at the calibrations. An excessive drift was found between the before- and after-survey calibrations.

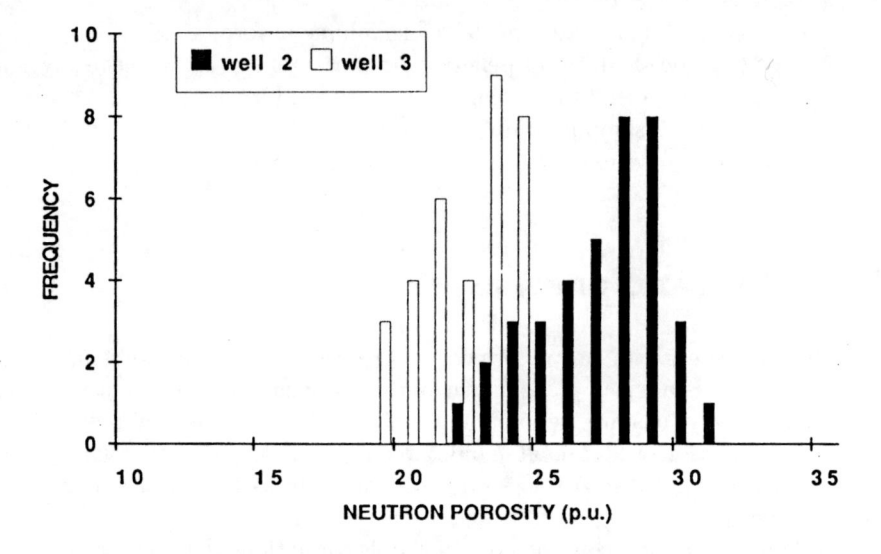

FIG. 27.4 : Neutron porosity in marker B of the same field.
Data from wells 2 and 3 are shown with different area codings. In well 2, the average porosity is 4 pu larger than in well 3 whose porosity is 0.5 pu larger than the field average.

---

[3]Coherence analysis is not a rigorous conformance check as it is impossible to verify the homogeneity of the rocks. Once a log has been rated "anomalous" with this method, other quality indicators need to be closely scrutinized. There are examples of wells with characteristics significantly different from the field average. Detailed crosschecks have proved that the logs run in these wells were correct. Geologists were able to create more sophisticated reservoir models to explain the facies changes.

TABLE 27.4
AVERAGE READINGS IN MARKERS A TO E

	Wells → Zones ↓	1	2	3	4	5	6	7	8	9	Field average
ϕ_N	A	27.69	32.91	27.67	28.24	26.96	29.59	32.11	35.65	27.30	29.79
	B	25.99	28.16	24.55	22.54	22.21	23.18	23.09	24.09	22.16	24.00
	C	38.07	40.59	37.33	37.61	38.27	37.37	38.79	42.29	38.71	38.78
	D	33.54	39.95	39.13	36.61	30.34	34.12	32.76	34.72	33.69	34.99
	E	32.84				39.34	33.26	37.71			35.79
ρ_b	A	2.293	2.219	2.321	2.293	2.313	2.295	2.301	2.239	2.315	2.288
	B	2.353	2.355	2.377	2.433	2.397	2.393	2.398	2.406	2.402	2.390
	C	2.253	2.239	2.250	2.262	2.233	2.311	2.308	2.312	2.305	2.275
	D	2.105	2.169	2.134	2.209	2.102	2.113	2.117	2.092	2.129	2.130
	E	2.056				2.073	2.062	2.063			2.063
Δt	A	95.3				94.4	94.6	95.5			95.4
	B	94.9				94.5	93.9	94.0			94.9
	C	94.2				94.3	95.2	94.8			94.5
	D	94.7				94.6	95.1	93.9			94.5
	E	94.8				94.9	94.1	94.8			94.5

TABLE 27.5
DIFFERENCES BETWEEN WELL READINGS AND FIELD AVERAGES

	Wells → Zones ↓	1	2	3	4	5	6	7	8	9
ϕ_N	A	-2.11	3.12	-2.12	-1.55	-2.83	-0.20	2.32	5.86	-2.49
	B	1.99	4.17	0.55	-1.46	-1.79	-0.82	-0.91	0.09	-1.83
	C	-0.72	1.81	-1.45	-1.18	-0.51	-1.41	0.01	3.5	-0.07
	D	-1.45	4.97	4.14	1.63	-4.64	-0.87	-2.22	-0.26	-1.29
	E	-2.94				3.55	-2.53	1.92		
	Average	-1.04	3.52	0.28	-0.64	-1.25	-1.16	0.23	2.3	-1.42
ρ_b	A	0.01	-0.07	0.03	0.01	0.03	0.01	0.01	-0.05	0.03
	B	-0.04	-0.04	-0.01	0.04	0.01	0.00	0.01	0.02	0.01
	C	0.02	-0.04	-0.02	-0.01	-0.04	0.04	0.03	0.04	0.03
	D	-0.02	0.04	0.00	0.08	-0.03	-0.02	-0.01	-0.04	0.00
	E	-0.01				0.01	0.00	0.00		
	Average	-0.02	-0.03	0.00	0.03	-0.01	0.01	0.01	-0.01	0.02
Δt	A	-0.09				-1.01	-0.76	0.11		
	B	-0.02				-0.43	-0.99	-0.88		
	C	-0.31				-0.19	0.67	0.36		
	D	0.27				0.12	0.60	-0.62		
	E	0.25				-0.24	-0.19	-0.16		
	Average	-0.02				-0.82	-0.17	-0.20		

27.4 Comparison with cores

Taking a core remains the only opportunity for the petroleum engineer or the geologist to physically examine a continuous interval of reservoir rock. This technique has been continuously improved to increase recovery and to keep the sample in conditions as close as possible to the ones observed downhole.

Three types of analysis [5] are performed depending on the size of the sample:

(a) **fullbore** analysis
(b) analysis of the **plugs** extracted from the fullbore core
(c) **sidewall cores**.

27.4.1 Comparison of logs with drilled cores

The comparison of core and log results is consuming large amounts of the petroleum engineer's and geologist's energy and time. There are a number of reasons why the parameters differ.

(a) **Physical configuration**

A fullbore core extracted from a hole of nominal bit size b would have a diameter approximately equal to $b/2$. The high-resolution logging tools, generally of pad type, have sensors directly against the borehole wall, while the outside of the core is several inches away from the borehole wall during drilling. This discrepancy causes a depth offset (*Fig. 27.5*).

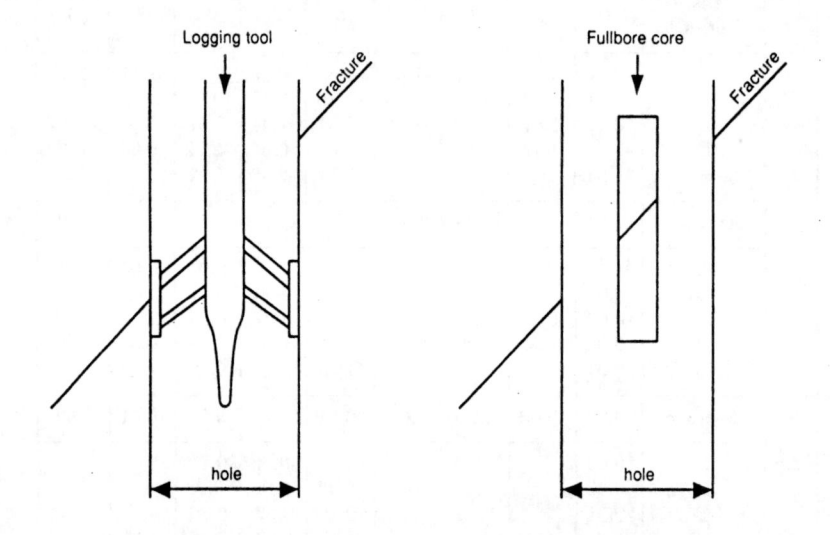

FIG. 27.5 : Depth offset detected by a pad logging tool and a fullbore core.
In this example, the offset is observed across a dipping fracture.
Courtesy of the Oilfield Review [1]

(b) **Depth mismatch**
Most laboratories match core depth with log depth by running a gamma ray survey over the core. However, when the recovery is poor and the core is short, the match is often unsatisfactory. It may even happen that the cores are wrongly labelled or put upside down in their boxes.

(c) **Change in fluid type and saturation**
When a conventional core is cut with a standard diamond or core barrel, it is partly flushed by the drilling fluid or mud filtrate. The core cut at downhole pressure and temperature contains a mixture of fluids including connate water, mud filtrate and residual oil or gas. When the core is brought to the surface, the pressure and temperature are reduced until they reach atmospheric conditions. The gas in solution in the oil is liberated. The free gas or solution gas expands, forcing mud filtrate, oil and possibly connate water out of the core. Consequently, a core taken from an oil-bearing zone contains gas, oil and a large amount of water (filtrate or connate). A core coming from a gas-bearing zone contains gas, no oil and a large amount of water.
These changes of phase may be reduced by special coring techniques using rubber or plastic sleeves. The cores are maintained at the bottomhole pressure at which they were cut until analyzed in the laboratory.

(d) **Difference of volume of investigation and of vertical resolution**
Logs average porosity over a depth sample, typically 0.3 m (1 ft). Core porosity may change by several pu within a few inches, as shown in *Table 27.6*. Core plug porosity measurements taken at 0.5 m (2 ft) intervals do not represent the average porosities every 0.5 m (2 ft).
In fractured or vuggy carbonates, the core laboratory analysts cut the plugs in homogeneous zones, and high-porosity zones are often missed.

(e) **Porosity**
Measured core porosity depends on the technique used (*Table 27.7*). For unconsolidated formations, the differences between porosities measured with and without overburden can be even larger.
In shaly sands, core porosity includes bound water volume and is equivalent to total porosity,[4] a parameter introduced in the Waxman-Smits and *Schlumberger* dual-water interpretation models. It could be significantly higher than log porosity, equivalent to effective porosity (*Fig. 27.6*).

(f) **Saturation exponent**
If careful procedures are not followed, the rock fabric of the core may be altered and the laboratory-derived saturation exponent may be erroneous [10].

---

[4]The reason for this is that the cores are often dried. Most of the bound water is evaporated. When the cores are dried under controlled humidity conditions, some of the bound water is preserved, but not all of it.

TABLE 27.6
VARIATIONS OF CORE PROPERTIES

Depth	Air	Helium porosity	Grain density
(ft)	(mD)	(pu)	(g/cm^3)
6023.6	91	17.9	2.64
6023.9	86	16.8	2.65
6024.1	75	17.4	2.64
6024.5	30	15.8	2.64
6024.7	56	17.2	2.65
6024.8	113	19.1	2.64
6032.8	28	16.0	2.64
6032.9	29	16.4	2.63
6033.1	68	18.3	2.64
6033.7	154	20.0	2.63
6033.9	56	17.4	2.64
6034.1	89	19.1	2.64
6065.9	0.04	5.8	2.64
6066.1	3.6	13.1	2.64
6066.2	9.2	13.9	2.65
6066.6	10	12.7	2.64
6066.9	14	13.3	2.64
6067.1	23	13.9	2.64
6115.1	30	15.5	2.64
6115.7	44	17.6	2.65
6115.9	58	17.4	2.64
6116.1	77	17.8	2.65
6116.3	65	16.7	2.64

Courtesy of the JPT [6]

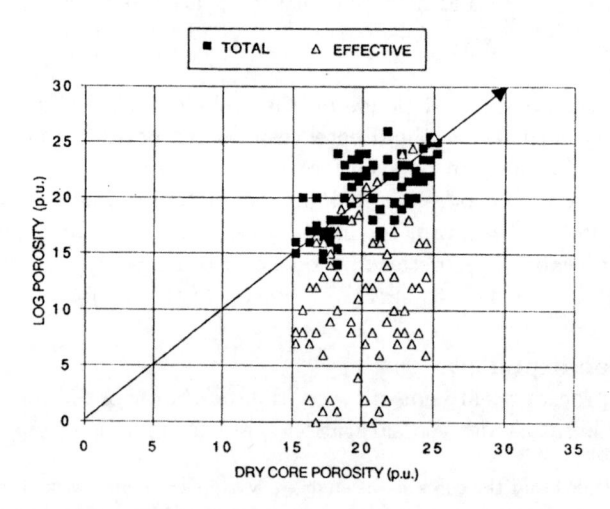

FIG. 27.6 : Core porosity versus log porosity.

TABLE 27.7
CORE POROSITY DERIVED FROM
DIFFERENT LABORATORY
TECHNIQUES

a. Summation of fluids.
b. Brine.
c. Helium, 200-psi overburden.
d. Helium, 3000-psi overburden.

Depth	Porosity (pu)			
(ft)	1^a	2^b	3^c	4^d
6023.33	20.2	13.1	15.4	14.6
6034.42	17.6	18.3	19.0	18.1
6040.42	11.6	13.8	15.7	14.9
6061.42	12.0	11.7	14.0	13.6
6066.33	14.9	11.5	13.7	12.1
6090.50	11.3	10.9	12.3	11.3
6109.42	14.9	9.6	10.2	9.3
6115.42	17.5	15.1	15.5	15.0
6124.42	8.7	8.1	8.2	7.3
6148.42	9.3	5.1	1.3	1.2
Average	13.8	11.7	12.5	11.7

Courtesy of JPT [6]

27.4.2 Comparison with sidewall cores

Sidewall samples are obtained by percussion. The petrophysical properties of the formation such as porosity and permeability cannot be evaluated because of compaction, mud invasion and shattering. Despite these limitations, sidewall cores are a useful complement to logs to evaluate lithology. Since the sidewall core and log information have different vertical resolutions, a potential difficulty in core/log comparison is depth matching.

Determination of depth shifts between and core and log data

The method is similar to the correlation of microresistivity dipmeter curves. It searches for an optimal displacement between the two parameters:

(a) Y = presence of hydrocarbon detected from core.
(b) Z = a function of the water saturation derived from logs.

Parameters Y and Z take values equalled to 1 and -1.

- $Y = 1$ if the core description indicates a hydrocarbon show.
- $Y = -1$ if the core description indicates *no show*.
- $Z = 1$ if S_w is larger than b, a threshold input by the analyst.[5] This corresponds to water-bearing zones.
- $Z = -1$ if S_w is smaller than b. This corresponds to hydrocarbon zones.

The coefficient of correlation $C(h)$ linking Y and Z is computed for different displacements, h, of Z as referred to Y:

$$C(h) = Y_d \times Z_{d+h}.$$

where Y_d is the value taken from the core description at depth d while Z_{d+h} is the value derived from the saturation at depth $d + h$. $C(h)$ is plotted versus h and the maximum of $C(h)$ defines the most probable displacement H. For a perfect sidewall core operation, H should be zero. If the correlation is perfect (when a show on a core always corresponds to a low water saturation and no show to a high water saturation), then the maximum of the correlogram is equal to the number of cores described:

$$C(h) = \sum_{\text{show cores}} (+1) \times (+1) + \sum_{\text{no-show cores}} (-1)(-1) = \text{number of cores}.$$

The maximum of the actual correlogram can be compared to this value. This gives an idea of the quality of the correlation. The sharpness of the correlogram indicates how precisely the displacement is known. *Figure 27.7* is an example of a log/sidewall core correlogram.

27.5 Comparison with production data

27.5.1 Capillary pressure curves

In a homogeneous formation, the **capillary pressure** curve gives the saturation profile (*Fig. 27.8*). The pressure is proportional to the height above the free water level, precisely defined as the depth where the capillary pressure is nil.[6] The threshold pressure is defined as the lowest capillary pressure at which water is displaced from the formation. It corresponds to the shallowest level with 100% water saturation computed from the logs. The curved section of the capillary pressure curve represents the transition zone. Corresponding to higher pressure, the zone above the transition displays the lowest water saturation values that reach an asymptotic minimum, the irreducible water saturation.

---

[5]The shape of the resulting correlogram does not vary substantially with b.

[6]On the pressure tester plot, the free water level depth coincides with the intersection of the water and hydrocarbon gradients.

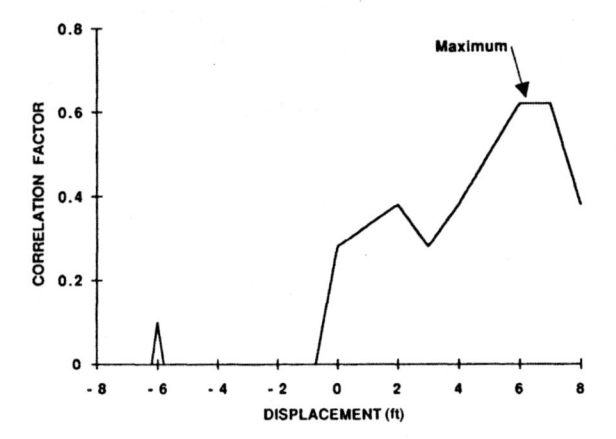

FIG. 27.7 : Example of correlogram.
Each value of the coefficient of correlation is normalized to the total number of cores. For instance, the value of the peak (0.62) represents the ratio of the coefficient of correlation divided by the total number of successful cores taken with the core gun. The displacement is 1.95 m (6.5 ft).

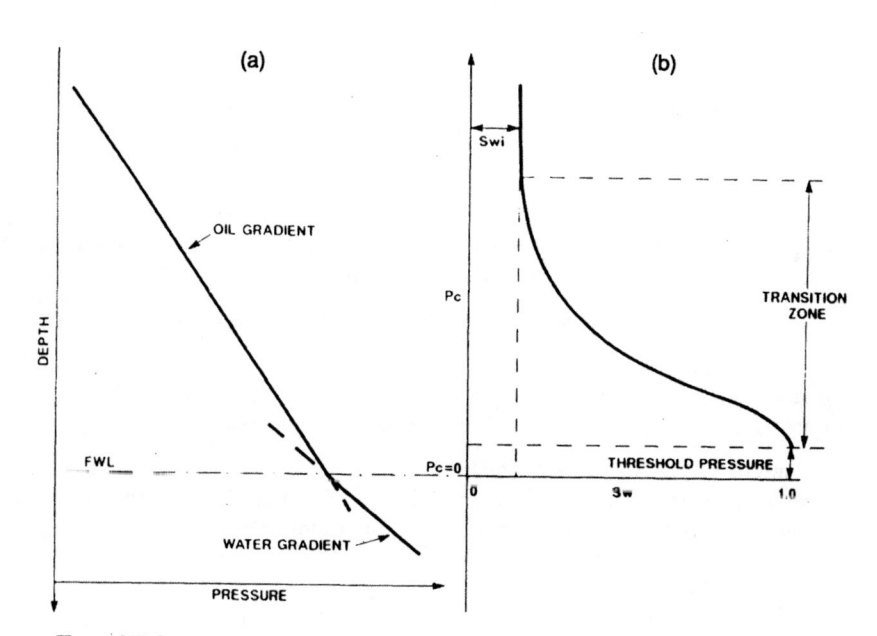

FIG. 27.8 : Pressure gradient and capillary pressure curve.

  a. Pressure versus depth plot. The pressure gradients intersect at the free water level.

  b. Corresponding capillary pressure curve. Zone A: 100% water saturation. Zone B: transition. Zone C: irreducible water saturation.

Courtesy of SPWLA [3]

Discrepancies between core- and log-derived saturations are mainly due to their respective resolution. Correct log values cannot be obtained from low-resolution tools in thin layers. In that case higher resolution logs, such as electromagnetic propagation logs [3 and 4] help resolve or explain the differences.

References

1 Basan, P., Hook, J., Hughes, K., Rathmell, J., Thomas, D., "Measuring porosity, saturation and permeability from cores, an appreciation of the difficulties," pp. 22-36, *The Technical Review*, Vol. 36, No. 4, 10-1988.

2 Ellis, D., Flaum, C., McKeon, D., Scott, H., Serra, O., Simmons, G., "Mineral logging parameters, nuclear and acoustic," pp. 38-52, *The Technical Review*, Vol. 36, No. 1, 1-1988.

3 Eriksen, S. H., "Use of EPT and capillary pressure to estimate saturation and permeability, a case study," paper L, *Trans.* SPWLA 11th European formation evaluation symposium, 1988.

4 Grimnes, J. P., "Brage, the invisible reservoir," paper C, *Trans.* SPWLA 11th European formation evaluation symposium, 1988.

5 Monicard, R. P., *Properties of reservoir rocks: core analysis*, Éditions Technip, Paris, 1980.

6 Neuman, C. H., "Logging measurement of residual oil, Rangely field, Co," SPE 8844, pp. 1735-1744, *Journal of Petroleum Technology*, Vol. 35, No. 9, 9-1983.

7 Prilliman, J., Bean, C. L., Hashem, M., Bratton, T., Fredette, M. A., Lovell, J. R.,"A comparison of wireline and LWD resistivity images in the Gulf of Mexico," paper DDD, *Trans.* SPWLA 38th annual logging symposium, Houston, 1997.

8 Serra, O., *Element mineral rock catalog*, Schlumberger, Paris, 1990.

9 Spross, R. L., Ball, M. S., Zannoni, S. A., "Case histories of MWD and wireline density log comparisons: an improved understanding of density log response," SPE 28430, SPE 69th annual technical conference and exhibition, New Orleans, 9-1994.

10 Worthington, P. F., Toussaint-Jackson, J. E., Pallat, N., "Effect of sample preparation upon saturation exponent in the Magnus field, UK North Sea," paper W, *Trans.* SPWLA 10th European formation evaluation annual symposium, 1986.

28

Optimum logging and uncertainty management

> We must improve in quantifying and reducing uncertainties in reservoir evaluation.
>
> *Farouk-Al-Kasim*

Identical logging programs are applied to vastly different conditions. Sometimes, the logging program has been written several years before the logging operation, at the beginning of a field development, and has not been modified despite the increased level of information at hand.

This approach does not allow the optimization of data collection and the minimization of costs. Data may be available in large quantities but could fail to be relevant. Conversely, it may be inexpensive but totally inadequate to solve the problems in formation evaluation or field development.

Optimal logging is an approach that detracts from the **ritual** logging program design.

28.1 Optimal logging

Considering the economical risk implied by the overestimation of reserves, it is possible to establish a ceiling for the uncertainty of a given formation parameter. Hence, by trial and error, the maximum uncertainty needed from the input parameters, the logging data, can be evaluated.

Once it is set, the logging program can be adapted to meet this requirement.

The elements of the logging program that can be modified:

(a) selection of the mode of conveyance
(b) selection of the type of logging tool that measures a given formation property (for instance, laterolog versus induction [5])
(c) logging speed/rate of penetration
(d) sampling rate
(e) improvements on the standard tool response
(f) improvements on the standard calibration procedure
(g) well environment (change of mud)
(h) choice of technology.

This approach is applied on a zone-by-zone basis. Zones of interest that demand quantitative evaluation require generally higher sampling rates, slower speed, and several runs over the same interval. Surface and intermediate zones require normal sampling rates, faster logging speed and possibly only one run. In these zones, depth is still critical and should be controlled accordingly.

28.1.1 Case study 1

Description of the problem

In a developed field, a reperforating campaign is planned. To select the zones to be perforated, the residual oil saturation (or, in a complementary manner, the water saturation), needs to be measured to ± 10 su.

The lithology is sandstone and the porosity around 25 pu. Formation water salinity is less than 10,000 ppm. The borehole diameter is 21.6 cm (8.5 in.). The casing outside diameter is 14 cm (5.5 in.). Casing weight is 14 lbm/ft. Water is present in the borehole.

The completion string is a tubing with a minimum restriction of 4.6 cm (1.83 in.). A workover operation costs several hundreds of thousands of dollars per well.

Analysis

The best adapted measurement of saturation in the given conditions is provided by an induced neutron spectroscopy tool, which gives the water saturation in any range of water salinity.

Through-tubing tools are available in two sizes, $1\frac{11}{16}$ in. and $2\frac{1}{2}$ in. Considering the minimum restriction in the production string, the $1\frac{11}{16}$-in. tool can be run. For the small-size tool, production must be stopped and the well shut in for the length of the acquisition. The tool must be run eccentered as the zone of interest is cased. The casing has an internal diameter of 12.73 cm (5.012 in.). If the tool were centralized, a potential 4.22 cm (1.66 in.) of casing fluid would surround the detectors, which would blur the information collected from the formation. In-line eccentralizers are available with the tool.

A job planner can be run before the acquisition. This software computes the uncertainty on the water saturation as a function of logging speed and porosity. In the given conditions of casing and lithology, it is found from that for a 10-su repeatability and a 68% confidence level, a logging speed of 19 m/h (62.1 ft/h) is required (*Figs 28.1* and *28.2*).

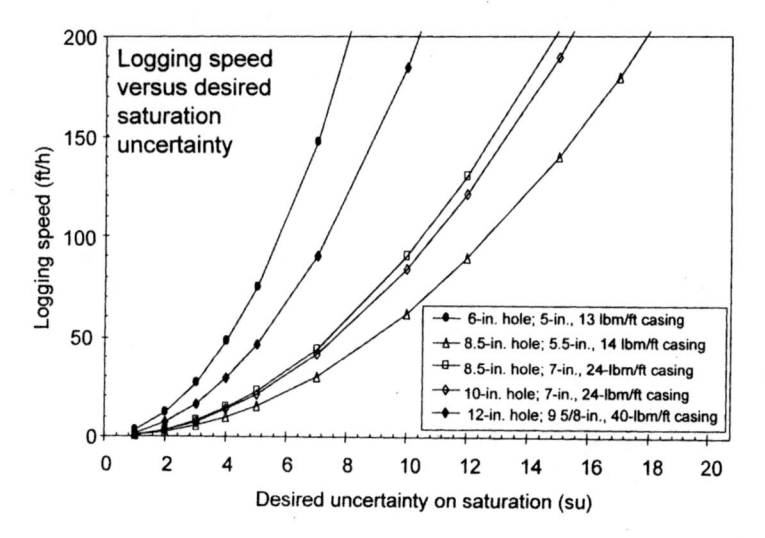

FIG. 28.1 : Required logging speed to obtain a desired precision on the values of S_w for water-filled cased holes of various sizes.
Courtesy of Schlumberger

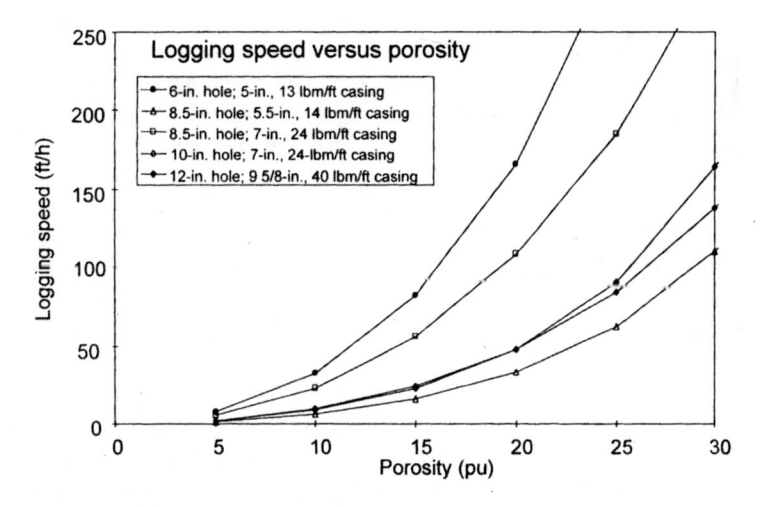

FIG. 28.2 : Required logging speed as a function of porosity and for various borehole configurations.
Courtesy of Schlumberger

The charts and numbers are valid for an alpha-processing length of 21 levels and a vertical averaging of 5 levels. If a better vertical resolution is required, less filtering can be selected, but a lower logging speed would be needed.

If a higher confidence level is required (for instance, 95%) then the logging speed must be reduced by a factor of 4.

In practical terms, logging speed can neither be less than 18.3 m/h (60 ft/h) nor higher than 36.6 m/h (120 ft/h). If the planned logging speed is below 18 m/h, several passes over the same interval can be performed. The passes are then merged after a check that the individual passes are in depth:

$$\text{Number of passes} = \frac{\text{Actual logging speed}}{\text{Proposed logging speed}}.$$

These values are valid if there is water in the borehole. If there is a mixture of oil and water over the interval to be logged, then the oil saturation uncertainty is dominated by the uncertainty on the proportions of oil and water in the mixture.

The operational procedure to log a cased-hole interval is decided after taking the expected uncertainty on saturation into consideration. The oil company objective is the key to the design of the job. The customer could also contribute to the collection of superior data by taking a decision on the fluid nature in the casing. Fresh water in the borehole is most favorable.

28.1.2 Case study 2

Definition of the problem

An absolute error of 1 pu on porosity and a relative error of 7% on water saturation are desired. The formation is made of laminations estimated to be around 20 to 30 cm (8 in. to 12 in.) thick, with a porosity varying between 20 pu and 30 pu. The lithology is mainly sand. Formation water resistivity is expected to be 0.1 Ω/m. The formations tend to deteriorate quickly during drilling because of shale swelling. Suggest the most appropriate logging program and procedures.

The following issues are covered:
(a) mud and drilling program
(b) repeat passes
(c) logging speed
(d) controlled environment
(e) calibrations.

The expectations in terms of porosity and saturation are quite ambitious. They require a tight control of tool uncertainty, of the environment and of the operational procedures. The presence of laminations further complicates the problem because then a higher sampling rate is needed.

Density logging

The 20-pu intervals correspond to the highest density. For a sandstone lithology, the density, ρ_b, is in the 2.32 to 2.34 g/cm^3 range depending on mud density. At 2.33 g/cm^3, at a logging speed of 9 m/min (1800 ft/h), without filtering and with a sampling rate of 15 cm (6 in.), the uncertainty is 0.015 g/cm^3.[1] Considering the thickness of the beds, a higher sampling rate (corresponding to a sampling length of 3 cm (1.2 in.) is recommended. This choice commands a logging speed five times slower, which will be considered in the zone of interest.

Neutron porosity

In the 20- to 30-pu range, the uncertainty is 2 pu for the standard sampling rate and logging speed. It is the same if the sampling length and the logging speed are decreased five-fold.

For the nuclear tools, accuracy can be preserved by running several repeat passes at the standard logging speed before stacking them, but depth-matching can be a source of problems.

Porosity determination

The uncertainties on density- and thermal neutron-derived porosities are, respectively, 1 pu and 2 pu. These inputs are combined with different weighting factors (40% for the neutron, 60% for the density).

Control of the well environment

In this example, formations are sensitive to water-base mud, which may affect the hole condition and would increase considerably the environment-related uncertainty on all measurements. Oil-base mud is recommended. It eliminates the need for a precise standoff for induction logging since mud resistivity is high. Borehole conductivity correction would be negligible.

Determination of saturation

Resistivities in the order of 60 Ω/m, equivalent to 16 mmho/m are expected in the reservoir. As oil-base mud is used, only induction type tools can be run. To obtain a sharp bed definition, a high frequency dielectric tool can be run in combination. At 16 mmho/m, the measurement uncertainty is 1.5 mmho/m for the standard induction and 0.5 mmho/m for the latest technology. Considering the constraint on the desired saturation uncertainty, the second type of equipment is necessary.

---

[1] The values indicated in this Section are taken from Chapters 7 and 21. For a specific test case, the precision/accuracy characteristics are to be checked with the logging company. The approach remains the same.

Calibrations

In the given conditions, the largest contribution in the induction tool uncertainty is its sonde error determination. The two-height calibration method[2] has to be used to reduce the calibration-related uncertainty to a fraction of mmho/m. Similarly, density and neutron calibrations are to be undertaken in controlled conditions to meet the challenging demands on porosity determination.

28.2 Quantification of uncertainties

Optimal logging methods provide the most appropriate program for given geological and petrophysical conditions. After data is acquired, it is necessary to check that the expectations have been met. This is achieved by quantifying the actual logging uncertainties.

A program quantifying uncertainties for simple logging programs [6] has been developed. The flowchart of the program in shown in *Fig. 28.3*.

Because the logging environment and procedures are so complex and diverse, a number of assumptions and approximations are linked to the program.

(a) Only resistivity, neutron porosity and density logs are handled by the program.
(b) The uncertainty caused by the environment is assumed to be a fixed percentage of the environmental correction.

The program has been successfully run in several fields and formations. An early application highlighted the importance of boundary determination in thin reservoirs, an intuitive conclusion whose quantification was made possible by the development of the program.

28.3 Using uncertainties in interpretation

Once an uncertainty has been assigned to a set of logging measurements, it is possible to combine them to obtain formation characteristics with the smallest error possible. The minimum propagated error for a linear combination of inputs is obtained if each contributing input is weighted with the inverse of the square of its own error.[3] Unfortunately this finding cannot be applied directly to petrophysics where the results are often nonlinear combinations of inputs. Still, a number of interpretation methods have been designed along the same lines of thought, whereby a logging measurement is weighted by a parameter directly related to the measurement uncertainty. Several papers describe these methods [1 to 4].

---

[2]Section 17.6.4.
[3]Section 7.4.3.

Input general information
(heading type)

Input individual log characteristics	
(for each log, R_{ILD}, ρ_b, ϕ_N, auxiliary logs)	
Logging speed Filtering Sampling interval Drift between after and before survey verifications Repeatability check	Average Standard deviation

Input average of sampled data per zone
Thickness
Log reading (R_{ILD}, ρ_b, ϕ_N)
Counting rates for nuclear tools

Input petrophysical parameters	
aR_w	Mean and σ
m	Mean and σ
ρ_{ma}	Mean and σ
ρ_{mf}	Mean and σ

Compute error linked to environmental corrections

Input general information on tools	
Induction	Uncertainty on sonde error Uncertainty on loop signal Minimum error Percentage error
Density and neutron ρ_b and ϕ_N	Coefficients of the tool response algorithm

Compute uncertainties			
R_{ILD}	ρ_b Random	ϕ_N Random	h
Systematic	Systematic	Systematic	

Merge uncertainties	
Computation of uncertainty on	ϕ_d ϕ_e S_w Hydrocarbon pore volume

FIG. 28.3 : Flowchart of the program quantifying logging uncertainties.

The concept common to all these methods is the search for an optimal solution, a set of **outputs** that minimizes an error function I, called **incoherence**.

The density log, for instance, gives a reading ρ_b. The theoretical value of the density, $\hat{\rho}_b$, recomputed from a set of formation parameters, S, is

$$\hat{\rho}_b = f(S) = f(\phi, S_w, S_{xo}, V_1, V_2, V_3, ..., V_k).$$

V_i is the volumetric percentage of mineral i in the formation. If σ_{ρ_b} is the uncertainty on ρ_b,[4] the incoherence related to S with respect to the density log is

$$I(\rho_b) = \frac{(\rho_b - \hat{\rho}_b)^2}{\sigma_{\rho_b}^2}.$$

The larger σ_{ρ_b}, the larger the difference between ρ_b and $\hat{\rho}_b$ could be. That σ_{ρ_b} is large means that input ρ_b does not control set S of formation parameters. The overall optimization process is controlled by the logs that have the smallest uncertainty.

For a complete set of logging measurements l_i, indexed from 1 to n, the total incoherence $\mathcal{I}$ is

$$\mathcal{I} = \sum_{i=1}^{i=n} \frac{(l_i - \hat{l}_i)^2}{\sigma_{l_i}^2}.$$

In the ideal and impossible case where the total incoherence $\mathcal{I}$ is nil, every contributing incoherence, positive or nil by definition, needs to be nil. A perfect match between theoretical and actual values is obtained.

References

1 Alberty, M., Hashmy, H. K., "Application of Ultra to log analysis," paper Z, *Trans.* SPWLA 25th annual logging symposium, 1984.

2 Freedman, R., Puffer, J. E., "Self-consistent log interpretation methods (SLIM) application to shaly sands," paper T, *Trans.* SPWLA 29th annual logging symposium, 1988.

3 Mayer, C., Sibbit, A., "Global, a new approach to computer processed log interpretation," SPE 9341, *Trans.* SPE 55th annual technical conference and exhibition, Dallas, 1980.

4 Quirein, J., Kimminau, S., Lavigne, J., Singer, J., Wendel, F., "A coherent framework for developing and applying multiple formation evaluation models," paper DD, *Trans.* SPWLA 27th annual logging symposium, Houston, 1986.

5 Souhaité, P., Misk, A., Poupon, A., "R_t determination in the eastern Hemisphere," paper LL, *Trans.* SPWLA 16th annual logging symposium, 1975.

6 Spurlin, J., Theys, P., "Quantification of the uncertainties on equivalent oil column height from the use of petrophysical data," report on project initiated by the Norwegian Petroleum Directorate, Stavanger, 1989.

---

[4]Some algorithms split this uncertainty in two terms: one linked to the measurement itself, the second related to the response equation.

Appendix 1: abbreviations, unit symbols and acronyms

alternating current	AC
barrels per day	B/D
bits per second	bps
bottomhole temperature	BHT
capture units	cu
centimeter	cm
centipoise	cP
counts per second	cps
cubic centimeter	cm^3
cubic foot	ft^3
cubic meter	m^3
darcy	D
degree (American Petroleum Institute)	°API
degree Celsius	°C
degree Fahrenheit	°F
direct current	DC
electron volt	eV
feet per minute	ft/min
feet per second	ft/s
feet per hour	ft/h
feet square	ft^2
foot	ft
gallon	gal
gram	g
hectare	ha
hertz	Hz
hour	h
inch	in.
inches per second	in./s
inside diameter	ID
kiloelectron volt	keV
kilobits per second	kbps
kilogram	kg
kilohertz	kHz
kilometer	km
kilowatt	kW
liter	L
logarithm	log
logarithm (natural)	ln
maximum	max
megahertz	MHz
meter	m
microsecond	µs
millidarcy, millidarcies	mD
millimeter	mm
millimho per meter	mmho/m
million (=10^6)	MM
million cubic feet	MMcf
million electron volt	MeV
milliseconds	ms
millivolt	mV
minimum	min
minutes	min
nanotesla	nT
ohm	Ω
ohm-meter	Ω/m
outside diameter	OD
parts per million	ppm
porosity units	pu
pound (mass)	lbm
pound (weight or force)	lbf

pound per gallon .... lbm/gal
pounds per square inch .... psi
seconds .... s
saturation units .... su
square centimeter .... cm^2
square foot .... ft^2
square inch .... in.^{2}
square meter .... m^2
tangent .... tan
thousand .... M
volt V

Data quality objectives .... DQO
Hydrocarbon pore volume .... HCPV
Logging/measurement-
while-drilling .... MWD
Nuclear magnetic
resonance .... NMR

Rate of penetration .... ROP
American Petroleum Institute ... API
Canadian Well Logging
Society .... CWLS
Environment Protection
Agency .... EPA
International Organization
of Standards .... ISO
Journal of Petroleum
Technology .... JPT
Log Characterization
Consortium LCC
Society of Exploration
Geophysicists .... SEG
Society of Petroleum
Engineers .... SPE
Society of Professional Well
Log Analysts .... SPWLA

Appendix 2: metrological definitions

The number between brackets refers to the reference document. The other numbers relate to the article in the said reference. A definition may originate from different reference documents.

Accuracy of the measurement [2] 3.6, [3] 3.05
The closeness of the agreement between the result of a measurement and the (conventional) true value of the measurand.
Notes: (1) Accuracy is a qualitative concept. (2) The use of the term "precision" for "accuracy" should be avoided.

Influence quantity [2] 3.5
A quantity which is not the subject of the measurement but which influences the value of the measurand or the indication of the measuring instrument.
Examples: ambient temperature; frequency of an alternating measured voltage.

Measurand [2] 3.4, [3] 2.09
A quantity subjected to measurement.

Precision [1] 3.14
The closeness of agreement between the results obtained by applying the experimental procedure several times on identical materials and under prescribed conditions. The smaller the random part of the experimental error, the more precise the procedure.

Reference conditions [2] 3.13
Conditions of use for a measuring instrument prescribed for performance testing, or to ensure valid intercomparison of results of measurements.
Note: The reference conditions generally specify "reference values" or "reference ranges" for the influence quantities affecting the measuring instrument.

Repeatability [1] 3.17
(a) Qualitatively: the closeness of agreement between independent results obtained in the normal and correct operation of the same method on identical test material,

in a short interval of time, and under the same test conditions (same operator, same apparatus, same laboratory).

The representative parameters of the dispersion of the population which may be associated with the results are qualified by the term "repeatability," for example repeatability standard deviation, repeatability variance.

(b) Quantitatively: the value equal to or below which the absolute difference between two single test results obtained in the above conditions may be expected to lie with a probability of 95%.

Reproducibility [1] 3.19

(a) Qualitatively: the closeness of agreement between individual results obtained in the normal and correct operation of the same method on identical test material, but under different test conditions (different operators, different apparatus, different laboratories).

The representative parameters of the dispersion of the population which may be associated with the results are qualified by the term "reproducibility," for example reproducibility standard deviation, reproducibility variance.

(b) Quantitatively: the value equal to or below which the absolute difference between two single test results on identical material obtained by operators in different laboratories, using the standardized test method, may be expected to lie with a probability of 95%.

Specified measuring range [2] 3.12

The set of values for a measurand for which the error of a measuring instrument is intended to lie within specified limits.

True value [1] 3.24

For practical purposes, the value towards which the average of single results obtained by n laboratories tends, as n tends towards infinity; consequently, such a true value is associated with the particular method of test.

Note: a different and idealized definition is given in ISO 3534, Statistics - Vocabulary and symbols.

Other definition [3] 1.18

The value of a measurand that is completely defined.

Notes: (1) This is the result that would be obtained by a perfect measurement. (2) True value is an idealized concept.

Uncertainty of measurement [2] 3.7

Result of the evaluation aimed at characterizing the range within which the true value of a measurand is estimated to lie, generally with a given likelihood.

References

1 International Organization for Standardization, *Petroleum products - Determination and application of precision data in relation to methods of test (ISO - 4259)*, Genève, Switzerland, 1992.

2 International Organization for Standardization, *Quality assurance requirements for measuring equipment (ISO - 10012-1:1992)*, Genève, Switzerland, 1992.

3 International Organization for Standardization, *VIM (1984) International vocabulary of metrology, basic and general terms in metrology, BIPM/CEI/ISO/OIML*, Genève, Switzerland, 1984.

Appendix 3: cumulative Gaussian probability

The values of the integral of the Gaussian distribution (*Fig. A3.1*) are listed as percentages versus $z = |s - \mu|/\sigma$ in *Tables A3.1* ($0.0 \leq z \leq 3.0$) and *A3.2* ($z \geq 3.0$). For instance, the cumulative probability for $z = 1.5$ is 87.885%.

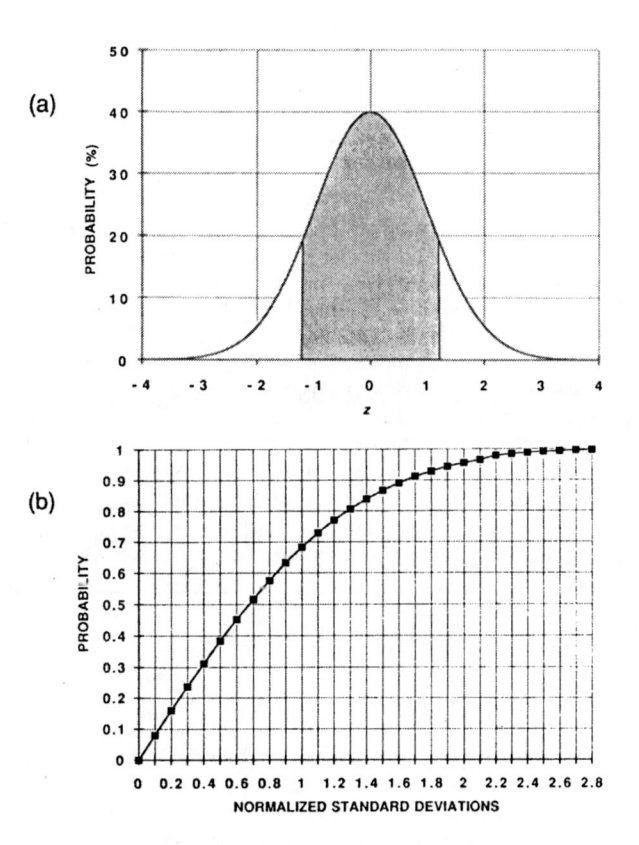

FIG. A3.1 : Gaussian distribution.
a. Probability versus z. **b.** Cumulative distribution.

Appendix 3: cumulative Gaussian probability

Table A3.1

z	0.00	0.01	0.02	0.03	0.04	0.05	0.06	0.07	0.08	0.09
0.0	0.000	0.798	1.596	2.393	3.191	3.988	4.784	5.581	6.376	7.171
0.1	7.966	8.759	9.552	10.343	11.134	11.924	12.712	13.499	14.285	15.069
0.2	15.852	16.633	17.413	18.191	18.967	19.741	20.514	21.284	22.052	22.818
0.3	23.582	24.344	25.103	25.860	26.614	27.366	28.115	28.862	29.605	30.346
0.4	31.084	31.819	32.551	33.280	34.006	34.729	35.448	36.164	36.877	37.587
0.5	38.292	38.995	39.694	40.389	41.080	41.768	42.452	43.132	43.809	44.481
0.6	45.149	45.814	46.474	47.131	47.783	48.431	49.075	49.714	50.350	50.981
0.7	51.607	52.230	52.847	53.461	54.070	54.674	55.274	55.870	56.461	57.047
0.8	57.629	58.206	58.778	59.346	59.909	60.467	61.021	61.570	62.114	62.653
0.9	63.188	63.718	64.243	64.763	65.278	65.789	66.294	66.795	67.291	67.783
1.0	68.269	68.750	69.227	69.699	70.166	70.628	71.085	71.538	71.985	72.428
1.1	72.866	73.300	73.728	74.152	74.571	74.985	75.395	75.799	76.199	76.595
1.2	76.985	77.371	77.753	78.130	78.502	78.869	79.232	79.591	79.945	80.284
1.3	80.639	80.980	81.316	81.647	81.975	82.298	82.616	82.930	83.240	83.546
1.4	83.848	84.145	84.438	84.727	85.012	85.293	85.570	85.843	86.112	86.377
1.5	86.638	86.895	87.148	87.397	87.643	87.885	88.123	88.358	88.588	88.816
1.6	89.039	89.259	89.476	89.689	89.898	90.105	90.308	90.507	90.703	90.896
1.7	91.086	91.272	91.456	91.636	91.813	91.987	92.158	92.326	92.491	92.654
1.8	92.813	92.969	93.123	93.274	93.422	93.568	93.711	93.851	93.988	94.123
1.9	94.256	94.386	94.513	94.638	94.761	94.882	95.000	95.115	95.229	95.340
2.0	95.449	95.556	95.661	95.764	95.864	95.963	96.059	96.154	96.247	96.338
2.1	96.426	96.513	96.599	96.682	96.764	96.844	96.922	96.999	97.074	97.147
2.2	97.219	97.289	97.358	97.425	97.490	97.555	97.617	97.679	97.739	97.797
2.3	97.855	97.911	97.965	98.019	98.071	98.122	98.172	98.221	98.268	98.315
2.4	98.360	98.404	98.448	98.490	98.531	98.571	98.610	98.648	98.686	98.722
2.5	98.758	98.792	98.826	98.859	98.891	98.922	98.953	98.983	99.012	99.040
2.6	99.067	99.094	99.120	99.146	99.171	99.195	99.218	99.241	99.264	99.285
2.7	99.306	99.327	99.347	99.366	99.385	99.404	99.422	99.439	99.456	99.473
2.8	99.489	99.504	99.520	99.534	99.549	99.563	99.576	99.589	99.602	99.615
2.9	99.627	99.638	99.650	99.661	99.672	99.682	99.692	99.702	99.712	99.721

Table A3.2

z	0.00	0.10	0.20	0.30	0.40
3.0	99.73002	99.80648	99.86257	99.903315	99.932614
3.5	99.953474	99.968178	99.978440	99.985530	99.9903805
4.0	99.9936656	99.9958684	99.9973308	99.9982920	99.9989174
4.5	99.99932043	99.99957748	99.99973982	99.99984132	99.999904149
5.0	99.999942657	99.999966024	99.999980061	99.999988410	99.999993327
5.5	99.999996193	99.999997847	99.999998793	99.999999328	99.999999627

Appendix 4: solution of exercises

Exercise 1. Gaussian distribution

Looking at *Table A3.1*, Appendix 3, 98% corresponds to:

$$z = \frac{x - \bar{x}}{\sigma} = 2.33.$$

$$x_{min} = \bar{x} - 2.33\sigma \qquad x_{max} = \bar{x} + 2.33\sigma.$$

Exercise 4. Geometrical factor

$$R_t = R_w/(\phi^2 S_w^2)$$

This gives $R_t = 444$ Ω/m for $S_w = 5$ su and $R_t = 197$ Ω/m for $S_w = 7.5$ su. $C_{xo} = C_{mf}\phi^2 S_{xo}^2 = 500$ mmho/m.

Borehole and invasion correction

$R_m = 0.1$ Ω/m	$C_m = 10,000$ mmho/m	$R_{mf} = 0.1$ Ω/m	$C_{mf} = 10,000$ mmho/m

TABLE A4.1
BOREHOLE AND INVASION EFFECTS

	Spherically focused log (mmho/m)	Medium induction (mmho/m)	Deep induction (mmho/m)
Borehole →		4	−2.00
Invasion diameter (in.) ↓			
15	90	3.8	−1.20
20	165	17.5	−1.75
25	215	40.0	−1.50
30	255	75.0	0.00

Table A4.2
Recomputed values for $R_t = 444$ Ω/m, $C_t = 2.25$ mmho/m

Invasion diameter (in.) ↓	Medium induction		Deep induction	
	Conductivity (mmho/m)	Resistivity (Ω/m)	Conductivity (mmho/m)	Resistivity (Ω/m)
15	10.0	100.0	−0.95	>2000
20	23.7	42.2	−1.50	>2000
25	46.2	21.6	−1.25	>2000
30	81.2	12.3	0.25	>2000

Table A4.3
Recomputed values for $R_t = 197$ Ω/m, $C_t = 5.06$ mmho/m

Invasion diameter (in.) ↓	Medium induction		Deep induction	
	Conductivity (mmho/m)	Resistivity (Ω/m)	Conductivity (mmho/m)	Resistivity (Ω/m)
15	12.8	78.0	1.9	538
20	26.6	37.7	1.3	763
25	49.1	20.4	1.6	641
30	84.1	11.9	3.1	327

Exercise 5. Reliability computation

Table A4.4
Triple combo mean time between failure

Tool	MTBF/h	Failure rate	System MTBF (h)
Telemetry/survey	800	0.001250	
Resistivity/GR	1200	0.000833	
Density/neutron	1200	0.000833	
Total system		0.002917	343

Exercise 6. Partial derivative 1

$$y = f(x, u) = 3x^2 + 4xu^3.$$

$$\frac{\partial y}{\partial x} = 6x + 4u^3 \quad \frac{\partial^2 y}{\partial x^2} = 6 \quad \frac{\partial y}{\partial u} = 12xu^2$$

$$\frac{\partial^2 y}{\partial u^2} = 24xu \quad \frac{\partial^3 y}{\partial u^3} = 24x \quad \frac{\partial^2 y}{\partial x \partial u} = 12u^2$$

$$\frac{\partial^3 y}{\partial x \partial u^2} = 24u \quad \frac{\partial^3 y}{\partial x \partial u^3} = 24 \quad \frac{\partial^3 y}{\partial u^2 \partial x} = 24u.$$

Note

$$\frac{\partial^2 y}{\partial x \partial u} = \frac{\partial^2 y}{\partial u \partial x}.$$

Exercise 7. Partial derivative 2

$$\frac{\partial y}{\partial u} = \frac{5}{u} \quad \frac{\partial y}{\partial v} = \frac{16}{v}$$

$$\frac{\partial^2 y}{\partial u^2} = -\frac{5}{u^2} \quad \frac{\partial^2 y}{\partial v^2} = -\frac{16}{v^2}$$

$$\frac{\partial^3 y}{\partial u^3} = \frac{10}{u^3} \quad \frac{\partial^3 y}{\partial v^3} = \frac{32}{v^3}$$

$$\frac{\partial^n y}{\partial u^n} = (-1)^{n-1} 5(n-1)! \frac{1}{u^n} \quad \frac{\partial^n y}{\partial v^n} = (-1)^{n-1} 16(n-1)! \frac{1}{v^n}.$$

Exercise 9. Density-derived porosity

$$\phi_D = \frac{\rho_{ma} - \rho_b}{\rho_{ma} - \rho_{mf}}$$

$$\sigma_{\phi_D}^2 = \left(\frac{\partial \phi_D}{\partial \rho_b}\right)^2 \times \sigma_{\rho_b}^2 + \left(\frac{\partial \phi_D}{\partial \rho_{ma}}\right)^2 \times \sigma_{\rho_{ma}}^2 + \left(\frac{\partial \phi_D}{\partial \rho_{mf}}\right)^2 \times \sigma_{\rho_{mf}}^2$$

$$\left|\frac{\partial \phi_D}{\partial \rho_b}\right| = \frac{1}{\rho_{ma} - \rho_{mf}} \quad \left|\frac{\partial \phi_D}{\partial \rho_{ma}}\right| = \frac{\rho_b - \rho_{mf}}{(\rho_{ma} - \rho_{mf})^2} \quad \left|\frac{\partial \phi_D}{\partial \rho_{mf}}\right| = \frac{\rho_{ma} - \rho_b}{(\rho_{ma} - \rho_{mf})^2}$$

$$\sigma_{\phi_D}^2 = \frac{\sigma_{\rho_b}^2}{(\rho_{ma} - \rho_{mf})^2} + \frac{(\rho_b - \rho_{mf})^2 \times \sigma_{\rho_{ma}}^2}{(\rho_{ma} - \rho_{mf})^4} + \frac{(\rho_{ma} - \rho_b)^2 \times \sigma_{\rho_{mf}}^2}{(\rho_{ma} - \rho_{mf})^4}$$

$$\sigma_{\phi_D} = \frac{1}{(\rho_{ma}-\rho_{mf})^2} \times$$
$$\sqrt{(\rho_{ma}-\rho_{mf})^2 \times \sigma^2_{\rho_b} + (\rho_b-\rho_{mf})^2 \times \sigma^2_{\rho_{ma}} + (\rho_{ma}-\rho_b)^2 \times \sigma^2_{\rho_{mf}}}$$

where ρ and σ_ρ are expressed in g/cm^3, so that σ_{ϕ_D} is expressed in percentage of total volume. With x defined as $(\rho_{ma}-\rho_{mf})^2$:

$$\sigma_{\phi_D} = \frac{1}{x} \times \sqrt{x \times \sigma^2_{\rho_b} + (\rho_b-\rho_{mf})^2 \times \sigma^2_{\rho_{ma}} + (\rho_{ma}-\rho_b)^2 \times \sigma^2_{\rho_{mf}}}.$$

Numerical application

$\sigma_{\phi_D} = 0.0144$, equivalent to 1.44 pu.

Exercise 10. Saturation uncertainty

$$S_w^2 = \frac{1}{\phi^2} \frac{R_w}{R_t}.$$

$$\left|\frac{\partial S_w}{\partial C_t}\right| = 0.5 \frac{R_w^{1/2}}{\phi} \frac{1}{C_t^{1/2}} \qquad \left|\frac{\partial S_w}{\partial \phi}\right| = \frac{1}{\phi^2} R_w^{1/2} C_t^{1/2}$$

$$\sigma_{S_w}^2 = \frac{R_w}{R_t} \frac{1}{\phi^4} \sigma_\phi^2 + 0.25 \frac{R_w}{\phi^2} \frac{1}{C_t} \sigma_{C_t}^2$$

$$\frac{\sigma_{S_w}}{S_w} = \sqrt{\frac{\sigma_\phi^2}{\phi^2} + \frac{1}{4}\frac{\sigma_{C_t}^2}{C_t^2}}.$$

Numerical application

TABLE A4.5
SATURATION UNCERTAINTY FOR DIFFERENT TECHNOLOGIES

$C_t = 100$ mmho/m		$S_w = 25.8$ su	
		A	B
σ_{ϕ_d}	(g/cm^3)	0.017	0.010
$\frac{\sigma_{C_t}}{C_t}$	(%)	3.0	2.0
$\frac{\sigma_{S_w}}{S_w}$	(%)	5.9	3.5

Exercise 11. Archie's equation

Archie's equation is simplified: $1/C_t = R_t = R_w/(\phi S_w)^m$ which can be rearranged:
$$S_w = \phi^{-1} e^{\frac{1}{m} \ln(R_w C_t)}.$$

Then:

$$\sigma_{S_w}^2 = [\sigma_m \frac{\ln(R_w C_t)}{m^2} \phi^{-1} e^{\frac{1}{m}\ln(R_w C_t)}]^2 + [\sigma_{R_w} \frac{1}{mR_w}\phi^{-1} e^{\frac{1}{m}\ln(R_w C_t)}]^2$$

$$+ [\sigma_{C_t}\frac{1}{mC_t}\phi^{-1}e^{\frac{1}{m}\ln(R_w C_t)}]^2 + [\sigma_\phi \frac{1}{\phi^2}e^{\frac{1}{m}\ln(R_w C_t)}]^2$$

$$[\frac{\sigma_{S_w}}{S_w}]^2 = [\frac{\ln(R_w C_t)\sigma_m}{m^2}]^2 + [\frac{\sigma_{R_w}}{mR_w}]^2 + [\frac{\sigma_{C_t}}{mC_t}]^2 + [\frac{\sigma_\phi}{\phi}]^2.$$

Exercise 12. Simandoux's equation

$$\frac{1}{R_t} = \frac{1}{\beta}S_w^2 + \gamma S_w.$$

$$\frac{1}{\beta} = \frac{\phi^2}{aR_w} \qquad \gamma = \frac{V_{sh}}{R_{sh}}$$

$$S_w = \frac{aR_w}{2\phi^2}[\sqrt{\frac{4\phi^2}{aR_w R_t} + (\frac{V_{sh}}{R_{sh}})^2} - \frac{V_{sh}}{R_{sh}}]$$

$$x = \sqrt{\frac{4\phi^2}{aR_w R_t} + (\frac{V_{sh}}{R_{sh}})^2} = \sqrt{\frac{4}{\beta R_t} + \gamma^2}$$

$$S_w = \frac{1}{2}\beta(x - \gamma)$$

$$\sigma_{S_w}^2 = (\frac{\partial S_w}{\partial \gamma})^2 \sigma_\gamma^2 + (\frac{\partial S_w}{\partial \beta})^2 \sigma_\beta^2 + (\frac{\partial S_w}{\partial R_t})^2 \sigma_{R_t}^2$$

$$\frac{\partial S_w}{\partial \gamma} = \frac{\beta}{2}[\frac{\partial x}{\partial \gamma} - 1] \qquad \text{with} \qquad \frac{\partial x}{\partial \gamma} = \frac{\gamma}{x}$$

$$\frac{\partial S_w}{\partial \beta} = \frac{1}{2}[(x - \gamma) + \beta\frac{\partial x}{\partial \beta}] \qquad \text{with} \qquad \frac{\partial x}{\partial \beta} = -\frac{2}{\beta^2 x R_t}$$

$$\frac{\partial S_w}{\partial R_t} = \frac{\beta}{2}\frac{\partial x}{\partial R_t} \qquad \text{with} \qquad \frac{\partial x}{\partial R_t} = -\frac{2}{\beta x R_t^2}$$

$$\sigma_\gamma^2 = (\frac{\partial \gamma}{\partial V_{sh}})^2 \sigma_{V_{sh}}^2 + (\frac{\partial \gamma}{\partial R_{sh}})^2 \sigma_{R_{sh}}^2$$

$$\sigma_\beta^2 = (\frac{\partial \beta}{\partial R_w})^2 \sigma_{R_w}^2 + (\frac{\partial \beta}{\partial \phi})^2 \sigma_\phi^2$$

$$\frac{\partial \gamma}{\partial V_{sh}} = \frac{1}{R_{sh}} \qquad \frac{\partial \gamma}{\partial R_{sh}} = -\frac{\gamma}{R_{sh}}$$

$$\frac{\partial \beta}{\partial R_w} = \frac{a}{\phi^2} \qquad \frac{\partial \beta}{\partial \phi} = -\frac{2\beta}{\phi}$$

Combining and dividing by S_w:

$$\frac{\sigma_{S_w}^2}{S_w^2} = (\frac{\gamma}{x})^2 [\frac{\sigma_{V_{sh}}^2}{V_{sh}^2} + \frac{\sigma_{R_{sh}}^2}{R_{sh}^2}] + (1 - \frac{2}{\beta x R_t(x-\gamma)})^2 [\frac{\sigma_{R_w}^2}{R_w^2} + 4\frac{\sigma_{\phi}^2}{\phi^2}] + (\frac{2}{\beta x R_t(x-\gamma)})^2 \frac{\sigma_{R_t}^2}{R_t^2}$$

$$e = \frac{2}{\beta x(x-\gamma) R_t} \qquad x^2 = \gamma^2 + \frac{4}{\beta R_t}$$

$$e = \frac{x^2 - \gamma^2}{2x(x-\gamma)} = \frac{x+\gamma}{2x} = \frac{1}{2} + \frac{\gamma}{2x} \qquad 1 - e = \frac{1}{2} - \frac{\gamma}{2x}$$

$$\frac{\sigma_{S_w}^2}{S_w^2} = (\frac{\gamma}{x})^2 [\frac{\sigma_{V_{sh}}^2}{V_{sh}^2} + \frac{\sigma_{R_{sh}}^2}{R_{sh}^2}] + (\frac{1}{2} - \frac{\gamma}{2x})^2 [\frac{\sigma_{R_w}^2}{R_w^2} + 4\frac{\sigma_{\phi}^2}{\phi^2}] + (\frac{1}{2} + \frac{\gamma}{2x})^2 \frac{\sigma_{R_t}^2}{R_t^2}.$$

Numerical application 1

$$V_{sh} = 0 \qquad \gamma = 0 \qquad x = \sqrt{\frac{4}{\beta R_t}} \qquad e = \frac{1}{2}$$

$$\frac{\sigma_{S_w}^2}{S_w^2} = \frac{\sigma_{R_w}^2}{4R_w^2} + \frac{\sigma_{\phi}^2}{\phi^2} + \frac{\sigma_{R_t}^2}{4R_t^2}.$$

When $V_{sh} = 0$, the error on S_w is reduced to the error shown in Exercise 9 (Archie's equation).

Numerical application 2

$$V_{sh} = 0.2 \qquad \gamma = 0.2 \qquad \beta = 2 \qquad x = 0.49 \qquad e = 0.70$$

$$\frac{\sigma_{S_w}^2}{S_w^2} = 0.1667 \times [\frac{\sigma_{V_{sh}}^2}{V_{sh}^2} + \frac{\sigma_{R_{sh}}^2}{R_{sh}^2}] + 0.088 \times [\frac{\sigma_{R_w}^2}{R_w^2} + 4\frac{\sigma_{\phi}^2}{\phi^2}] + 0.50 \times \frac{\sigma_{R_t}^2}{R_t^2}.$$

Input → All other $\sigma s = 0$	$\frac{\sigma_{V_{sh}}}{V_{sh}} = 1$	$\frac{\sigma_{R_{sh}}}{R_{sh}} = 1$	$\frac{\sigma_{\phi}}{\phi} = 1$	$\frac{\sigma_{R_t}}{R_t} = 1$	$\frac{\sigma_{R_w}}{R_w} = 1$
$\frac{\sigma_{S_w}}{S_w}$	0.41	0.41	0.59	0.70	0.30

Appendix 4: solution of exercises

Exercise 13. Derivation of saturation by a numerical method

Nine different values of the saturation can be computed by taking the different input combinations. In increasing order, they are:

0.127, 0.137, 0.151, 0.154, 0.162, 0.166, 0.176, 0.182, 0.192.

Each value has a 1/9 (11.1%) probability (*Fig. A4.1*).

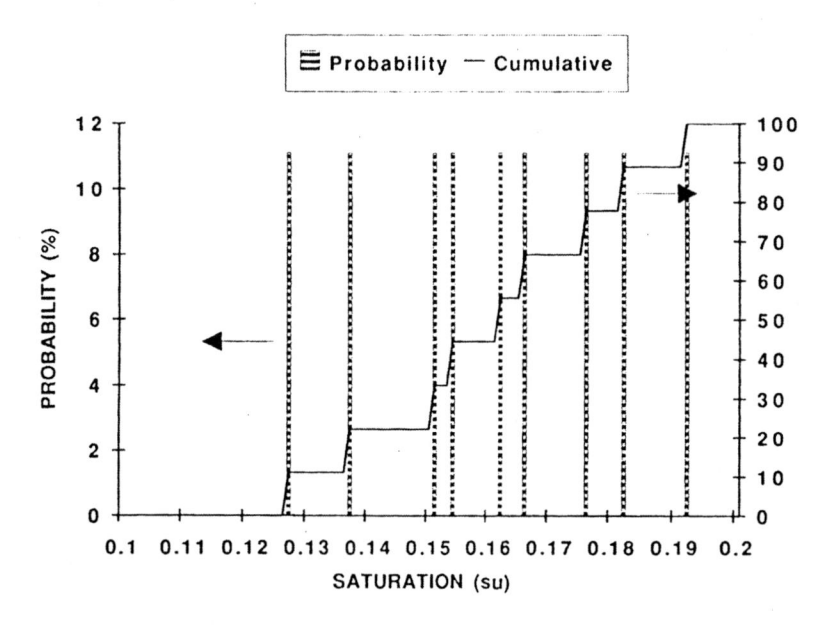

FIG. A4.1 : Histogram of the saturation distribution.

Exercise 15. Rate of penetration for an expected precision

The difference of ROP is equal to the square of the ratio of the desired precision over the reference precision (at a ROP of 100 ft/h), or 178 ft/h.

Exercise 17. Thermal neutron environmental correction

Since the effect of temperature only is considered, the contributing error reduces to:

$$\sigma_{\phi_N} = \frac{\partial f}{\partial T} \times \sigma_T$$

Over the range 50°F to 300°F, the correction is 10 pu. Therefore:

$$\frac{\partial f}{\partial T} = 40 \times 10^{-3}$$

With σ_T equal to 20, $\sigma_{\phi_N} = 0.8$ pu.

Bibliography

I Anderson, B., "The analysis of some unsolved induction interpretation problems using computer modeling," paper II, *Trans.* SPWLA 27th annual logging symposium, 1986.

II Bateman, R., *Log quality control*, IHRDC, Boston, 1985.

III Bevington, P. R., *Data reduction and error analysis for the physical sciences*, McGraw-Hill, New York, 1969.

IV Cerr, M., *Instrumentation industrielle*, Technique et Documentation, Paris, 1980.

V Ellis, D. V., *Well logging for earth scientists*, Elsevier Science Publishing Company, New York, 1987.

VI *Formation evaluation conference Indonesia*, Schlumberger, Jakarta, 1986.

VII Jorro, R. V. J., "A new calibration and quality control system for neutron logging instruments," pp. 129-138, *The Log Analyst*, Vol. 30, No. 3, 6-1989.

VIII Knoll, G. F., *Radiation detection and measurement*, John Wiley and Sons, New York, 1979.

IX Olesen, J. R., "A new nuclear tool calibration, wellsite verification and log quality control system," *Trans.* SPWLA 13th European formation evaluation symposium, 1990.

X Roscoe, B. A., Grau, J. A., Wraight, P. D., "Statistical precision of the neutron-induced gamma-ray spectroscopy measurements," paper CC, *Trans.* SPWLA 27th annual logging symposium, 1986.

XI *Schlumberger wellsite products and mnemonics guide (MAXIS* units OP6.0 and Cyber Service* Units CP 40)*, Schlumberger wireline and testing, Houston, 1993.

XII *Schlumberger log interpretation charts*, Schlumberger Oilfield Services, Houston, 1996.

XIII *Schlumberger log interpretation principles/applications*, Schlumberger Educational Services, Houston, 1989.

XIV Serra, O., *Diagraphies différées. Bases de l'interprétation. Tome 1. Acquisition des données diagraphiques*, Elf Aquitaine, Pau, 1983.

XV Serra, O., *Fundamentals of well log interpretation*, Elsevier, Amsterdam, 1984.

XVI Serra, O., *Diagraphies différées. Bases de l'interprétation. Tome 2. Interprétation des données diagraphiques*, Elf Aquitaine, Pau, 1985.

XVII Slider, H. C., *Slip, Worldwide practical petroleum reservoir engineering methods*, PennWell, Tulsa, 1983.

XVIII Theys, P., "Log quality control and error analysis, a prerequisite to accurate formation evaluation," paper V, *Trans.* SPWLA 11th European annual symposium, 1988.

XIX Tittman, J., *Geophysical well logging*, Academic Press, Orlando, Florida, 1986.

Index

Accelerometers, 113
Access,
  downhole –, 76
Accuracy, 5, 42-43, 105, 433
Acoustic waves, 168
Alpha processing, 154
Anhydrite, 411
Anisotropy, 201
Annulus, 207
  anti –, 207
Anti-Delaware, 386
Antisqueeze, 387
Archie equation, 80, 89, 443
Artificial formations, 161-164, 222
ASCII, 350
Audit, 317
Azimuth, 299
Azimuthal,
  averaging, 144-145
  density, 227

Barite, 221, 224-225
BGO detector, 71-72, 110
Big-hole logging, 384
Bimetallism, 376
Blunders, 34, 103, 300
Borehole,
  alteration, 198
  compensation, 377
  correction, 226
  environment, 192-201
  rugosity, 228

  shape, 192-195
  trajectory, 199
Boron, 91, 393

Cable, 120
Calibration, 233-258
  caliper –, 249
  density –, 256
  gamma ray –, 250
  induction –, 251-252
  n-ary –, 234-235
  neutron –, 252-256
  presentation of –, 242
Caliper,
  accuracy, 186, 409
Capillary curve, 420
Capture cross section, 392-393
Casing, 407-408
C/O logging, 424
Cement bond log, 27, 410
Channel, 19
Chlorine, 392
Coherence analysis, 272
Compton scattering, 217
Conductive anomalies, 390
Confidence interval, 38
Conformance, 310
Consonant sensors, 68
Core, 10, 416
  analysis, 417
  measurements, 417
  plugs, 10

Core,
  porosity, 417
  sidewall –, 419
Cycle-skipping, 377
Cycles on induction, 391

Data,
  formats, 349
  frame, 350
  management, 343-357
  models, 348
  quality,
    plan, 310
  storage, 346
  transfers, 344
  transmission, 120
Dead time, 74, 113
Decision-making, 23
Deconvolution, 153
Deep-electromagnetics, 11
Delaware, 386
Density, 37, 41, 87, 114, 217-231
  256-257, 265, 381-382, 427

Depth, 5, 275-297
  absolute –, 275
  control, 276, 294
  correlated –, 276
  cranking, 277, 339
  – derived compensation, 290
  differential –, 276
  driller's –, 284
  matching, 103, 293-294
  mismatch, 286-292
    from dip, 290
  MWD –, 284
  policy, 295
  reference, 327
  of investigation, 11, 18, 56-60
Derivative, 79-80
  partial –, 80-89 , 441-444
Destickage, 288-289
Detector resolution, 113, 265

Dielectric
  constant, 211
Dip,
  effect of –, 200
Dipmeter, 378
Directional surveys, 299-306
DLIS, 351
Dogleg severity, 76
Dolomite, 222-223
DQO method, 25-26

Earth's
  gravity field, 109
  magnetic field, 109, 302
Eccentering, 196, 253-254, 380-382
Electric diameter, 60
Electromagnetic
  propagation, 117
    transmission, 123
  transmitters, 108
Enhancement of resolution, 154-157
Enhanced resolution induction, 156
Environment, 5,
  of deposition, 21
  real –, 191-216
Environmental corrections, 175-190
Error, 33
  causes of –, 103
  from calibration, 248
  from caliper, 186-187
  in directional surveys, 300, 304
  on horizontal position, 103
  random –, 34, 37, 103, 300
  systematic –, 34, 36, 103, 300
  propagation of –, 5, 79-98
Experimental measurements, 161

Filtering, 5, 133-145
  active –, 142-143
  optimal –, 135
Floppy disks, 102
Formation MicroScanner, 395
Fresnel radius, 15

Fullbore coring, 10

Gadolinium, 392
Gamma ray,
   log, 35-36, 250, 374
   effect of – on logs, 72-73
   natural –, 109
Gaussian distribution, see statistical
Geiger-Müller detector, 74-75, 109
Geometrical factor, 48- , 439
   2D –, 51
   elemental, 50
   integrated, 52
   radial, 52
Geosteering, 30
Gravimetry, 11
Groningen effect, 387-389
GSO detector, 111
Gyroscope surveys, 304

Heading, 329
Helium detectors, 111
High resolution, 148
   density, 154-155
   sensors, 151
   sonic, 151-153
Horizontal resolution, 15
Hydraulic,
   continuity, 28
   separation, 27
Hydrocarbon
   pore volume, 7, 81, 147-149
Hysteresis, 73

Image, 395-397
Incoherence, 430
Induction, 117, 156, 182, 196
Interrupt, 124-125
Invasion, 206-211

Job planning, 311

LAS, 350
Laterolog, 10, 117, 197, 386-388
LIS, 350
Logging speed, 359-360, 426
Log quality control, 6, 314-321
   database, 320
   form, 317-319
Lost circulation material, 205
LSO detectors, 111
LWD, see MWD

Magnetic surveys, 301
Magnetism, effect of –, 72
Magnetometers, 113
Markers, 413
Mathematical modeling, 167-173
   nuclear, 170
   resistivity, 169
   sonic, 168
Measurement,
   chain, 104-105
   duration, 119
   sequence,
Median averaging, 140
Memory dump, 123
Metrology, 33-78
Microlog, 147
Microresistivity, 117
Micro-Spherical Focused Log, 139, 319
Mirror images, 378
Monte Carlo, 90, 172
MTBF, 77
Mud,
   data, 331
   filtrate, 207
   heavy – algorithm, 221
   homogeneity, 205
   segregation, 206
   silicate –, 204
   solids, 204
Mudcake, 224-225
Mud-pulse transmission, 121
Multigroup, 170
Multiple pulsing, 264

Multiwell, 413
MWD, 5, 76, 187, 227

Net-to-gross ratio, 7
Neutron porosity log, 11, 37, 185, 196,
    252-254, 264, 414, 427, 445
Noise, 376-377
Nonconformity, 316
Nuclear
    magnetic resonance, 398
        measurement duration, 119, 132
        quality control, 403-405
        relaxation times, 398
    measurements, 118

Oil-base mud, 203
    pseudo -, 204
Optimal,
    filtering, 135
    logging, 423
Overburden, 417

Permeability, 14,
Photoelectric effect, 219-221
Photomultiplier, 110
Physical simulation, 160
Piezoelectric transducer, 111-112
Platform express, 227
Porosity, 7, 80, 87, 88, 427, 441
    worth of a - unit, 97
Potassium, 266
Precision, 5, 42-43, 104, 117, 132-134,
    433
Pressure, 329, 386
    effect of -, 171
    gauges, 112,
Print, 353-356
Pseudogeometric factor, 53-56
Pulse-echo spacing, 400
Pyrite, 393

Quality,
    curves, 336
    flags, 336

Radius of investigation, 58
Random error, see error
Rate of penetration, 125-127, 130, 336,
    359, 445
Recovery factor, 150
Relaxation time, 74, 398
Relative bearing, 299
Reliability, 77, 440
Relief well logging, 12
Repeat formation tester, 412
Repeatability, 45, 364-371, 434
Repeat section, 5, 337, 396
Reproducibility, 45, 434
Resistor network, 160
Resolution matching, 67-68

Sampling,
    interval, 137-139
    rate, 130-131
Scaled experiments, 164-166
Scintillation detectors, 110
Seismics, 14-18
    borehole -, 16
    crosswell -, 16
    surface -, 14
Semiconductor detectors, 110
Sensitivity analysis, 95-96
Shocks,
    effect of -, 76
Short-axis logging, 380-381
Shoulder-bed effect, 177-178
σ, see standard deviation
Signal processing, 5, 133, 153
Signal transmission, 120-121
Simandoux equation, 84, 443-444
Sliding indicator, 336
Sodium iodide detectors, 71, 110

Sonde error, 251
Sonic,
  imaging, 11
  log, 117, 164, 196, 377, 409
  transducers, 111
Sources,
  nuclear –, 107
  sound –, 108
Specifications, 312
Spectral gamma ray, 142-143
Speed correction, 292,-293, 396
Spine and ribs, 114
Spontaneous potential, 109, 374
Squeeze, 387
Stability, 45
Stacking, 140-141
Standard deviation, 38
Standards, 314
Standoff, 182-183
Statistical distribution,
    binomial –, 39
    Gaussian –, 40, 437-439
    Poissonian –, 39
Sticking indicator, 336
Stoneley waves, 164
Stretch correction, 279-280
Surface system alignment, 236
Systematic error, see error

Tandem logging, 382
Tape, 5
Telemetry, 122
Telluric currents, 377
Temperature, 191, 332, 386
  effect, 71
Tension, 279-281, 336
Thermal decay time, 46
Thorium, 266
TIF, 352
Time,
  after bit, 336
  -indexed data, 334
  marks, 337
  recovery –, 74

Time,
  response, 73
  – to depth conversion, 124
Tomography, 11
Tool,
  configuration, 362, 396
  description, 333
  face, 299
  positioning, 196
  response, 5, 159-174
  rotation, 361
True value, 33, 434
Turbidite, 19-20
Two-height calibration, 252

Ultra-long spaced electric log, 11
Ultrasonic, 117
Uncertainty, 24, 46, 423, 434
Units, 325-326
Uranium, 266

Variance, 83
Verification, 259-261
Vertical resolution, 10, 14, 16, 18, 61-67,
  enhancement of –, 67, 147-157
  reduction of –, 67, 135
Vibrations,
  effect of –, 76

Wait time (NMR), 399
Water,
  bound –, 417
  saturation, 7, 89, 94, 427, 442
Waxman-Smits, 417
Well testing, 12

X-signal, 251

Yo-yo, 281

Printed on June, 2009
by Imprimerie Maury S.A.S.
Z.I. des Ondes – 12100 Millau
(F09/43614L)

Printed in France